Green Energy and Technology

For further volumes:
http://www.springer.com/series/8059

Andrea Cipollina · Giorgio Micale · Lucio Rizzuti
Editors

Seawater Desalination

Conventional and Renewable Energy Processes

With 143 Figures and 55 Tables

Springer

Editors
Andrea Cipollina, Giorgio Micale, Lucio Rizzuti
Dipartimento di Ingegneria Chimica
dei Processi e dei Materiali
Università degli Studi di Palermo
Viale delle Scienze, Ed. 6
90128 Palermo
Italy
e-mail: micale@dicpm.unipa.it

ISSN 1865-3529 e-ISSN 1865-3537
ISBN 978-3-642-26059-9 e-ISBN 978-3-642-01150-4
DOI 10.1007/978-3-642-01150-4
Springer Heidelberg Dordrecht London New York

Cover design: WMX Design GmbH, Heidelberg

Printed on acid-free paper

Springer is part of Springer Science+Business Media (www.springer.com)

. . . Water, water, everywhere,
Nor any drop to drink. . .

The Rime of the Ancient Mariner
by Samuel Taylor Coleridge

The Editors' royalties of this book will be donated to the European Desalination Society, as a small sign to support EDS extensive activities in the field of Desalination, with particular regard to the widespread of knowledge and awareness at all levels of the global community.

Foreword

The seas around us harbors over 97% of water in the world. Yet water stress and scarcity loom large as population grows.

Desalination technology has come to harness the water treasure in the seas. Desalination plants growing in number, size and efficiency are supplying more and more water with lower energy requirements, more attention to environment and lower cost. At the same time, great efforts are being made to promote water conservation by more efficient use. Yet we have a long way to go.

Lowering energy requirements and hence cost is the major challenge of desalination as it is for all energy consuming technologies. We now look to sun, wind and wave and we once looked to their Gods. If today we have not yet reached efficient ways to unlock this mighty source of energy and some skeptics lurk in the wings, research has even brought us to the moon. So it is essential to pursue research on solar energy coupled to desalination as well as wind and wave sources with determination to link them to more and more efficient clean green sources for desalination at lower costs.

The editors and authors of chapters in this book have been dedicated to this objective and should be congratulated for their efforts in explaining desalination and links to renewable energy in a reader-friendly style.

Miriam Balaban
European Desalination Society

Preface

by the Editors

The idea of a novel book on seawater desalination has been for quite a long time in the minds of the Editors, who have been living in the Italian island of Sicily and thus well acquainted with the overwhelming potential offered by the sea and the sun.

It was not until the Summer of 2007 that the Editors came to the final decision to start the preparation of the present book, thanks to the insightful support of Dr. Christoph Baumann, Engineering Editor at Springer Heidelberg.

This book is an attempt to write an original comprehensive yet concise introduction to Seawater Desalination processes, with the specific aim to abridge the gap between conventional technologies and novel sustainable renewable energy processes.

The first section of this book presents, in a technical but reader-friendly way, an overview of currently-used desalination processes, from thermal to membrane processes, highlighting the relevant technical features, and development potential. It also gives a rapid insight into the economic aspects of fresh water production from seawater.

The second section of the book presents novel processes which use Renewable Energies for fresh water production. From the first solar still evaporators, which artificially reproduced the natural cycle of water, technology has progressed to develop complex systems to harness energy from the sun, wind, tides, waves, etc. and then to use this energy to power conventional or novel desalination processes. Most of these processes are still at a preliminary stage of development, but some are already being cited as examples in remote areas, where they are proving to be valuable in solving the problems of water scarcity.

The book actually fills the need for a wide coverage systematic description of a remarkable range of alternative desalination process and plant configurations, thus helping the reader to understand the key technological advantages and disadvantages and the relevant suitability of a given configuration for the specific site of installation.

Worldwide acknowledged leading experts in the various fields of desalination processes have contributed writing those chapters of the book relevant to their specific field of expertise.

Perhaps the most immediate value of this book resides in providing valuable accessible and up-to-date information for all readers who may be interested in the field of Desalination within the context of present and future sustainability: engineering professionals, academics and scientists, technicians, managers, private and public institutions, students and common citizens.

The Editors wish to express their most heartfelt thanks to the Authors of the chapters for their efforts in contributing to such a challenging task, and to Mya for performing the language revision of the book.

Last but not least the Editors wish to thank Alice, Cristina, Giuseppe, Laura, Pietro and Serena for the invaluable support throughout the time taken for the preparation of the manuscript; without their encouragement and understanding this book would not be completed today.

Palermo, July 2009

Andrea Cipollina
Giorgio Micale
Lucio Rizzuti

Contents

Contributors

M. T. Chaibi National Research Institute for Rural Engineering Water and Forestry, B. P 10, 2080 Ariana, Tunisia, chaibi.medthameur@iresa.agrinet.tn

Andrea Cipollina Dipartimento di Ingegneria Chimica dei Processi e dei Materiali, Università di Palermo, Viale delle Scienze, Ed. 6, 90128 Palermo, Italy

Efrem Curcio Department of Chemical Engineering and Materials, University of Calabria Via P. Bucci CUBO 44A – 87030 Rende (CS), Italy

Enrico Drioli Institute on Membrane Technology ITM-CNR, c/o University of Calabria Via P. Bucci CUBO 17C – 87030 Rende (CS), Italy

Ali M. El-Nashar 22 Ahmed Gharbo Street, Apt. 703, Zizinia, Alexandria, Egypt, elnashar100@hotmail.com

Hisham Ettouney Department of Chemical Engineering, College of Engineering and Petroleum, Kuwait University, P.O. Box 5969, Safat 13060, Kuwait, ettouney@hotmail.com

Dieter Geyer Center for Solar Energy and Hydrogen Research Baden-Württemberg, Industriestr. 6, D-70565 Stuttgart, Germany, dieter.geyer@zsw-bw.de

I. Khamis Nuclear Power Technology Development Section, Division of Nuclear Power, IAEA, Vienna, Austria, I.Khamis@iaea.org

Joachim Koschikowski Fraunhofer Institute for Solar Energy Systems ISE, Freiburg, Germany, PSE AG – Projects in Solar Energy, Freiburg, Germany, Joachim.koschikowski@ise.fraunhofer.de

Sabine Lattemann Institute for Chemistry and Biology of the Marine Environment, Carl von Ossietzky University of Oldenburg, Postfach 2503 26111 Oldenburg, Germany, sabine.lattemann@icbm.de

Dimitris Manolakos Agricultural University of Athens, Dept. of Natural Resources and Agricultural Engineering Iera Odos Street 75, Athens 11855, Greece

Giorgio Micale Dipartimento di Ingegneria Chimica dei Processi e dei Materiali, Università di Palermo, Viale delle Scienze, Ed.6, 90128 Palermo, Italy, micale@dicpm.unipa.it

B.M. Misra 501 Emerald Heights, 32 Union Park, Chembur, Mumbai 400071, India, bmmisra@gmail.com

Essam Sh. Mohamed Agricultural University of Athens, Dept. of Natural Resources and Agricultural Engineering Iera Odos Street 75, Athens 11855, Greece

George Papadakis Agricultural University of Athens, Dept. of Natural Resources and Agricultural Engineering Iera Odos Street 75, Athens 11855, Greece, gpap@aua.gr

Michael Papapetrou WIP-Renewable Energies, Sylvensteinstr. 2, 81369, Munich, Germany, michael@papape.com

Baltasar Peñate Canary Islands Institute of Technology (ITC), Pozo Izquierdo, Gran Canaria, Canary Islands, Spain, agua@itccanarias.org

Jürgen Rheinländer Center for Solar Energy and Hydrogen Research Baden-Württemberg, Industriestr. 6, D-70565 Stuttgart, Germany, juergen.rheinlaender@zsw-bw.de

Lucio Rizzuti Dipartimento di Ingegneria Chimica dei Processi e dei Materiali, Università di Palermo, Viale delle Scienze, Ed.6, 90128 Palermo, Italy

Matthias Rommel Institute for Solar Technology SPF University of Applied Sciences Rapperswil HSR Oberseestr. 10, CH-8460 Rapperswil – Switzerland

Vicente J. Subiela Canary Islands Institute of Technology (ITC), Pozo Izquierdo, Gran Canaria, Canary Islands, Spain

Eftihia Tzen Centre for Renewable Energy Sources (CRES), Wind Energy Department, 19th km Marathonos Ave, 19009 Pikermi, Greece, etzen@cres.gr

Marcel Wieghaus Fraunhofer Institute for Solar Energy Systems ISE, Freiburg, Germany; PSE AG – Projects in Solar Energy, Freiburg, Germany

Mark Wilf Membrane Technology, Tetra Tech, Inc. 10815 Rancho Bernardo Road, San Diego, CA 92127, USA, mark.wilf@tetratech.com

Chapter 1
Seawater Desalination for Freshwater Production

Giorgio Micale, Andrea Cipollina, and Lucio Rizzuti

Abstract In the last decades more and more countries have experienced water scarcity problems, thus pointing at alternative non-conventional sources of fresh water. Seawater desalination has proven to be a reliable and economically sustainable water resource since the second half of the 20th Century. A number of well proven technologies already exist, with advantages and disadvantages making each of them more suitable in specific sites. Moreover, quite recently, coupling the use of renewable energy to the production of fresh water from seawater results in novel technologies, able to minimise the environmental impact that desalination processes can create due to their intense energy consumptions.

On the above basis, the chapter opens the book providing a general overview on the world water scarcity problem, followed by a classification of the main desalination technologies based on different principles. Also an overview on the coupling of desalination processes with RE technologies is presented, thus introducing on the topics discussed in details in the second part of the book.

1.1 Introduction

Water means life. Every day human beings perform a remarkable variety of activities which directly or indirectly involve the use of water, often in very large quantities. Water is needed in all industrial activities, in agriculture and for domestic purposes. Pro capita consumptions may vary quite considerably depending on the specific geographical area of the world, with obvious large differences between, for example, the Saharan Region of Africa and the Great Lakes Region in North America. Average water consumptions reach values of up to 400 l per person per day in the USA, while they may drop to 150 l in other western countries, where actions have

G. Micale (✉)
Dipartimento di Ingegneria Chimica dei Processi e dei Materiali, Università di Palermo, 90128 Palermo, Italy
e-mail: micale@dicpm.unipa.it

A. Cipollina et al. (eds.), *Seawater Desalination*, Green Energy and Technology,
DOI 10.1007/978-3-642-01150-4_1, © Springer-Verlag Berlin Heidelberg 2009

been successfully implemented to reduce freshwater demand. In contrast, in some African countries, where serious water shortages are experienced, pro capita freshwater consumptions are in the range of 20 l per day. The World Health Organisation (WHO) recommends a lower limit for survival of 15–20 l per person per day, which is able to guarantee only basic needs such as drinking, food preparation, personal hygiene and laundry. This minimum amount needs to be increased significantly, up to 50 l per person per day, in order to guarantee needs beyond individual ones, such as those connected to hospitals, schools, basic infrastructures, etc.

There are several factors which have contributed to the steadily increasing world freshwater consumptions, among them being the world demographic increase and the general improvement in the quality of life at all levels, with consequential increase in freshwater demand for a wide variety of uses.

Based on the above, it is expected that by the year 2040 the world demand for freshwater will be greater than the amount available. A future scenario such as this, of course, requires the utmost consideration and all possible efforts should be made in order to ensure constant freshwater demand fulfilment.

To this end considerable effort has been made by scientists and technicians to optimise the management of conventional water resources in order to achieve optimum efficiency of water utilization. Notwithstanding the improvements made in this respect, it is now evident that alternative sources of freshwater are necessary to meet the current and future trend in freshwater demand.

In the past, freshwater was considered as a naturally available resource, unmanufactured, and at most to be treated by means of some simple physico-chemical process in order to achieve the characteristics required for its final use.

Only recently has freshwater started to be considered as a product that can be manufactured, with quality standards depending on the specific use. As such, a non conventional, yet sustainable route for freshwater production could be that offered by seawater desalination.

Desalination processes make freshwater from the separation of salt from seawater or brackish water. As detailed in the following paragraphs, seawater, in abundance, constitutes a reliable sustainable source of freshwater, with high potential to fulfil the continuously increasing freshwater demand of the future.

1.2 Water Resources and Data

Water is one of the most abundant substances present on Earth; an estimate of the total amount of water provides a figure of about $1.4 \cdot 10^9$ km^3. Seawater is about 97.5% of the total available water, while the remaining 2.5% (i.e. $3.5 \cdot 10^7$ km^3) is constituted by underground and surface waters. A remarkable 80% of the latter is frozen water in glaciers, so that only 0.5% of the total amount available is to be found in lakes, rivers and aquifers.

The water cycle makes the total amount of water on Earth constant over time. Water evaporates, from the seas and surface waters, into the atmosphere and accumulates in clouds from which rainfall originates. Precipitated water in turn feeds into underground and surface waters and seas. Thus, as far as water is concerned, Earth behaves like a terrarium, which is a self-sufficient closed system with no need for external supply.

Of course, the local availability of freshwater depends on the level of precipitation. In areas where rainfall is abundant, freshwater is also abundant. Conversely in arid lands, where rainfall is scarce, freshwater is scarce too.

Freshwaters differ substantially from seawaters by the relative amount of salts found in them. Table 1.1 shows a very simple classification of natural waters on the basis of their saline content.

Freshwaters may have salinity up to 1,500 ppm, brackish waters exhibit salinity in the range of 3,000–10,000 ppm, while the salinity of seawater typically ranges from 10,000 ppm (as in the case of the Baltic Sea) up to 45,000 ppm (as in the Arabian Gulf). Low salinity may be as a result of the presence of inflow from rivers and melting icecaps, as well as due to the abundance of precipitation. High salinity may be a result of the remoteness from land and of high temperatures promoting evaporation.

The reference average salinity of seawater is taken as 35,000 ppm. Typical seawater composition is given in Table 1.2.

The main chemical constituents found in seawater are essentially sodium (30%) and chloride (55%), i.e. the components of common table salt, which account for 85% of the total content of dissolved solids of seawater. Sulphate and magnesium

Table 1.1 Water classification based on salinity content

Type	Total dissolved solids (TDS)	Note
Freshwater	Up to 1,500	Variable chemical composition
Brackish water	1,500–10,000	Variable chemical composition
Salt water	> 10,000	Variable chemical composition
Seawater	10,000–45,000	Fixed chemical composition
Standard seawater	35,000	Fixed chemical composition

Table 1.2 Standard seawater composition

Chemical Ion	Concentration [ppm]	Percentage of total salt content [%]
Chloride Cl^-	19,345	55.0
Sodium Na^+	10,752	30.6
Sulfate SO_4^{2-}	2,701	7.6
Magnesium Mg^{2+}	1,295	3.7
Calcium Ca^{2+}	416	1.2
Potassium K^+	390	1.1
Bicarbonate HCO_3^-	145	0.4
Bromide Br^-	66	0.2
Borate BO_3^{3-}	27	0.08
Strontium Sr^{2+}	13	0.04
Fluoride F^-	1	0.003

are also abundant, with a percentage of almost 8 and 4%, respectively. It should be noted that although salinity of seawater may vary depending on the specific region of the world, the percentage composition of seawater is essentially constant throughout the world (i.e. the proportions of the major constituents are constant).

The composition of seawater, described above, is quite different from that of a typical river water, shown in Table 1.3.

Comparison with seawater highlights the fact that sodium chloride content in river water is far lower, representing less than 16%; conversely calcium and bicarbonate content is far higher (i.e. about 17 and 32%, respectively) accounting for nearly 50% of the total dissolved solids. Silica is another major constituent of river water, but is negligibly present in seawater.

As a result of the high content of total dissolved solids the physical properties of seawater are very different from those of freshwaters. Table 1.4 shows main thermodynamic properties in the case of standard seawater composition, i.e. 35,000 ppm salinity at 20°C.

It is worth noting that both Osmotic Pressure and Boiling Point Elevation are fundamental properties in the design and operation of membrane and thermal desalination processes. Osmotic pressure is related to the quantity of ions dissolved in

Table 1.3 Comparison between seawater and river water

Chemical ion	Percentage of total salt content in river water [%]	Percentage of total salt content in seawater [%]
Chloride Cl^-	8.6	55.0
Sodium Na^+	6.9	30.6
Sulfate SO_4^{2-}	12.4	7.6
Magnesium Mg^{2+}	4.6	3.7
Calcium Ca^{2+}	16.6	1.2
Potassium K^+	2.6	1.1
Bicarbonate HCO_3^-	31.9	0.4
Bromide Br^-	–	0.2
Borate BO_3^{3-}	–	0.08
Strontium Sr^{2+}	–	0.04
Fluoride F^-	–	0.003
Silica SiO_2	14.6	–
Iron Fe^{2+}	0.7	–
Nitrate NO_3^-	1.1	–

Table 1.4 Thermodynamic properties of seawater with 35,000 ppm salinity at 20°C

Density [kg/m^3]	1,024
Viscosity [kg/ms]	$1.074 \cdot 10^{-3}$
Specific Heat [kJ/kg°C]	3.998
Osmotic Pressure [bar]	27
Boiling Point Elevation, at 20°C [°C]	0.32
Boiling Point Elevation, at 100°C [°C]	0.51

the salt water and strongly influences the tendency of water to pass through semi-permeable membranes, for the separation of freshwater from a salty stream. Boiling Point Elevation represents the increase in boiling temperature of a solution. It is strongly related to the concentration of salts in the solution itself, but is only weakly dependent on its temperature. This parameter largely affects seawater evaporation phenomena, thus influencing the overall performance of evaporative separation processes.

1.3 An Overview of Desalination Processes

In order to produce freshwater from salt water a suitable separation process must be devised. The task can be accomplished in many different ways, all of them based on the principle that water and salts do not separate spontaneously, and thus require some kind of energy source to power the separation process. Historically, the first desalination units used for freshwater production were based on the evaporation of pure water via the addition of heat provided by the sun or by combustion processes. This principle, though implemented via complex and highly energy-integrated processes, still applies to current desalination technologies based on thermal separation processes.

In recent years, the development of modern polymer materials has led to the production of membranes which allow the selective passage of water (semi-permeable membranes) or ions (ionic exchange membranes), thus providing the basis for membrane desalination processes.

In general, desalination technologies can be classified according to three criteria: (1) *what is extracted* from the seawater; (2) the *type of separation process* adopted; (3) the *type of energy* used.

Within the first classification desalination process technologies can be divided into two main groups: (i) processes in which water is removed from the main stream, thus producing a salt-free product; (ii) processes in which salts are removed from the main stream, thus leaving the latter salt-free.

Since seawater salt content is generally lower than 4%, it may seem more convenient to select processes where salts are removed from seawater, rather than extracting pure water from seawater.

However due to technological constraints, up to now, processes which remove water give better performances and are thus preferred. In particular, suitable design of highly energy-integrated configurations allows the production of freshwater, with relatively low energy consumptions.

Among these processes, some accomplish the separation with phase change of the produced freshwater, typically in the form of vapour. These processes are usually indicated as evaporative (or thermal) processes and require a significant amount of energy, given the high heat of vaporisation of liquid water. Today, two main types of evaporative desalination processes are used worldwide, Multiple Effect Distillation (MED) and Multi Stage Flash (MSF) desalination. In MSF, seawater evaporates in flashing chambers, in which the pressure is kept below the feed vapour pressure. This leads to a flash of seawater, resulting in the production of vapour and the cooling of the remaining brine. An arrangement of flashing stages connected in series, with decreasing pressure, allows the production of large amounts of vapour which is then re-condensed on the external surface of a tube bundle. Condensation heat in the condensing bundle is used for preheating the seawater feed stream, thus reducing thermal energy requirements for heating the feed to its Top Brine Temperature, the temperature at which the feed enters the first flashing stage. Thus, the only thermal energy input is in the Brine Heater, where pre-heated seawater is further heated by means of low-pressure steam to its Top Brine Temperature. Standard MSF units require an average of 1 kg of low-pressure steam for every 8–10 kg of distillate produced.

In MED, evaporation occurs on the external surface of a tube bundle which is heated by motive steam condensing inside the tubes. Vapour produced in one effect is then used as motive steam in the following effect, which operates at lower pressure and, therefore, at a lower boiling temperature. Such a heat integration arrangement allows the attainment of very high energy efficiency for the process, resulting in a performance ratio up to 10–12 kg of distillate per kg of motive steam fed into the first effect. In order to enhance the process energy efficiency further, MED units can also be coupled to vapour recovery devices. Thermal Vapour Compression (TVC) and Mechanical Vapour Compression (MVC) are the most common systems. In TVC part of the vapour from the last effect is recompressed by a steam ejector (powered by mid-pressure steam), and then used as motive steam for the first effect, thus increasing the performance ratio up to 15–16 kg of distillate per kg of mid-pressure motive steam fed into the ejector. In MVC all the vapour exiting from the last effect is directly compressed by a mechanical compressor, and then used as motive steam in the first effect. In this case only mechanical (or electrical) energy is used to power the process. Other two minor processes which remove vapour from salt water are the Humidification-Dehumidification and the Membrane Distillation processes. Both of them will be described in the relevant chapters highlighting their coupling with Renewable Energies.

The production of frozen desalted water, via the removal of heat from a salt solution, is also a possibility. The energy requirements are significantly lower than the previous case, as the heat of fusion of water ice is much lower than the heat of vaporisation. Nevertheless, both the technology of refrigeration cycles, and the management of the relevant desalination process, presently prevent a successful economical application of such an option.

Other processes accomplish the separation of freshwater from salt water without phase change, as in the case of pressure-driven membrane separation processes. Among these processes, Reverse Osmosis is the most widely adopted. It achieves the separation of freshwater through the use of a semi-permeable membrane which allows the passage of water whilst preventing the passage of salts into the permeate stream.

Salts can be removed from salty waters by using devices able to capture the ionic content of the stream, either by ionic exchange membranes or by ionic exchange resins. In the first case (Electro Dialyses process) ions are forced to pass through

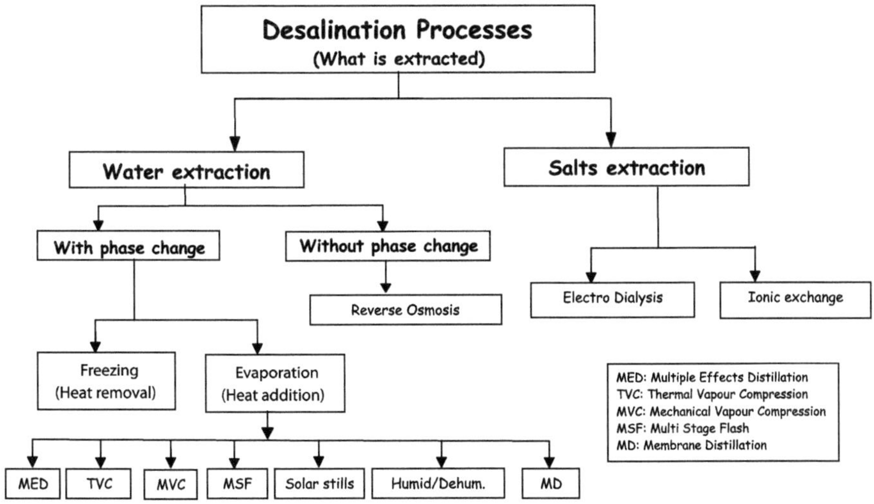

Fig. 1.1 Desalination technologies classification based on what is extracted from the feed stream

the membranes and thus separated from the main stream. In the second case (Ionic Exchange) ions are captured within the solid matrix of the resin. Both these processes, however, are limited to brackish waters only. Figure 1.1 illustrates how conventional processes can be classified according to the above classification criteria.

The second classification is made on the basis of the separation process adopted. The first group is made up of membrane processes where separation occurs by means of selective membranes. When semi-permeable membranes are used, water can pass through the membrane into the permeate stream, while salts are rejected. This is the case of the Reverse Osmosis process, where the driving force for the separation is a pressure difference between the faces of the membrane itself. Conversely, in the Electro Dialysis process, ionic exchange membranes are employed, allowing the selective passage of positive or negative ions. The driving force for the passage is a very large difference in electrical potential between two electrodes positioned on the external sides of a stack of channels consisting of alternate anionic and cationic membranes. Anions and cations are forced to move towards positive and negative electrodes respectively, flowing through anionic and cationic membranes. When the anions eventually encounter cationic membranes, and the cations anionic ones, they are stopped and trapped in channels where they accumulate in a concentrated stream. On the other side, streams from which ions are drawn remain, with very low salt content, and thus result in dilute streams.

The second group is made up of thermal processes, where separation occurs by adding or removing heat to obtain pure water from the saline solution. Most commonly evaporation is adopted to produce pure water in the form of a vapour from liquid salty water. The vapour is then condensed, and the condensation heat is usu-

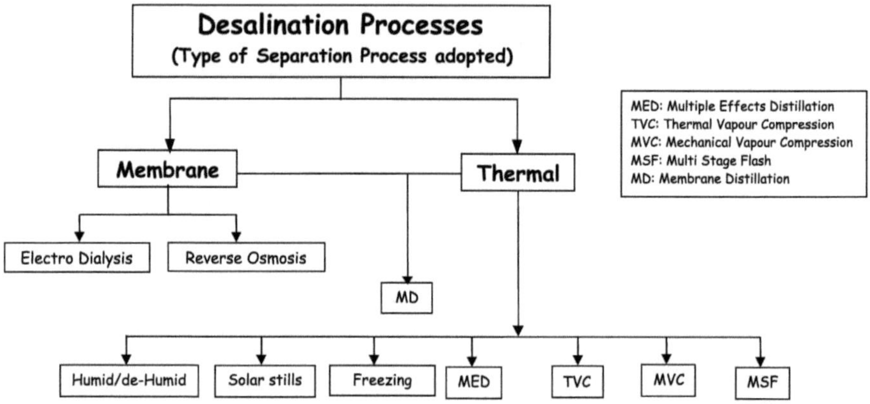

Fig. 1.2 Desalination technologies classification based on the separation process adopted

ally recovered to preheat the feed stream or to evaporate further vapour. Figure 1.2 shows the technologies classified according to the separation process adopted.

The third criterion classifies desalination technologies on the basis of the type of energy used, i.e. thermal, mechanical or electrical. Most evaporative processes need not only thermal energy, but also mechanical and electrical energy to power circulation pumps and auxiliary units. The present classification only accounts for thermal energy which is the prime mover for thermal technologies.

It is worth noting that the classification criterion based on energy input requirements is of paramount importance when considering the possibility of coupling conventional desalination technologies with alternative sources of energy, for example in order to determine the coupling potential of desalination with solar thermal col-

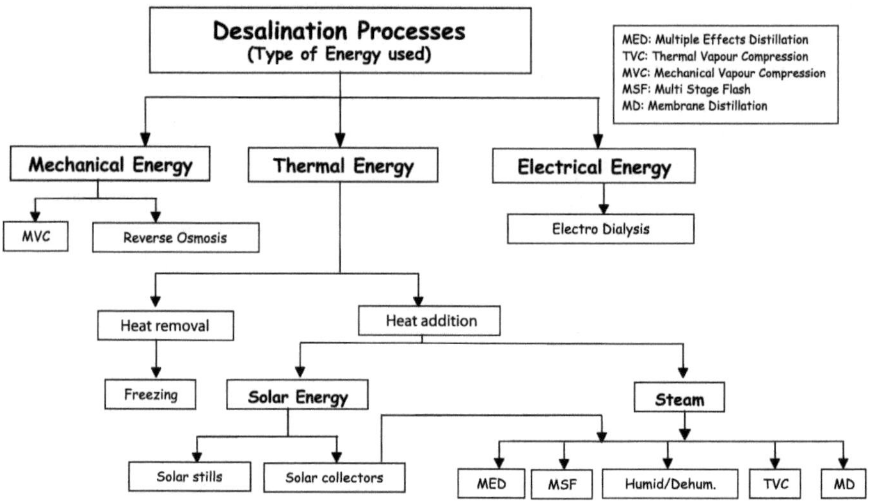

Fig. 1.3. Desalination technologies classification based on the type of energy used

lectors, solar photovoltaic panels, wind turbines, etc. A detailed description of all possible couplings is presented in the second part of this book, in Chaps. 6, 7, 8, 9, and 10. Figure 1.3 shows the classification based on the type of energy used.

1.4 Energy Sustainability of Desalination Processes

As discussed above, desalination of salt water often requires significant amounts of energy to separate the salts from the water. Such energy can be provided as heat, in the case of thermal processes, or as mechanical or electrical energy, as in the case of membrane processes. An exception is the MVC process which requires mechanical energy, although it can be classified as a thermal process.

Energy consumption varies from process to process. The most energy-demanding processes are thermal ones, which normally produce from 8 to 12 kg of freshwater per kg of heating steam (the equivalent thermal energy, expressed in thermal kWh, would be about 50–70 kWh_{th}/m^3 of distilled water). Reverse Osmosis requires from 3 to 6 kWh_{el} to produce 1 m^3 of permeate water, while Electro Dialysis consumptions are lower, but always limited to the treatment of low salinity brackish water.

A cost analysis of desalination processes indicates that a very large proportion of the cost of water is related to the cost of energy. Considering also the dramatic increase in energy costs of recent years, and the likely rising trend, it appears imperative to find suitable alternative solutions for powering the production of desalinated water.

Several different couplings between alternative energy sources and desalination processes have been proposed so far. Nuclear energy can be used for the production of desalinated water, either using nuclear plant waste heat to power thermal units, or through the use of electrical energy generated by the nuclear power station, for powering Reverse Osmosis units in off-peak hours.

The energy sustainability of desalination processes could be improved by the coupling of desalination technologies with Renewable Energy (RE) sources. There is a wide range of well established technologies for the exploitation of renewable energy. Solar energy can be used for the collection of heat, with very high efficiencies, through the use of modern thermal solar collectors, or for the production of electrical energy, through the use of fast-developing photovoltaic technologies. Wind energy is also a well established source of RE, with wind turbines able to produce electrical energy for any electrically-powered desalination process. More recently, wave and tidal energy has also been demonstrated to be a promising and large-capacity energy source, although the technology is still in start up phase. Other minor RE sources can also be used, such as geothermal, biomass, etc., although they are usually limited to specific sites and their use for freshwater production through desalination has yet to be demonstrated.

One very important aspect is that often the demand for an alternative supply of freshwater is combined with the abundance of one of the aforementioned renewable energy sources. In fact, for example, the sunniest countries are those where water scarcity has led to the use of desalination as an alternative freshwater source. Wind energy is often easily exploitable close to the coast or even offshore, where seawater availability is one of the main features of the site. This is obviously also true for wave and tidal energy. All these reasons have led researchers, companies and institutions to focus their research efforts on the study of producing desalinated water using RE.

Some technological problems should also be noted. The first one is related to the cost of RE exploitation, which is still higher than for conventional energy sources. This prevents the use of RE-desalination coupling for large scale installations, where conventional technologies are still more competitive. The other technological limitation is related to the inconsistent nature of most RE sources. This limitation can be overcome in two ways: (1) by buffering the energy via a storage unit (a battery for electrical energy or a thermal buffer tank for thermal energy); (2) by designing a suitable desalination process, able to operate and perform well in transient state.

Nowadays, the use of RE-desalination is already a viable and convenient option in small installations for remote sites, where the cost of conventional energy, as well as conventional freshwater supply, is very high. Moreover, given the rapid improvement in both RE and desalination technologies, it is important to underline the great potential which such new technologies, compared to traditional processes, present.

Several examples of operating units can be found in relevant literature and research is currently being carried out on the development of more efficient and technologically viable novel processes, as presented in the second part of this book.

To give an outline of the possibilities of coupling Renewable Energies with desalination technologies, Fig. 1.4 is a block diagram indicating how the most common RE sources can be used for the production of desalinated freshwater.

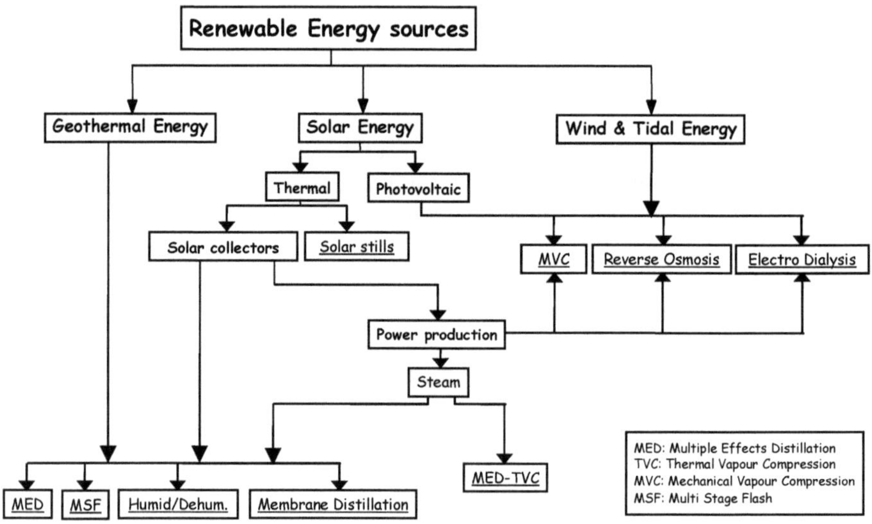

Fig. 1.4 Coupling potentials between renewable energies and desalination technologies

1.5 Pre- and Post- treatments for Desalinated Water Production

An introduction to desalination technologies must take into account two crucial aspects of the desalination process, these being the pretreatment of salty water and the post-treatment of freshwater produced, in order to meet the standards required for its final use.

Pretreatment is often necessary to guarantee proper operation of desalination units. It depends strictly on the type of desalination process adopted.

Thermal processes usually require very little pretreatment as they are intrinsically more robust than membrane processes. Typical pretreatment performed on the seawater feed of MED and MSF units are:

- grid filtration and suspended solids settling, in order to reduce the concentration of suspended matter in the feed stream to values suitable for operation;
- disinfection, usually through the addition of common disinfectants such as hypochlorite, chlorine, etc., in order to reduce the formation of algae and bio-fouling especially in cold parts of the unit (feed ducts, filters, etc.);
- de-aeration, with the double aim of (i) reducing the quantity of CO_2, bi-carbonates and carbonates which can lead to scaling, and (ii) reducing non-condensable gases which could then prevent the achievement and keeping of vacuum conditions within evaporation stages;
- addition of anti-scaling chemicals, usually consisting of dimeric or polymeric organic acids or chelating agents, able to reduce the formation of Calcium Carbonate and Sulphate;
- addition of anti-foaming chemicals, for example consisting of polyglycols, which reduce the formation of foam during the evaporation process;

Effects of scaling in MED units are much more damaging than in MSF, due to evaporation on the hot external surface of tubes, which is very hard to clean. However MSF units usually operate at higher Top Brine Temperatures, which in turn leads to a higher potential for scaling inside the tubes of the condensing bundles. Thus the addition of anti-scaling agents is fundamental in both processes. Anti-foaming agents are more frequently used in MSF units where the flashing phenomenon can easily give rise to foaming, which can increase the entrainment of salt droplets into the distillate, and thus reducing the product quality.

Membrane processes, and particularly Reverse Osmosis, require more substantial pretreatment, due to the sensitivity of membranes to fouling problems. Typical pretreatments are described in the chapters relevant to membrane processes. A standard pretreatment of feed water for a membrane desalination unit consists of:

- disinfection, usually through the addition of common disinfectants such as hypochlorite, chlorine, etc., in order to reduce the formation of algae and bio-fouling in the most critical components of the plant, such as the filters and the membranes themselves. As many types of membranes (e.g. polyamide membranes) are very sensitive to oxidants, the addition of a reducing agent is often also required (e.g. phosphite salts) which neutralises the residual oxidizing agent before the feed enters the membrane assembly.

- filtration, performed by media filters or, more recently, by micro- and ultra-filtration, is necessary to reduce the feed Silt Density Index (see Chap. 3) to values below 2–3, thus guaranteeing a longer membrane life and less frequent maintenance problems and cleaning of the membrane assembly;
- addition of anti-scaling chemicals, similar to those used in thermal units, able to reduce, mainly, the formation of Calcium Carbonate;

Pretreatments in reverse osmosis plants can often influence the overall water costs by up to 30–40%, thus highlighting the importance of such a step in the overall economics of the process. Where feed seawater is of very good quality, with low SDI and turbidity (for example when intake is from very deep waters with rocky floors), pretreatments become less important. However they are still necessary to guarantee correct and long term operation of the plants.

In some cases pre-filtration may not even be necessary, for example when beach wells are used for feed water intake. However, this option is only feasible for small capacities and cannot always guarantee high feed water quality or availability for very long periods.

Once desalinated water has been produced, its suitability for potable, civil or industrial use must always be checked. Indeed, post-treatments are often necessary to guarantee the achievement of the standards required for the final use of the product water. The type and intensity of post-treatments depend strictly on both the process adopted to desalinate water and the final use of the product water itself.

Thermal plants produce almost distilled water, with very low salt content (usually not more than 10–20 ppm). Thus thermally desalinated water is suitable for industrial applications in which pure water is needed (e.g. in boilers for steam production, or in heat exchange circuits etc.), however it must be re-mineralised for domestic and drinking uses. Re-mineralisation is necessary to increase the salinity up to a few hundred ppm, and especially to increase the hardness of the water, without which the water would be too aggressive for domestic distribution and not potable according to World Health Organisation standards. Post-treatment steps may vary according to the scale of the thermal plant. In very large plants they usually involve the absorption of CO_2 within a gas-liquid contactor, followed by the addition of lime ($Ca(OH)_2$), or the dissolution of limestone ($CaCO_3$), thus increasing the content of calcium and carbonates, and also controlling the final pH of the product. Finally some disinfected seawater is added to increase the overall salinity to values which may vary between 200 and 400 ppm. Sodium Hydroxide (NaOH) can be used for final pH adjustment. It is worth mentioning that the use of CO_2 is very convenient in large installations. This gas is usually recovered from the de-aeration process and stored so that it can be used for the post-treatment of product water without any additional cost. In small installations, chemicals in solid form can be used, e.g. limestone ($CaCO_3$), Sodium Carbonate (Na_2CO_3), Calcium Chloride ($CaCl_2$). This choice avoids the need for a gas-liquid contactor or the storage of CO_2.

Membrane plants for seawater desalination usually produce freshwater with salinities varying from 200 to 500 ppm. However bivalent ions, such as Calcium and Magnesium, are completely removed by reverse osmosis membranes, thus leaving the produced water very soft, and therefore, very aggressive. In order to increase water hardness, Calcium Chloride ($CaCl_2$) plus carbonate salts (e.g. $NaHCO_3$ or Na_2CO_3) can be added, or percolation through a limestone-packed bed can be adopted. Further pH adjustment may also be necessary before the final distribution of freshwater. If membrane desalinated water is to be used for industrial purposes in which high purity water is needed, then a further deionising step may be required. This can be performed by a second RO stack, or by a column of ionic exchange resins.

In all cases, water to be distributed for domestic uses, requires a disinfectant agent with residual effect (like chlorine or hypochlorite) to guarantee the disinfection of water during distribution.

Specific pre- and post-treatment may be necessary in specific situations (for example removal of silica or boron from waters rich in such constituents), but a comprehensive outline of these processes is outside the scope of this book.

More details are given in Chaps. 3 and 4, however for a more in-depth analysis relevant quoted literature should be consulted.

1.6 Outline of Chapters

In the first part of this book (Chaps. 2, 3 and 4), conventional desalination technologies are described. These focus on technological and operating aspects in order to give a complete overview of the current state of desalination processes for the production of freshwater and the potential for future developments and improvements.

Chapter 2 is focused on the description of the main conventional thermal processes, namely Multiple Effect Distillation (MED), stand-alone or coupled with Thermal or Mechanical Vapour Compression (TVC or MVC), and Multi Stage Flash evaporation (MSF). A brief description of each process is given, along with some process and plant design features typical of industrial units. Moreover information on performance parameters and typical operating conditions for each process is provided.

In Chap. 3 the fundamentals of conventional membrane processes for water treatment and seawater desalination are described. An overview of membrane technology and its application to brackish and seawater desalination is presented, along with some insight into the fundamental laws which regulate such processes. The process of Reverse Osmosis is certainly the most important of the membrane processes, but a short description of other processes, i.e. Electro Dialysis and Micro-, Ultra-, Nano-Filtration (often proposed as very efficient pretreatment units in seawater Reverse Osmosis plants), is also given in order to provide a complete overview of membrane water treatment processes available for the various steps of a seawater desalination process.

Chapter 4 presents some real data on the industrial processes described in the first two chapters. Information on historical production trends of each technology and on typical performances of industrial units is presented. Completing the outline of industrial units is an overview of costs, of both the plant and produced water, for the different technologies. This chapter ends the first part of the book, devoted to conventional technologies and providing a concise yet complete overview of the most frequently adopted desalination processes worldwide.

In the second part of the book, novel coupling schemes between desalination and non-conventional energy sources are discussed.

Chapter 5 presents an overview of possible couplings of desalination processes with nuclear energy, through the use of waste heat for thermal processes or electrical energy for powering membrane processes. Nuclear energy is now a well established energy source, providing energy with a relatively low environmental impact with respect to pollutant gas emission, especially when compared to conventional thermal power plants. Nevertheless, the disposal of radioactive waste still constitutes a problem, which influences public opinion and environmentalists to consider nuclear power as an environmentally unfriendly energy source . Yet nuclear power could be an essential energy source in the mid-term, acting as an effective bridge between old conventional power generation technologies and novel renewable energy ones. In this respect Chap. 5 successfully addresses alternative sources of energy for power desalination installations. This leads into the second part of the book which is devoted to the use of Renewable Energy coupled to desalination processes for the production of freshwater.

Chapters 6, 7, 8, 9 and 10 of this book are in fact all devoted to the description of the wide field of desalination powered by renewable energy, ranging from the simplest of units to the newest and most technologically advanced ones.

In Chap. 6, desalination technologies powered by solar thermal energy are presented. The chapter opens with an overview of simple solar still units, which provide freshwater purely by "sun evaporation" and vapour condensation on a glass surface, thus replicating exactly the natural water cycle. Though not very efficient, these units have been proposed as simple stand-alone units to be used, for example, in combination with greenhouses, where glass surfaces are commonly used as roofs. The rest of the chapter is devoted to the coupling of solar thermal energy collectors for the production of heat to power thermal desalination units. An overview of several commercial devices for solar heat collection, together with different coupling schemes, is presented. An example of a solar powered MED unit, with a long operational life, is critically discussed in detail.

Membrane Distillation (MD), a novel technology coupled to solar thermal energy, is presented in Chap. 7. A process description, operating and performance parameters, and possible configurations are presented in the first part of the chapter, while the second part presents some data on successfully operating Solar MD units.

Chapter 8 closes the overview of solar energy powered processes, by presenting potential couplings of desalination technologies with photovoltaic solar energy. The chapter focuses on the analysis of different coupling schemes, taking into account several possible scenarios of isolated sites where freshwater production is a primary need, together with electrical energy production. Thus the use of photovoltaic panels can provide both these primary products, using different coupling schemes suitably adapted to the different needs of local populations.

Chapter 9 introduces the use of wind and wave energy for freshwater production. The most viable schemes provide for the use of Reverse Osmosis units powered by wind or wave energy, giving the highest efficiency in terms of freshwater production. A brief description of wind and wave technologies is given, followed by an outline of possible coupling schemes. Each technology description is also accompanied by several examples of pilot scale operating units, which, especially in the case of Wind-RO, have shown to be reliable and efficient.

The part devoted to Renewable Energy is closed by Chap. 10 where a general overview of all RE-desalination coupling schemes is presented. Minor technologies are also mentioned for completeness, although indications are given on the trends of the most promising processes. A final section on the basic economics of RE-desalination is also provided, with some rough data and indications on cost analysis and potential for improvements in the main RE-desalination technologies which have been presented in this book.

Finally, Chap. 11 outlines the main environmental aspects relating to the production of freshwater via desalination. Focusing mainly on the marine environments affected by the presence of the desalination industry (mainly closed seas such as the Mediterranean Sea, Arabian/Persian Gulf and the Red Sea), this chapter highlights the main aspects of desalination processes which endanger marine life, e.g. by the disposal of concentrated brines containing chemicals, metal ions and high concentration salts. Different scenarios are considered and information on the potential effects of such activities on marine flora and fauna is presented.

Chapter 2
Conventional Thermal Processes

Hisham Ettouney

Abstract Thermal desalination processes account for about 50% of the entire desalination market. The remaining market share is dominated by the reverse osmosis (RO) process. The main thermal desalination processes include multi-stage flash desalination (MSF), multiple-effect distillation (MED), and mechanical vapor compression (MVC). Other thermal desalination processes, e.g., solar stills, humidification dehumidification, freezing, etc., are only found on a pilot or experimental scale. Thermal desalination processes consume a larger amount of energy than RO; approximately the equivalent of 10–15 kWh/m^3 for thermal processes versus 5 kWh/m^{3}for RO. Irrespective of this, the reliability and massive field experience in thermal desalination keeps its production cost competitive compared to the RO process. Also, the large scale production capacity for a single MSF unit, approximately 75,000 m^3/day, is sufficient to provide potable water for 300,000 inhabitants. An increase in production capacity for the MED system has been realised recently, with unit production capacities of up to 30,000 m^3/day. This chapter covers various aspects of thermal desalination processes. It includes a review of design, operating, and performance parameters. The analysis for each process includes a brief review of some of the recent literature studies, process descriptions, process models, and an illustration of system design and performance analysis. The chapter is divided into two parts, the first is on evaporation processes, which includes MED and MVC, and the second is on flashing processes, which include MSF. Each section starts with a description and analysis of the individual stage, for either evaporation or flashing. This is to simplify the explanation of the main processes that takes place during evaporation or flashing. Each division gives a complete description of each desalination process, together with the main modelling equations. Performance charts are presented for each system and explained in terms of main design and operating conditions.

H. Ettouney (✉)
Department of Chemical Engineering, College of Engineering and Petroleum, Kuwait University, P.O. Box 5969 – Safat 13060, Kuwait
e-mail: ettouney@hotmail.com

A. Cipollina et al. (eds.), *Seawater Desalination*, Green Energy and Technology,
DOI 10.1007/978-3-642-01150-4_2, © Springer-Verlag Berlin Heidelberg 2009

2.1 Conventional Thermal Desalination Processes

Multi-stage flash (MSF), multiple-effect distillation (MED) and mechanical vapour compression (MVC) are the main thermal desalination processes. The market shares of these three processes are 87.3, 12.5, and 0.2% for MSF, MED and MVC, respectively [1]. Other types of thermal desalination processes, i.e. solar stills, humidification-dehumidification and freezing, are not found on a commercial scale and are limited to either experimental types or conceptual designs [2].

MSF and MED systems are often constructed in cogeneration plants where power and water are produced simultaneously. This is convenient because both systems require low pressure heating steam which can be easily extracted from the power plant at fairly low cost. The MVC system is operated solely on electric power.

The material presented in this chapter focuses on description, modelling and analysis of the main desalination processes. The discussion starts by explaining the elements of the evaporation and flashing processes. This is followed by a description of the entire flow diagram for each of the three processes. The main performance charts are presented for each process. Discussion of the evaporation systems includes multiple effect distillation combined with thermal and mechanical vapour compression, as well as single effect mechanical vapour compression. The section on the MSF process includes a description of the flashing stage elements, as well as a description and modelling of the MSF process.

2.2 Evaporators

Thermal evaporation is the principal mechanism in generating fresh water vapour from seawater. The evaporation process is based on creating a hot surface using heating steam; the heating steam condenses on one side and vapour is formed on the other. Evaporators include submerged tube, falling films and plates [3].

Figure 2.1 shows a schematic of a two-effect submerged tube evaporator. As shown, heating steam condenses on the wall of the tube in the first effect, and releases its latent heat to a thin layer of liquid surrounding the outside surfaces of the tubes. This results in the formation and release of vapour bubbles, which rise through the liquid and are released into the vapour space. The formed vapour is routed to the second effect, where it condenses on the wall of the tube and results in the formation of a smaller quantity of vapour. The vapour released in the second effect can be either routed to another effect, or condensed against the feed seawater.

Submerged tube evaporators are used in household humidifiers and electric kettles. They were also used in small industrial desalination units during the first half of the twentieth century. These early units were plagued with rapid fouling and scaling of the outside surface of the tubes. This required lengthy and expensive cleaning procedures of the tube bundle.

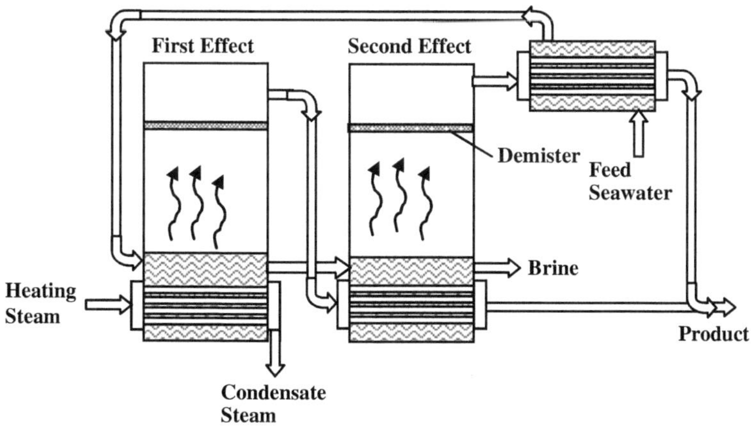

Fig. 2.1 Elements of two-effect submerged evaporator

Another drawback of the submerged tube evaporator is the reduction in the overall heat transfer coefficient, caused by the static head of liquid surrounding the outside surface of the tube. This hinders the formation, growth and release of vapour bubbles.

The falling film configuration eliminates the drawbacks of the submerged tube evaporator [4]. As shown in Figs. 2.2 and 2.3, there are two arrangements for the falling film system which include horizontal or vertical tubes. The horizontal falling film evaporator (Fig. 2.2) eliminates the static pressure effect on the evaporating surface, and as a result, higher overall heat transfer coefficients are obtained. However, the horizontal falling film arrangement necessitates operation at temperatures below 70°C, to limit the scaling rate of the outside surface of the tubes and to reduce the frequency of chemical cleaning. The horizontal falling film configuration is the industry standard and is used in most MED and MVC systems.

Figure 2.3 shows a schematic of a vertical tube falling film evaporator. As shown, the feed seawater forms a thin falling film on the inside surfaces of the tubes. Film formation is more difficult to maintain and control than with horizontal falling film. As a result, dry patches may form and result in a high scaling rate and uneven tube expansion. Although the vertical tube arrangement allows for use of on-line ball

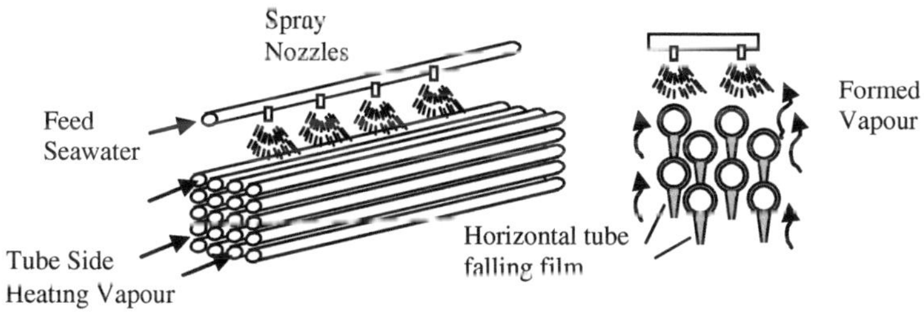

Fig. 2.2 Horizontal tube falling film evaporator

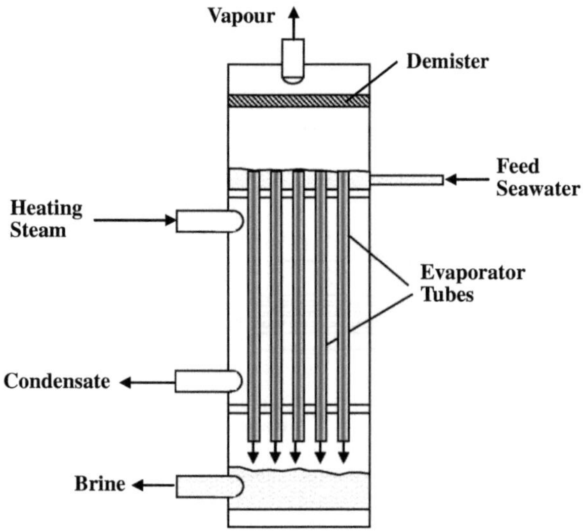

Fig. 2.3 Vertical tube falling film evaporator

cleaning, which would considerably reduce the scaling effects and allow for higher temperature operation, it is used on a very limited scale in the desalination industry.

Plate evaporators have been developed and tested on limited scale [5]. A schematic for the plate evaporator is shown in Fig. 2.4, where the heating steam condenses on one side of the plate, and water evaporates on the other. Plate evaporators can be manufactured using metal, plastic or polymer-coated metal. These plate heat exchangers have lower hold-up volumes, closer temperature approaches, lighter weights, smaller space requirements, higher heat transfer coefficients and lower fouling resistances. Irrespective of the many attractive features of plate evaporators they remain limited to experimental and prototype units [3].

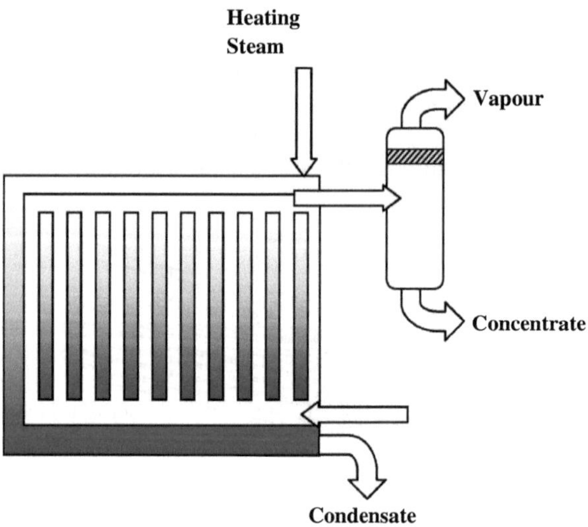

Fig. 2.4 Plate evaporator

2.3 Single-Effect Evaporation

The single-effect evaporation system for seawater desalination has no practical use on an industrial scale. This is because the system has a thermal performance ratio of less than 1, i.e. the mass of water produced is less than the mass of heating steam used to operate the system. However, the understanding of this process is essential as it is a constituent of other single-effect vapour compression systems, as well as multiple-effect evaporation processes [2].

Figure 2.5 shows a schematic diagram for a horizontal tube, falling film, single-effect evaporation system. The main components of the unit are the evaporator, feed preheater or down condenser, the vacuum system and the pumping units. As shown, the intake seawater (M_{cw}+M_f), at temperature T_{cw} and with salt concentration X_f, is introduced into the tube side of the preheater, where its temperature increases to T_f. The cooling water (M_{cw}) is released back into the sea. The function of the cooling water in the condenser is to remove excess heat added in the evaporator by the heating steam. This implies that the evaporator does not consume all the supplied heat, instead, it degrades its quality. The heating of the feed seawater (M_f) in the condenser tubes from T_{cw} to T_f is essential to increase the thermal performance of

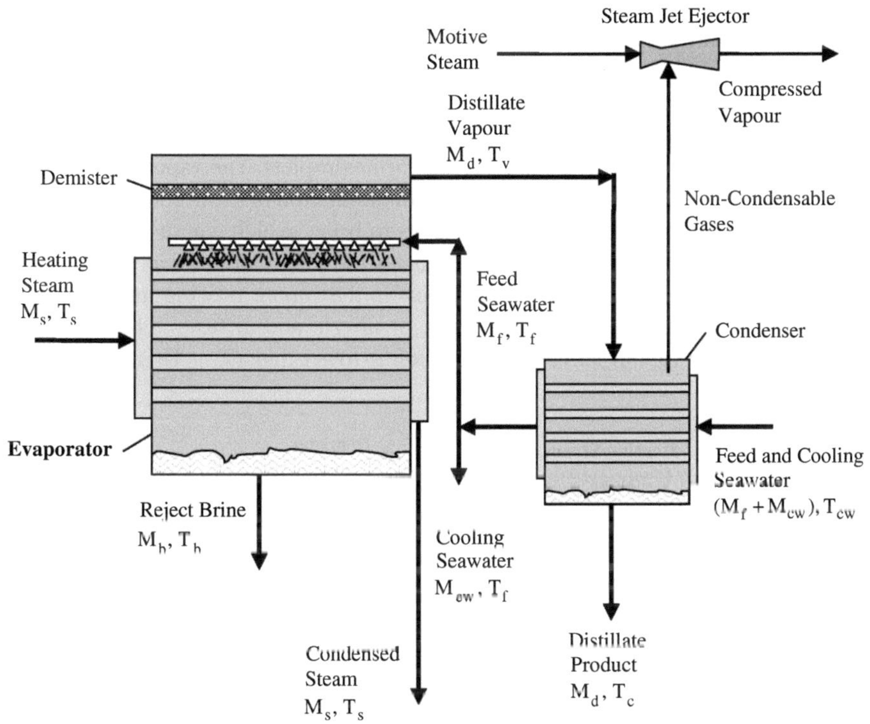

Fig. 2.5 Single-effect evaporation desalination process

the process. The heat needed to warm the seawater inside the condenser tubes is supplied by condensing the vapour formed in the evaporator (M_d).

The vapour condensation temperature and consequently the pressure in the vapour space, for both the evaporator and the condenser, is controlled by the cooling water flow rate (M_{cw}), the feed water temperature (T_{cw}), the available heat transfer area in the condenser (A_c), the overall heat transfer coefficient between the condensing vapour and the circulating seawater (U_c). Accordingly, the condenser has three functions: (1) to remove the excess heat from the system, (2) to improve the process performance ratio, and (3) to adjust the boiling temperature inside the evaporator.

The feed seawater (M_f) is chemically treated and de-aerated before being pumped into the evaporator. This chemical treatment is needed to prevent foaming and the tendency of scale formation in the evaporator. Both factors may seriously impair unit operation. Once inside the evaporator, the feed water is sprayed from the top; it falls in the form of a thin film through the rows of tubes arranged horizontally below. Condensation of the saturated heating steam, and release of its latent heat, provides the required sensible and latent heat for water evaporation from the feed seawater. As a result, the feed water temperature (T_f) is raised to boiling temperature (T_b). The value of $T_b$is mainly dictated by the type of chemicals used to control scale formation and the state of the heating steam. The vapour formed by boiling, with a flow rate of M_d, is completely free of salt. Figure 2.6 shows that the temperature of the generated vapour (T_v) is less than boiling temperature due to boiling-point elevation (BPE). Similarly, the temperature of the condensed vapour (T_d) is lower than the temperature of the generated vapour due to heat losses caused by the demister, transmission lines and condensation.

The generated vapour flows through a knitted wire mist separator, known as the wire mesh demister, to remove the entrained brine droplets. The vapour needs to be completely free of brine droplets to prevent contamination of the product water. This also prevents exposure of the condenser tubes to brine, which can result in scaling, surface corrosion and reduction of heat transfer rates. In thermal vapour compression, the presence of entrained water droplets in the vapour flowing into the steam jet ejector can result in erosion of the ejector nozzle and diffuser. The saturation temperature of the vapour leaving the demister is lower than (T_v). This temperature

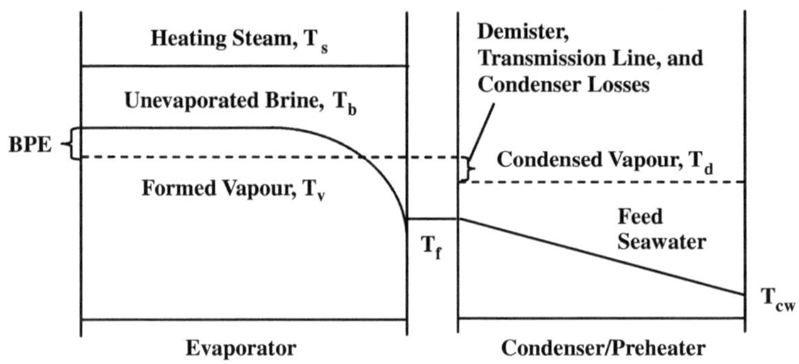

Fig. 2.6 Temperature profiles in evaporator and condenser of the single-effect evaporation system

reduction is caused by the frictional pressure loss in the demister. Further drops in pressure take place during vapour transfer between the evaporator and pre-heater and during vapour condensation. This further decreases the vapour condensation temperature.

The non-condensable gases in the vapour space of the condenser must be continuously vented to avoid downgrading of the heat transfer capacity of the condenser. The blanket of non-condensable gases masks some of the heat transfer area from the condensing vapour. In addition, the non-condensable gases reduce the partial pressure of the condensing vapours. As a result, condensation takes place at a lower temperature. This reduces process efficiency because of the decrease in the net driving force for heat transfer (i.e. the temperature difference between the condensing vapour and the stream, M_f), and consequently reduces the feed seawater temperature (T_f). Removal of these gases is made at points in the system where the temperature approaches its lowest value (i.e. that of cooling water entering the tubes). This permits the cooling of the non-condensable gases to the minimum possible temperature, thereby minimising the amount of vapour which escapes with the gases and decreasing the volume of pumped gases. In addition, it is possible to operate a counter-current condenser so that the exit water is within 3–5°C of the condensation temperature of the saturated vapour. This improves the thermal performance of the unit and minimises the mass flow rate of the cooling water.

2.3.1 Modelling of the Single-Effect Evaporator

Modelling and analysis of the evaporator can be based on a detailed set of equations that utilise a number of correlations to determine the heat transfer coefficient, thermodynamic losses, and physical properties of water and vapour. Solution of the detailed model requires an iterative procedure because of the non-linearity of the model equations and correlations. The model can be simplified by assuming constant physical properties, negligible heat losses to the surroundings, constant thermodynamic losses and a constant overall heat transfer coefficient. This assumption reduces the model to a set of material and energy balance equations, which can be used for design or simulation [6]. The evaporator energy balance is:

$$M_s \lambda_s = M_d \lambda_v + M_f C_p (T_b - T_f) \tag{2.1}$$

The left side of the equation defines the thermal load of the heating steam and the right hand side gives the latent heat of the formed vapour and the sensible heat of the feed seawater. The thermal load of the heating steam is also used to determine the required heat transfer area of the evaporator. This relationship is illustrated by:

$$A_e = \frac{M_s \lambda_s}{U_e \, (T_s - T_b)} \tag{2.2}$$

A similar set of equations can also be defined for the condenser and includes the condenser energy balance and the heat transfer equations:

$$M_d\lambda_d = (M_{cw} + M_f)C_p(T_f - T_{cw}) \quad (2.3)$$

$$A_c = \frac{M_d\lambda_d}{U_c\ LMTD_c} \quad (2.4)$$

In Eq. (2.4), the logarithmic mean temperature difference ($LMTD_c$) is given by the following relation:

$$LMTD_c = (T_{cw} - T_f)/\ln((T_d - T_f)/(T_d - T_{cw})) \quad (2.5)$$

Where it is worth noting that the distillate temperature Td is lower than the brine temperature due to thermodynamic losses. The remaining model equations define the overall mass balance and the salt mass balance of the entire system:

$$M_f = M_d + M_b \quad (2.6)$$

$$X_f M_f = X_b M_b \quad (2.7)$$

Other important design aspects include the tube wetting rate, the velocity of the vapour inside the evaporator tubes and the water velocity inside the condenser tubes [7]. The tube wetting rate (WR) is defined as the feed flow rate per unit length of the top row of the tubes. This relationship is defined by:

$$WR = M_f/(n_{rt}L_t) \quad (2.8)$$

where n_{rt} is the number of tube rows, and L_t is the tube length.

The wetting rate should remain within a range of 0.03–0.14 kg/(m s); lower values result in formation of dry spots and higher values affect the thermal characteristics of the system. The vapour velocity inside the evaporator tubes is given by:

$$V_s = \frac{M_s \cdot v_s}{n_e \cdot \frac{\pi d_e^2}{4}} \quad (2.9)$$

where v_s is the steam specific volume, n_e is the total number of tubes in the evaporator and d_e is the tube diameter.

The vapour velocity may vary within a range of 20–50 m/s. A similar relationship is used to determine the water velocity inside the condenser tubes:

$$V_f = \frac{(M_f + M_{cw})}{\rho_{cw} \cdot n_c \cdot \frac{\pi d_c^2}{4}} \quad (2.10)$$

where ρ_{cw} is the cooling water density, n_c is the total number of tubes in the condenser and d_c is the tube diameter. The water velocity may vary within a range of 1–3 m/s. Lower velocities can result in increased rates of fouling and scaling because of the increase in the contact time between the water and the tube wall. The upper limit on water velocity is set by the capacity of the feed pumps.

2.4 Multiple-Effect Distillation (MED)

The multiple-effect distillation process can be found in various industries, i.e. sugar, paper and pulp, dairy, textiles, acids and desalination. Small MED plants with capacities of less than 500 m^3/day were introduced to the desalination industry in the 1960 s. Subsequent developments lead to the increase in unit production capacity. In 2006, MED capacity increased to a value of 36,000 m^3/day [1]. Most MED processes operate at low temperatures, of less than 70°C. This is because the evaporators adopt a horizontal film configuration, where the feed seawater is sprayed on the outside surface of the tubes. Therefore low temperature operation limits the rate of scale formation on the outside surface of the evaporator tubes. Moreover, operation at low temperatures allows for efficient combination with thermal or mechanical vapour compression. The vapour compression process has been developed to improve the process performance ratio (kg product/kg heating steam) to values close to 16 for a twelve-effect system. The performance ratio drops to a value of 8 if the system operates with no vapour compression [8]. Prototype units of MED, combined with lithium bromide absorption vapour compression, give a performance ratio of more than 20 [9]. On a commercial scale, most MED systems are designed to operate in either standalone mode or in combination with thermal vapour compressors (MED/TVC). Mechanical vapour compression systems (MED/MVC) are found on a much more limited scale.

Ophir and Lokiec [10] presented a recent evaluation of the MED process and how its economics are superior to other desalination processes. As mentioned before, low temperature operation allows for highly efficient thermal vapour compression. It also allows for use of low grade energy. Another advantage is the use of relatively inexpensive construction material, which includes aluminium alloys for the heat transfer tubes, as well as carbon steel epoxy coated shells, for the evaporator shells. Ophir and Lokiec [10] reported water cost of \$0.54/m^3 for a plant of 5 units, each producing 20,000 m^3/day.

2.4.1 MED Process Description

Industrial MED systems include up to 12 evaporation effects, where evaporation in the first effect is driven by heat steam extracted from cogeneration boilers. The vapour formed in the first effect is used to drive evaporation in the second effect. This process continues in subsequent effects until the vapour temperature drops to about 30–40°C. Most industrial MED systems are designed to operate in dual mode, i.e. standalone, where it is driven by heating steam from the boiler, or in a thermal vapour compression mode, where part of the vapour formed in the last

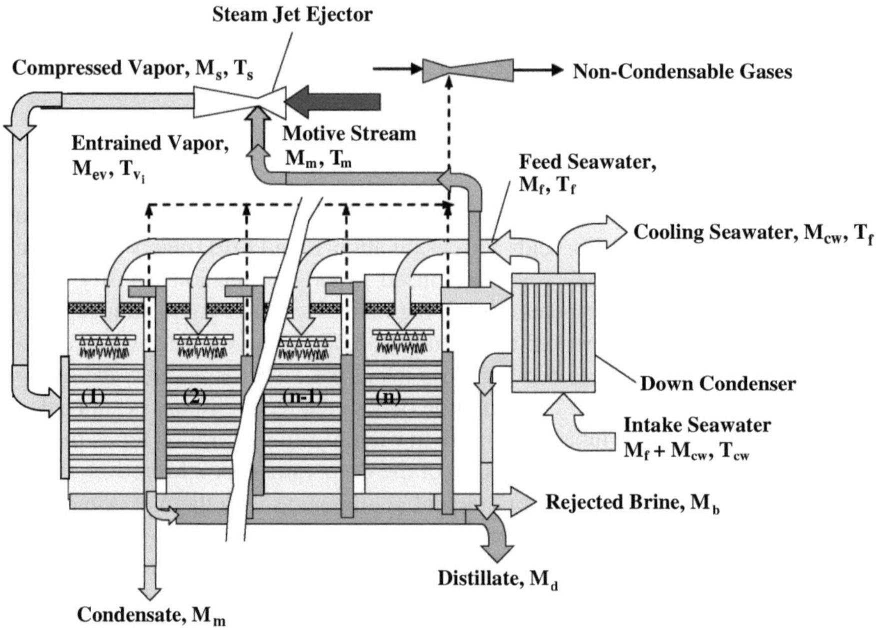

Fig. 2.7 Schematic of MED/TVC with thermal vapour compression

effect is compressed to the desired temperature and used to drive evaporation in the first effect. Figure 2.7 shows a schematic for a thermal vapour compression system (MED/TVC).

A small proportion of the MED system utilises a mechanical vapour compression configuration, where the entire vapour formed in the last effect is compressed mechanically to the desired temperature, and is used to drive evaporation in the first effect. Limitations in the compression capacity of the mechanical vapour compressor limit the number of effects in the system to less than 6.

Figure 2.8 shows a schematic for an MED/MVC process. MED/MVC systems, as shown in Fig. 2.8, have similar layouts to thermal vapour compression processes. The main differences are the absence of a down condenser and the use of a mechanical compressor, to compress the entire vapour formed in the last effect to the desired heating steam temperature. In addition, the outlet brine and distillate streams exchange heat with the feed stream in two pre-heaters.

2.4.2 Modelling and Design of MED and Vapour Compression

The MED model is based on the same set of equations used to model the single-effect evaporation process. The following analysis utilises the detailed model, which includes detailed correlations for heat transfer coefficients, physical properties and thermodynamic losses [7]. The model is used to design 30,000 m^3/day MED and

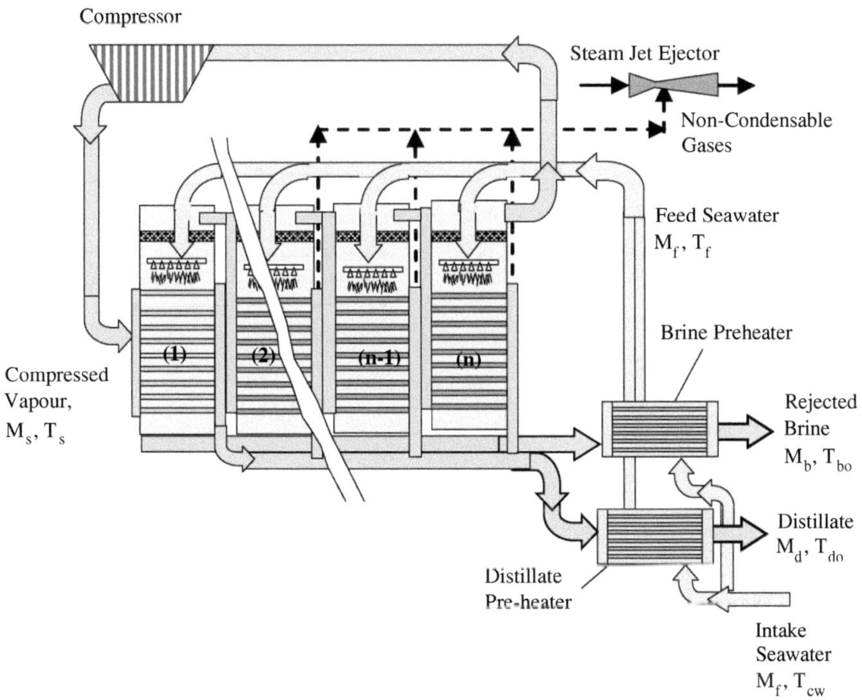

Fig. 2.8 Schematic of MED/MVC with mechanical vapour compression

MED/TVC systems. The calculations are performed for 8, 10, and 12 effects and heating steam temperatures of 65, 70 and 75°C. The system performance ratios and specific heat transfer areas are shown in Figs. 2.9 and 2.10, respectively.

Figure 2.9 shows the increase in performance ratio as the number of effects increases. This is caused by the constant production capacity used in all configurations, which results in a reduction in the amount of vapour formed in each effect, as the number of effects is increased. Therefore, the amount of heating steam required to drive evaporation in the first, reduces with the amount of vapour formed in each effect.

Figure 2.10 shows that the specific heat transfer area increases with the increase in the number of effects, and decreases with the increase in heating steam temperature. The increase in the specific heat transfer area, with the increase in the number of effects, is caused by a reduction in the temperature drop from one effect to the next, which reduces the driving force for heat transfer. This is because the temperature difference between the first and last effect is kept constant in all calculations. Increasing the heating steam temperature reduces the specific heat transfer area as the brine blow down temperature is kept constant in all calculations. Therefore the temperature drop from one effect to the next increases and results in an increase in the driving force for heat transfer in each effect.

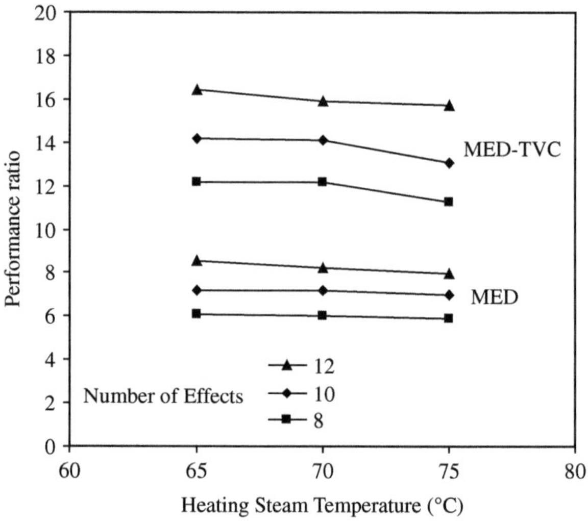

Fig. 2.9 Variations in performance ratio as a function of heating steam temperature and number of effects

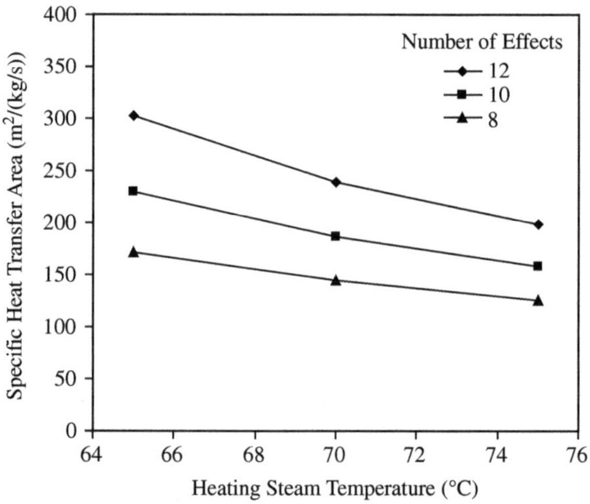

Fig. 2.10 Variations in specific heat transfer area as a function of heating steam temperature and number of effects

2.5 Single-Effect Mechanical Vapour Compression (MVC)

The MVC system was introduced in the 1980 s. Early studies by Matz and Fisher [11], Lucas and Tabourier [12] and Matz and Zimerman [13] were motivated by the need to develop a thermal desalination process driven solely by electrical power. The MVC process was pursued as a competitor to the newly introduced RO technology. However, the test of time has shown the dominance of the RO process and a very limited growth in the reliability of MSF and MED processes.

MVC unit capacity is very small in comparison with MSF and MED. Current unit capacity is below 5,000 m^3/day. However, recent conceptual designs by Kronenberg and Lokiec [14] demonstrate the feasibility of constructing units with 10,000 m^3/day capacity. The small size, and the fact that only electrical power is required, makes it feasible to operate using various forms of renewable energy, i.e. wind, photovoltaics, etc. On average MVC consumes 10–14 kWh/ $m^3$of electrical power, to operate the system and other associated equipment including pumps, controls and auxiliaries. A 5,000 m^3/day MVC would thus require 2–3 MW of electrical power, which can be easily provided by available renewable energy technologies.

2.5.1 Description of the MVC Process

Schematics for an MVC process are shown in Figs. 2.11 and 2.12. As shown in Fig. 2.11, the system contains a horizontal tube evaporator, spray nozzles, a vapour compressor, a recycle pump and plate preheaters. A cross section of the tube arrangement, spray nozzles, demister and compressor intake are shown in Fig. 2.12. As shown in Fig. 2.11, the heating steam or compressed vapour flows within the tubes and the brine is sprayed on the outside surface of the tubes. To limit the rate of scale formation on the outside surface of the tubes, the maximum possible saturation temperature of the compressed vapour is 70°C. As shown in Fig. 2.12, the tube bank is divided into two groups on either sides of the demister, which is placed in

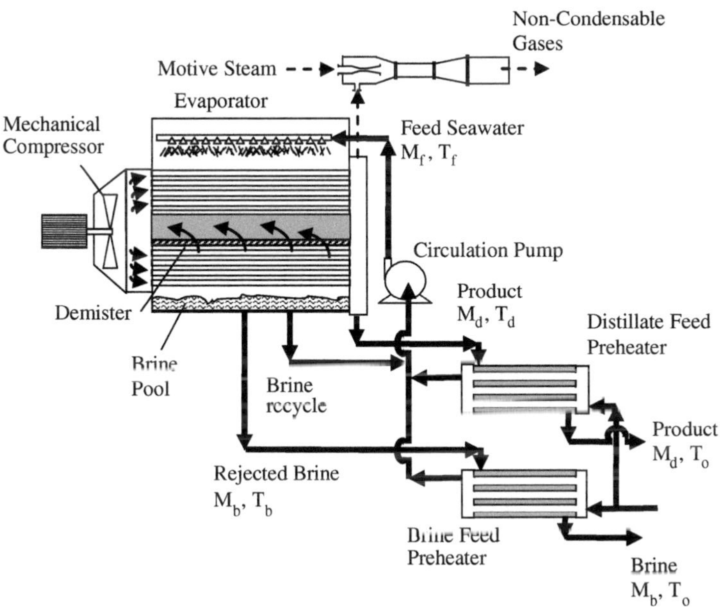

Fig. 2.11 Single-effect mechanical vapour compression

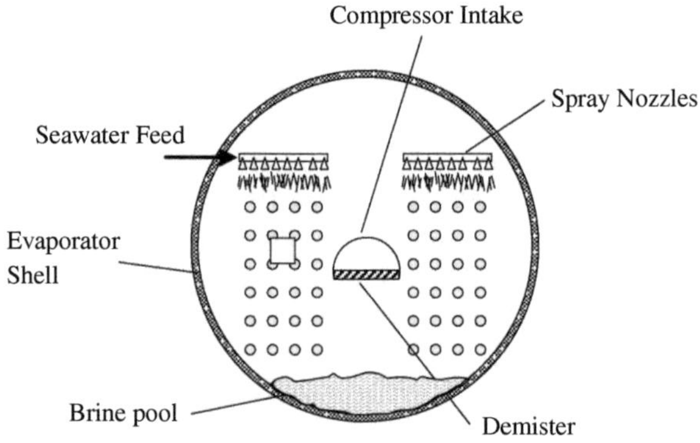

Fig. 2.12 Cross section of the evaporator showing two tube banks with square pitch, spray nozzles, demister, compressor intake and brine pool

the centre of the evaporator. The demister arrangement is connected directly to the compressor intake, where the formed vapour is compressed and superheated to the desired temperature. The feed water entering the system is pre-heated in two plate exchangers using the distillate condensate and the brine reject. Plate preheaters are compact and allow for a small approach temperature. Brine circulation takes place within the evaporator to achieve the desired wetting rate, i.e. the mass of sprayed water per unit length of top row tubes should be within a range of 0.03–0.14 kg/(m s). This is necessary to prevent formation of dry spots or flooding within the system [15]. Accumulation of non-condensable gases within the evaporator is controlled by the use of a jet ejector. Pumping units used in the system include pumps for the feed, distillate, brine reject and brine recycle.

2.5.2 Modelling and Design of MVC

MVC modelling focuses on the determination of the required heat transfer area for the evaporator and feed preheaters, power capacity of the compressor, brine circulation flow rate, tube arrangement, shell diameter, pumping power and capacity of the venting system. Commonly used assumptions in model development are: steady state conditions, distillate is salt-free, negligible heat losses to the surroundings and negligible vapour losses in the venting line [16]. Details of the model equations, correlations and solution method can be found in various studies by Darwish [17], El-Dessouky and Ettouney [2], Ettouney [16], Ettouney et al. [7].

The performance of MVC is illustrated in terms of variations in the specific heat transfer area and the specific power consumption, and as a function of brine temperature and the temperature difference between distillate condensate and brine (Figs. 2.13 and 2.14). As shown in Fig. 2.13, the specific heat transfer area has a negligible dependence on brine temperature. This effect is caused by the limited variation in the overall heat transfer coefficient that occurs upon varying the brine temperature. However, an increase in the temperature difference between the distil-

Fig. 2.13 Variations in specific heat transfer area of the MVC system as a function of brine temperature and the temperature difference between distillate brine

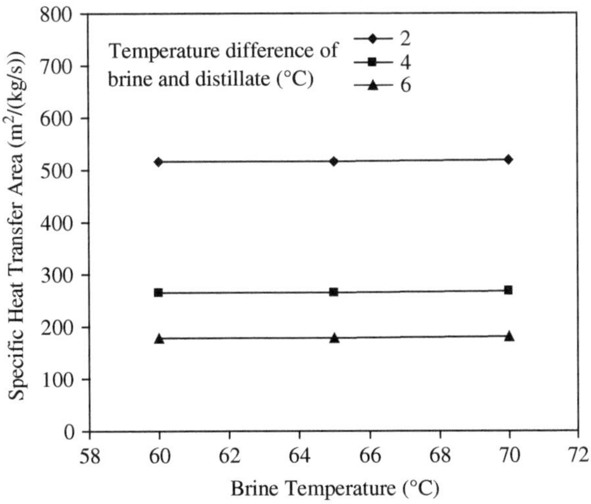

late and brine stream considerably reduces the heat transfer area. This is because of the increase in the driving force for heat transfer between the condensed vapour and the evaporating brine.

Variations in the specific power consumption for the system are shown in Fig. 2.14. The specific power consumption has little dependence on brine temperature, where the specific power consumption decreases with increases in brine temperature. This is caused by the decrease in the specific volume of the intake vapour, which in turn reduces the required compression power. The specific power consumption has a much greater dependence, however, on the temperature difference between distillate and brine. This is caused by the increase in the compression ratio of the vapour.

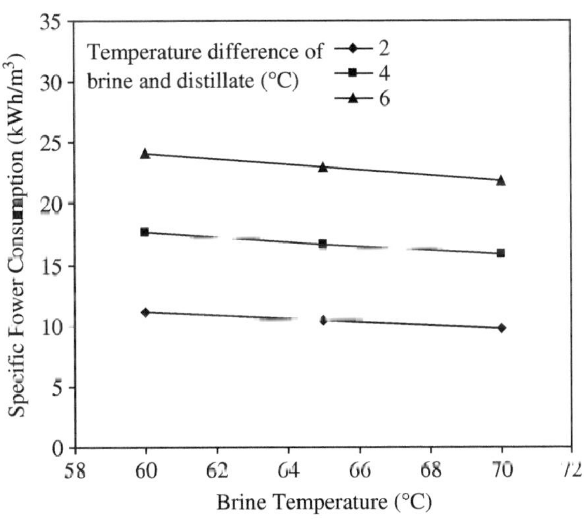

Fig. 2.14 Variations in specific power consumption for the MVC system, as a function of brine temperature and temperature difference between distillate and brine

A review of the data shown in Figs. 2.13 and 2.14 shows the need to optimise the temperature difference between brine and distillate. Operation at small differentials would result in large heat transfer areas; on the other hand, increases in the temperature differentials cause large increases in specific power consumption. The heat transfer area and the consumed power have significant effects on the product cost.

2.6 Multi-Stage Flash Desalination (MSF)

The MSF desalination process was introduced in the early 1950 s. In 1957, Silver patented the multi-stage flash desalination process [18]. The patent optimised the number of flashing stages and the heat transfer area. Since then, the MSF process has gone through several dramatic modifications and improvements, which have lead to a massive increase in unit capacity, from 500 m^3/day in the 1960 s to 75,000 m^3/day in the 1990 s [19]. Other developments include the use of demisters in all flashing stages, which limits entrainment rates of brine by the flashed-off vapour. As a result, product salinity is maintained below 10 ppm. Also, development of the on-line ball cleaning system has resulted in less frequent use of acid cleaning and plant shutdown. Currently, MSF plants can be operated for periods varying from 2 to 5 years before a major overhaul is necessary [20]. Recent field experience shows that a large number of old MSF plants are being rehabilitated to improve performance and extend service life [21, 22].

2.6.1 MSF Flashing Chamber

The main element in the MSF process is the flashing chamber. A schematic of the MSF flashing stage is shown in Fig. 2.15 and includes the following items.

A large brine pool with a similar width and length to the flashing stage and with a depth of 0.2–0.5 m.

A brine transfer device, comprising a weir and splash plate combination between the stages, is designed to seal the vapour space between the stages and to enhance turbulence and mixing of the inlet brine stream. This device promotes flashing by controlling the formation of vapour bubbles, their growth and subsequent release.

A demister formed of wire mesh layers and supporting system. The demister function is to remove the entrained brine droplets from the flashed-off vapour. This is essential to prevent an increase in the salinity of the product water and scale formation on the outer surface of the condenser tubes.

A tube bundle of condenser/pre-heater tubes, where the flashed-off vapour condenses on the outer surface of the tubes. The released latent heat of condensation results in heating of the brine recycle stream flowing inside the tubes. This energy recovery is essential to maintain high system performance.

A distillate tray, where the condensed distillate product is collected and cascades through the stages. The distillate product is withdrawn from the tray in the last stage.

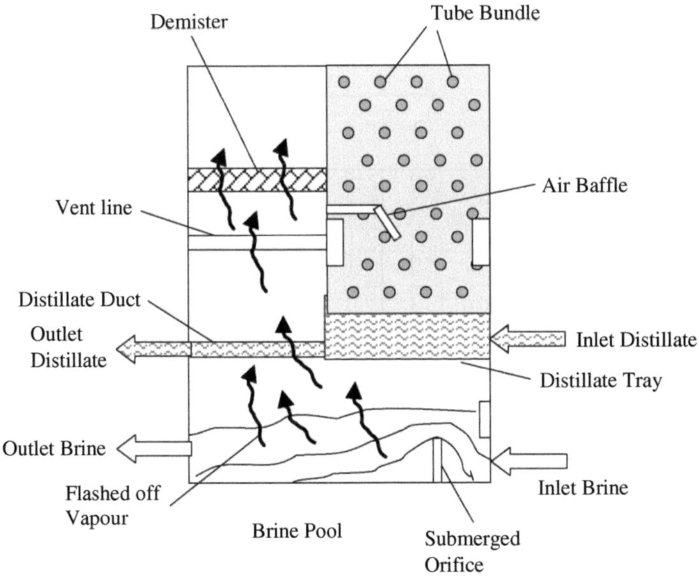

Fig. 2.15 MSF flashing stage

Water boxes at both ends of the tube bundle to transfer the brine recycle stream between adjacent stages.

Connections for the venting system, which remove non-condensable gases (O_2, N_2 and CO_2), which are dissolved in the feed seawater, even after de-aeration. CO_2 can also be generated during decomposition of bicarbonate compounds in the high temperature stages. Another major source of non-condensable gases is air in-leakage from the ambient surroundings into flashing stages operating at temperatures below 100°C, which correspond to vacuum conditions.

Instrumentation, which includes thermocouples, a level sensor and a conductivity meter, is placed in the first and last flashing stages. The data measured at these stages is used by the process control system. Accordingly, and subject to disturbances in the system parameters, i.e. feed seawater temperature, increase in fouling thermal resistance, available steam, etc., adjustments are made in the controllers to maintain the desired operating conditions. The magnitude of these adjustments depends on the measurements made in the first and last stages.

2.6.2 *MSF Processes*

There are two main layouts for the MSF process. The first is the once-through system and the second is the brine circulation system. The brine circulation system is to be found on a larger scale than the once-through system. Figure 2.16 shows a schematic for the MSF once-through process. As shown, the system includes brine heater, flashing stages, vacuum ejector, chemical addition pumps and feed screens [23].

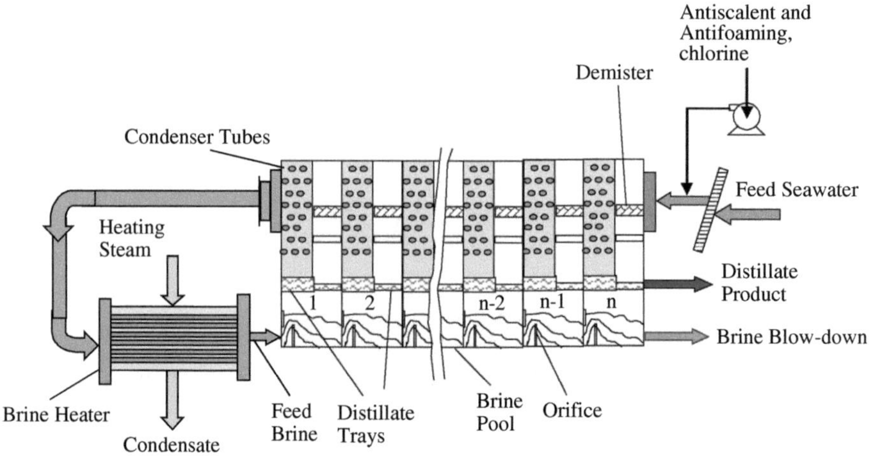

Fig. 2.16 Once-through multi-stage flash process

In the MSF process the separation is achieved by evaporating some of the feed in each stage by flashing. In fact, the hot feed seawater entering each stage encounters a lower pressure than its own vapour pressure; it then flashes off producing vapour on one side and so cooling the brine flowing on the other side. This results in a decrease in brine temperature and an increase in its salinity, stage by stage. Such a flashing process is repeated, stage after stage, due to the continuous decrease in stage pressure, afforded by the drops in pressure of the brine flowing through the brine orifices. Vapour produced at each stage passes through a demister, thus dramatically reducing the entrainment of brine droplets, and condenses on the external surface of the tube bundle. Latent heat of condensation is transferred to the feed seawater, which flows through the condenser tubes from stage *n* to stage 1. The feed seawater temperature increases upon condensation of the flashed-off vapour on the outside surface of the condenser tubes. A further increase in the temperature of feed seawater exiting from the 1st stage condensing tubes takes place in the brine heater. Here, heating takes place by the use of saturated steam, which is usually extracted from the low pressure turbines of a power generation system.

The brine circulation system is shown in Fig. 2.17. As shown, the flashing stages are grouped in two sections, i.e. the heat recovery and heat rejection sections. The heat recovery section is almost identical to the flashing stages of the once-through system. The heat rejection section usually contains only two to three stages and is designed to control the temperature of the intake seawater and reject the excess heat added by the brine heater. As is shown, the flow rate of the intake seawater is equal to the sum of the flow rates of the feed seawater and the cooling seawater. During winter operation, part of the rejected cooling seawater is recycled and is mixed with the intake seawater. This is to control the intake seawater temperature. This practice is common in most brine circulation MSF plants, especially when the winter season is long. Control of the feed seawater temperature prevents a reduction in the temperature in the last system stage. This would otherwise result in an increase

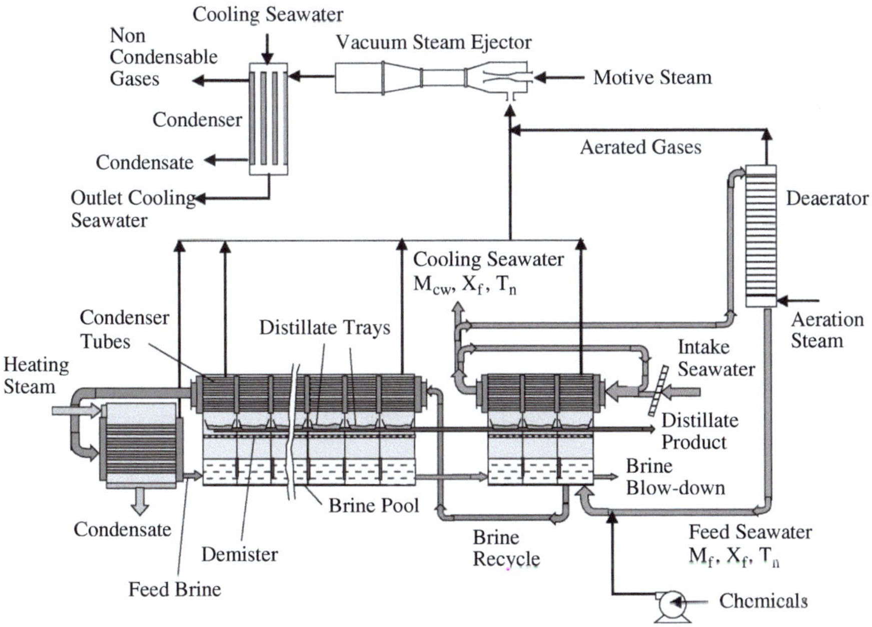

Fig. 2.17 Brine circulation multi-stage flash process

of the specific volume of the flashed-off vapour and a subsequent increase in vapour velocity, as well as an increase in the amount of entrained brine. This in turn would cause an increase in product salinity and the product stream might not be suitable for use as makeup boiler water. Instead, part of the distillate product in the first stages should be used.

De-aeration of the feed seawater is an essential element in the MSF brine circulation system. It removes dissolved gases from the feed stream, i.e. oxygen, nitrogen, and carbon dioxide. If these gases are not removed, they will be released in the flashing stages. The released gases have low thermal conductivity and would reduce the heat transfer rate around the condenser tubes. Carbon dioxide and oxygen may also promote corrosion reactions in various locations in the flashing stages. The deaerator may have a vertical or horizontal configuration equipped with spray nozzles or trays. De-aeration is performed by heating steam, which results in an increase in the feed temperature and as a result reduces gas solubility in the feed water.

2.6.3 Modelling and Design of MSF

The MSF model assumptions are similar to those for the MVC and MED systems. The system model can be simplified or made more complex depending on the assumptions used to define the heat transfer coefficient, thermodynamic losses and physical properties. Basic model equations have been developed to describe the material and energy balance in each flashing stage.

These equations include:

- global mass balance,
- salt mass balance
- energy balance for the flashing brine,
- energy balance for the condenser tubes,
- heat transfer equation for the condenser tubes.

They can be described as follows:

$$B_{j-1} = B_j + D_j \tag{2.11}$$

$$X_{b_{j-1}} \cdot B_{j-1} = X_{b_j} \cdot B_j \tag{2.12}$$

$$D_j \cdot \lambda_j = B_{j-1} \cdot C_{pb_j} \cdot \left(T_{b_{j-1}} - T_{b_j}\right) \tag{2.13}$$

$$D_j \cdot \lambda_{c,j} + C_{pd} \cdot \left(T_{c_{j-1}} - T_{c_j}\right) \sum_{k=1}^{j-1} D_k = M_f \cdot C_{pf_j} \cdot \left(T_{f_j} - T_{f_{j+1}}\right) \tag{2.14}$$

$$M_f \cdot C_{pf_j} \cdot \left(T_{f_j} - T_{f_{j+1}}\right) = U_{c_j} \cdot A_c \cdot (LMTD)_{c_j} \tag{2.15}$$

Other model equations for the brine heater, which include energy balance and heat transfer equations, are as follows:

$$M_s \cdot \lambda_s = U_h \cdot A_h \cdot (LMTD)_h \tag{2.16}$$

$$(LMTD)_h = \frac{\left(T_s - T_{b_0}\right) - \left(T_s - T_{f_1}\right)}{\ln\left(\frac{\left(T_s - T_{b_0}\right)}{\left(T_s - T_{f_1}\right)}\right)} = \frac{\left(T_{f_1} - T_{b_0}\right)}{\ln\left(\frac{\left(T_s - T_{b_0}\right)}{\left(T_s - T_{f_1}\right)}\right)} \tag{2.17}$$

$$M_s \cdot \lambda_s = M_f \cdot C_{p_h} \cdot \left(T_{b_0} - T_{f_1}\right) \tag{2.18}$$

Further model details can be found in studies by Abdel-Jabbar, et al. [24], El-Dessouky and Ettouney [2], and Ettouney et al. [23]. These studies give variations in the system performance ratio, specific heat transfer area, and the specific flow rates of cooling water and brine recycle. The analysis is performed as a function of maximum brine temperature (temperature of brine entering the first stage), or Top Brine Temperature (TBT), and number of stages.

The design and analysis presented here is limited to variations in the specific heat transfer area as a function of system production capacity and maximum brine temperature. Additional design data and performance analysis can be found in the studies which have been previously cited. These studies include evaluation of variations in the system performance ratio and specific flow rates of the cooling seawater and brine recycle, as well as stage dimensions. In addition, there are other models and

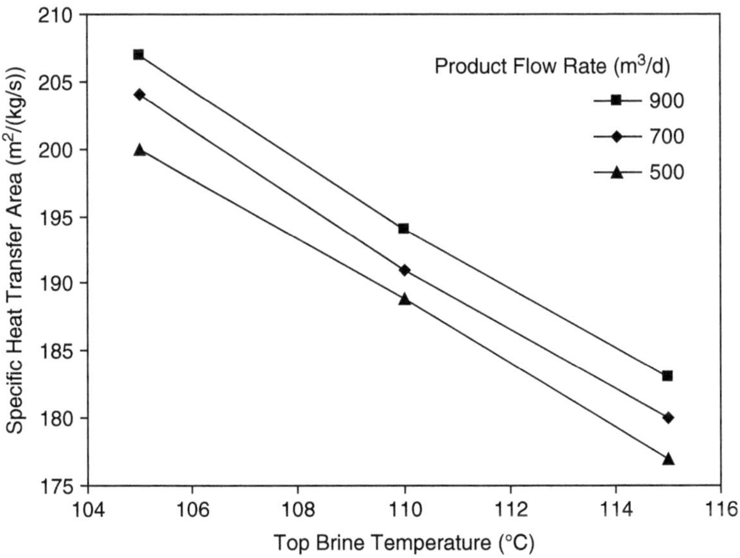

Fig. 2.18 Variations in specific heat transfer area as a function of top brine temperature and product flow rate

analyses which focus on thermo-economic optimisation [25, 26], system dynamics [27] and control [28].

Variations in the specific heat transfer area are shown in Fig. 2.18. As shown, the specific heat transfer area decreases with increase in the maximum brine temperature. This is because of the increase in flashing range and the temperature drop between one stage and the next, which increases the driving force for heat transfer. Also, the specific heat transfer area increases with the increase in production capacity. This is due to the increase in the thermal load of the system.

Nomenclature

A	Heat transfer surface area, m^2
B	Brine flow rate, kg/s
Cp	Specific heat at constant pressure, kJ/kg °C
d	Diameter, m.
D	Flow rate of distillate formed by evaporation, in the ith effect, kg/s
L	Length, width or thickness, m
LMTD	Logarithmic mean temperature difference, °C
M	Mass flow rate, kg/s
n	Number of tubes
T	Temperature, °C
U	Overall heat transfer coefficient, kW/m2 °C

v	Specific volume, m3/kg
V	Velocity, m/s
WR	Wetting rate, kg/(m s)
X	Salinity, ppm

Greek Symbols

λ	Latent heat of vaporization, kJ/kg
ρ	Density, kg/m3

Subscripts

b	Brine
c	Condenser or condensate
cw	Cooling water or intake seawater
d	Distillate
e	Evaporator
f	Feed seawater
h	Brine heater
j	jth stage
o	Brine leaving the brine heater
rt	Number of tube rows
s	Heating steam
t	tube
v	Saturated vapour

Abbreviations

IDA	International Desalination Association
LMTD	Logarithmic Mean Temperature Difference
MED	Multiple-Effect Distillation
MSF	Multi-stage Flash
MVC	Mechanical Vapour Compression
TBT	Top Brine Temperature
TVC	Thermal Vapour Compression

References

1. IDA, Int. Desalination Association, Worldwide Desalting Plants Inventory, 2006.
2. El-Dessouky, H.T., and Ettouney, H.M., Fundamentals of Salt Water Desalination, Elsevier, New York, USA, 2002.

3. Tonner, J.B., Hinge, S., and Legorreta, C., Plates – the next breakthrough in thermal desalination, Desalination, 134 (2001) 205–211.
4. Yang, L., and Shen, S., Experimental study of falling film evaporation heat transfer outside horizontal tubes, Desalination, 220 (2008) 654–660.
5. Renaudin, V., Kafi, F., Alonso, D., and Andreoli, A., Performances of a three-effect plate desalination process, Desalination, 182 (2005) 165–173.
6. El-Dessouky, H.T., Alatiqi, I., and Ettouney, H.M., Process synthesis: The multi-stage flash desalination system, Desalination, 115(1998) 155–179.
7. Ettouney, H.M., Abdel-Jabbar, N., Mjalli, F.S., and Qiblawy, H., Development of Web Based Computer Package for Simulation of Thermal and Membrane Desalination Processes, final report, Project ID MEDRC-04-AS-001, MEDRC, Muscat, Oman, 2008.
8. Temstet, C., Canton, G., Laborie, J., and Durante, A., A large high-performance MED plant in Sicily, Desalination, 105 (1996) 109–114.
9. de Gunzbourg, J., and Larger, D., Cogeneration applied to very high efficiency thermal seawater desalination plants, Desalination 125 (1999) 203–208.
10. Ophir, A., and Lokiec, F. Advanced MED process for most economical sea water Desalination, Desalination 182 (2005) 181–192.
11. Matz, R., and Fisher, U., A comparison of the relative economics of sea water desalination by vapor compression and reverse osmosis for small to medium capacity plants, Desalination, 36 (1981) 137–151.
12. Lucas, M., and Tabourier, B., The mechanical vapour compression process applied to seawater desalination: a 1500 ton/day unit installed in the Nuclear Power Plant of Flamanville, France, Desalination, 52 (1985) 123–133.
13. Matz, R., and Zimerman, Z., Low-temperature vapour compression and multi-effect distillation of seawater. Effects of design on operation and economics. Desalination, 52 (1985) 201–216.
14. Kronenberg, G., and Lokiec, F., Low-temperature distillation processes in single- and dual-purpose plants, Desalination, 136 (2001) 189–197.
15. Han, J., and Fletcher, L., Falling film evaporation and boiling in circumferential and axial grooves on horizontal tubes, Ind. Eng. Chem. Process Des. Dev., 24 (1985) 570–597.
16. Ettouney, H.M., Design of single effect mechanical vapor compression, Desalination, 190 (2006) 1–15.
17. Darwish, M.A., Thermal analysis of vapor compression desalination system, Desalination, 69 (1988) 275–295.
18. Silver, R.S., Multi-stage flash distillation – the first 10 years, 3rd Int. Sym. On Fresh Water from the Sea, Athens, Greece, 1 (1970) 191–206.
19. Borsani, R., and Rebagliati, S., Fundamentals and costing of MSF desalination plants and comparison with other technologies, Desalination 182 (2005) 29–37.
20. Al-Falah, E., Al-Shuaib, A., Ettouney, H.M., and El-Dessouky, H.T., On-Site training program in desalination plants, Eu. J. Eng. Edu., 26 (2001) 407–418.
21. Thirumeni, C., Deutsche Babcock Rehabilitation and uprating of Ras Abu Fontas MSF, desalination units: process optimisation and life extension, Desalination 182 (2005) 63–67.
22. Helal, A.M., Uprating of Umm Al Nar East 4-6 MSF desalination plants, Desalination 159 (2003) 43–60.
23. Ettouney, H.M., El-Dessouky, H.T., and Al-Juwayhel, F., Performance of the once through multistage flash desalination, Proc. Inst. Mech. Eng. Part A, Power and Energy, 216 (2002) 229–242.
24. Abdel-Jabbar, N.M., Qiblawey, H.M., Mjalli, F.S., and Ettouney, H., Simulation of Large Capacity MSF Brine Circulation Plants, Desalination, 204 (2007) 501–514.
25. Fiorini, P., and Sciubba, E., Thermoeconomic analysis of a MSF desalination plant, Desalination, 182 (2005) 39–51
26. Cipollina, A., Micale, G., and Rizzuti, L., Investigation of flashing phenomena in MSF chambers, Desalination, 216 (2007) 183–195

27. Bogle D., Cipollina A., and Micale G., Dynamic modelling tools for solar powered desalination processes during transient operations, in Solar Desalination for the 21st Century (Eds. L. Rizzuti, H.M. Ettouney, A. Cipollina), Springer. ISBN: 978-1-4020-5506-5 (2007).
28. Tarifa, E.E., Domínguez, S.F., Humana, D., Martínez, S.L., Nunez, A.F., and Scenna, N.J., Faults analysis for MSF plants, Desalination, 182 (2005) 131–142.

Chapter 3
Membranes for Desalination

Efrem Curcio and Enrico Drioli

Abstract This chapter provides a general overview of the fundamentals of membrane processes currently under operation in desalination plants, including Microfiltration (MF), Ultrafiltration (UF), Nanofiltration (NF), Reverse Osmosis (RO) and Electrodialysis (ED). The first part provides a description of the preparation techniques for porous, dense or thin film composite membranes made from polymeric materials, along with a technical comparison of the different membrane modules used in filtration processes. The basic transport phenomena relating to mass transfer for the different membrane separation units are also illustrated. In the second part, attention is focussed on analysis of the critical operational issues, with emphasis on fouling and related actions to prevent or control it by appropriate cleaning methods or suitable pretreatment (conventional or membrane-based).

3.1 Membrane Desalination Technology

Today the growing need for water sources is considered a key element in sustainable development, and possible solutions to this holistic problem can be found only by rational integration and implementation of new industrial, economical, environmental and social strategies. Membrane operations, with their intrinsic characteristics of efficiency and operational simplicity, high selectivity and permeability for the transport of specific components, compatibility between different membrane operations in integrated systems, low energy requirements, good stability under operating conditions, environment-compatibility, easy control and scale-up and great flexibility, can offer an interesting answer for the rationalisation of desalination processes [1].

E. Curcio (✉)
Department of Chemical Engineering and Materials – University of Calabria, Via P. Bucci CUBO 44A 87030 Rende (CS) Italy; Institute on Membrane Technology ITM-CNR, c/o University of Calabria, Via P. Bucci CUBO 17C – 87030 Rende (CS) Italy
e-mail: e.curcio@unical.it

A. Cipollina et al. (eds.), *Seawater Desalination*, Green Energy and Technology,
DOI 10.1007/978-3-642-01150-4_3, © Springer-Verlag Berlin Heidelberg 2009

Currently membranes are frequently used in water desalination; large scale desalination plants are under construction, or will be realised in the coming years, making pressure-driven membrane systems the leading technology in this strategic area. Nowadays, around the world, there are about 12,500 sea and/or brackish water desalination plants; installations based on Reverse Osmosis (RO) technology account for around 50% of the total desalination capacity. However, RO is not the only technology much used in this field; Microfiltration, Ultrafiltration, Nanofiltration and Membrane Contactors will be more and more utilised in the pre- and post- treatment phases, so realising integrated membrane systems. These are expected to satisfy the goal of process intensification and to develop sustainability pathways, thus providing reliable options for industrial growth and environmental protection. Starting with basic definitions, a synthetic membrane can be considered as an interphase that restricts the transport of components in a specific manner. In most cases (membrane filtration systems, electrodialysis) the membrane shows an intrinsic selectivity due to peculiar physico-chemical characteristics, such as pore size, charged surface, etc. In other cases (for example, membrane distillation) the membrane is a non-selective barrier used merely to separate and contact two adjacent phases; mass and energy transfers occur according to the principles of thermodynamic equilibrium.

Separation is the result of differences in the transport rate of various species through the membrane. The transport rate is mainly determined by the nature of forces acting on each individual component (usually called "driving force"), and by its mobility through the membrane. According to these basic concepts, membranes are usually classified on the basis of the applied driving force(s). Focussing only on membrane operations used in desalination, the following can be defined:

1. Pressure-driven membrane processes, such as Reverse Osmosis (RO), Nanofiltration (NF), Ultrafiltration (UF) and Microfiltration (MF).
2. Electrical potential-driven membrane processes, such as Electrodialysis (ED).
3. Temperature-driven membrane processes, such as Membrane Distillation (MD) and Membrane Crystallisation (MCr).

3.2 Membrane Materials and Modules

3.2.1 Preparation Techniques

The microstructure of a membrane is a critical subject, and is strictly dependent on the preparation procedures. Commonly, membranes are classified according to their morphology (Fig. 3.1): dense homogeneous polymer membranes, porous membranes and (thin film) composite membranes, consisting of a dense top layer on a porous structure of a different material. The latter membranes are today widely used in sea- and brackish-water RO desalination. Porous membranes can be also symmetric or asymmetric. Symmetric membranes may have straight or sponge-like pores.

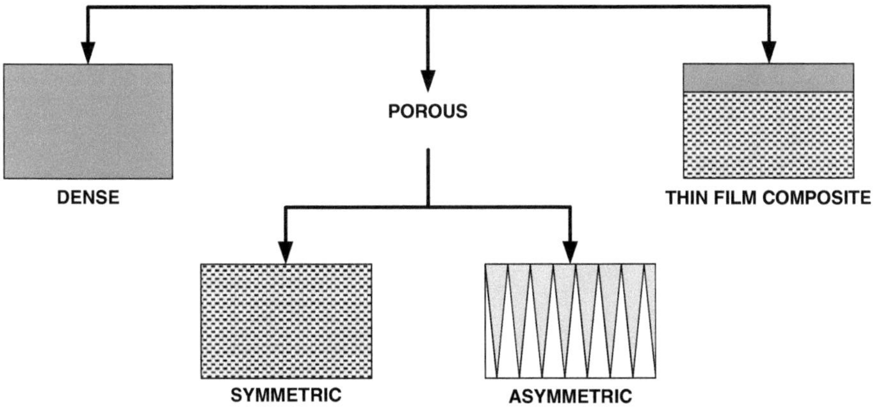

Fig. 3.1 Membrane classification based on morphology

Asymmetric membranes display a thin dense skin layer, with or without pores, on top of a deep porous sublayer: the thickness of the selective skin offers the advantage of low resistance to transport through the membrane.

Membranes are generally manufactured by sintering, stretching, track-etching and, most commonly, by phase inversion.

Sintering is a simple technique: a powder of polymeric particles is pressed into a film or plate and sintered to just below the melting point [2]. The process yields a microporous structure having a porosity in the range of 10–40% and rather irregular pore sizes, ranging from 0.2 to 20 μm (Fig. 3.2a).

Microporous membranes can be also prepared by stretching a homogeneous polymer film made from a partially crystalline material [3]. Films are obtained by extrusion of a polymeric powder at temperature close to the melting point, coupled with a rapid drawdown. Crystallites in the polymers are aligned in the direction of drawing. After annealing and cooling, mechanical stress is applied perpendicularly to the direction of drawing. This manufacturing process gives a relatively uniform porous structure with pore size distribution in the range of 0.2–20 μm and porosity of about 90% (Fig. 3.2b).

Microporous membranes, with uniform and perfectly round cylindrical pores, can be obtained by track-etching [4]. Homogeneous thin films, usually with thicknesses of 5–15 μm, are exposed to irradiation with collimated charged particles, with an energy of about 1 MeV. These particles damage the polymeric matrix. The film is then immersed in an acid or alkaline bath, where the polymeric material is etched away through the tracks so leaving perfect pores with a narrow size distribution. Typical pore size ranges between 0.02 and 10 μm and the surface porosity generally is below 10% (Fig. 3.2c).

A large number of membranes for RO and NF are prepared using a phase inversion technique with polymers that are soluble at a specific temperature in an appropriate solvent or solvent mixture, and that can be precipitated in a continuous phase by changing temperature and/or composition of the system (Fig. 3.2d). These

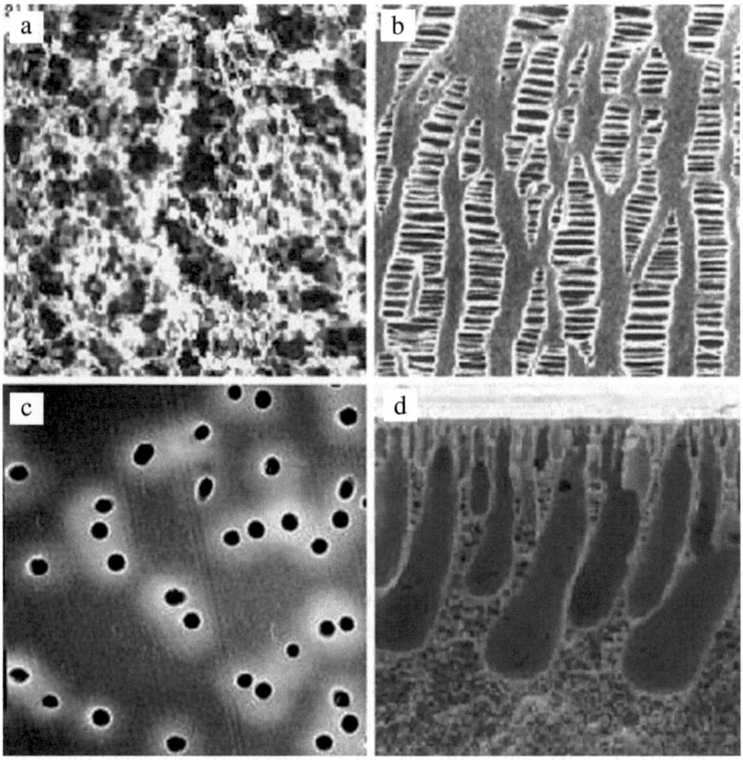

Fig. 3.2 Cross-section micrographs of (**a**) sintered PTFE membrane; (**b**) stretched PP membrane; (**c**) track-etched PCA membrane; (**d**) PVDF membrane prepared by phase inversion

changes aim to create a miscibility gap in the system for a given temperature and composition; from a thermodynamic point of view the free energy of mixing of the system then becomes positive [5].

The formation of two different phases, i.e. a solid phase forming the polymeric structure (symmetric, with porosity almost uniform throughout the membrane cross-section, or asymmetric, with a selective thin skin on a sub-layer) and a liquid phase generating the pores of the membrane, is created by a few and conceptually simple actions:

1. by changing the temperature of the system (cooling a homogeneous polymer solution which separates in two phases) – Temperature-Induced Phase Separation technique (TIPS);
2. by adding non-solvent or a non-solvent mixture to a homogeneous solution – Diffusion-Induced Phase Separation (DIPS);
3. by evaporating a volatile solvent from a homogeneous polymer solution prepared using solvents with different dissolution capacity.

Polymeric membrane formation by phase inversion is a complex process which depends on a great number of parameters, such as the type and concentration of polymer used, the type of solvent, non-solvent and additives used in the casting solution and/or coagulation bath, the casting temperature, etc. These factors determine whether the membrane will become dense or porous, symmetric or asymmetric, whether it will have a dense or porous skin, whether it will exhibit macrovoids or not, and whether it will have a finger-like, sponge-like, cellular or particulate morphology [6].

3.2.2 Membrane Materials

3.2.2.1 Membranes for Microfiltration

Microporous hydrophobic membranes for Microfiltration have a pore size typically in the range of 0.1–10 μm, and are commonly prepared by stretching and phase inversion. The typology and main characteristics of the polymers frequently used as material for microporous membranes are given in Table 3.1.

Table 3.1 Common polymeric materials used in the preparation of MF membranes

Polymer	Formula	Properties
Polytetrafluoroethylene (PTFE)	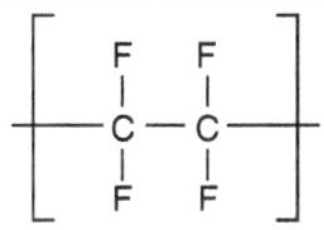	High temperature and chemical (acid) resistance; cannot be irradiated; inherently hydrophobic
Polyvinylidenefluoride (PVDF)	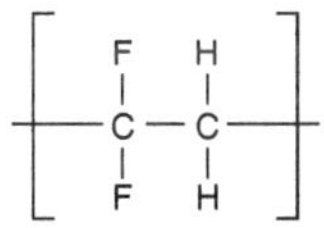	High temperature resistance; inherently hydrophobic
Polypropylene (PP)	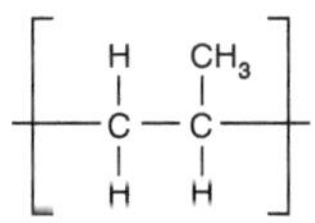	Chemically resistant; hydrophobic
Polycarbonate (PCA)	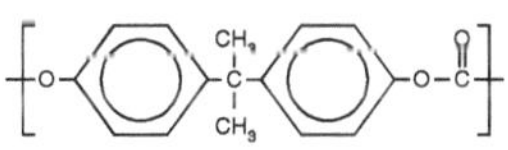	High wet/dry strength; mechanical properties suitable for track-etching preparation method
Polyethyleneterephtalate (PET)	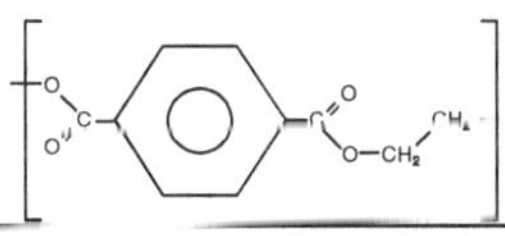	Mechanical properties suitable for track-etching preparation method

Common examples of commercial membranes are: the Celgard® membrane made of PP produced by stretching, the Gore-Tex PTFE membrane prepared by stretching and the Accurel® PP membrane prepared by thermal-induced phase separation. Dense films of polycarbonate (PCA) or polyethyleneterephtalate (PET) can be transformed, by track-etching, into porous Microfiltration membranes with a fine pore size distribution.

3.2.2.2 Membranes for Ultrafiltration

The availability of well-established, chemically resistant membranes able to work in harsh process conditions is of significant relevance to ultrafiltration operations. UF membranes are usually prepared by phase inversion. The most widely used polymers are listed in Table 3.2.

Both polysulfone (PSU) and polyethersulfone (PES) are easily prepared by conventional phase inversion, because of their good solubility in chloroform and dimethylformamide. One of the main disadvantages of these materials is their high non-specific adsorption ability as a result of their hydrophobicity. This property increases the risk of fouling, leading to a reduction of membrane performance. Common strategies aimed at increasing the hydrophilic character of the membrane include: control of sulphonation, blending with hydrophilic polymers such as

Table 3.2 Common polymeric materials used in the preparation of chemically resistant UF membranes

Polymer	Formula	Properties
Polysulfone (PSU)	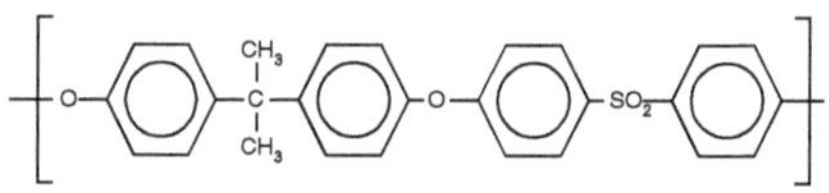	pH and temperature resistant; poor hydrocarbon resistance
Polyethersulfone (PES)	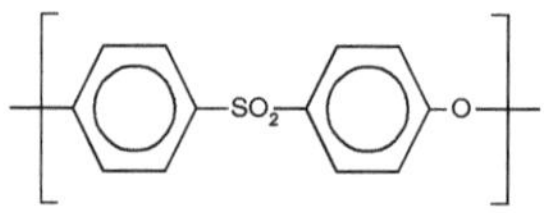	High thermal and chemical stability
Polyimide (PI)	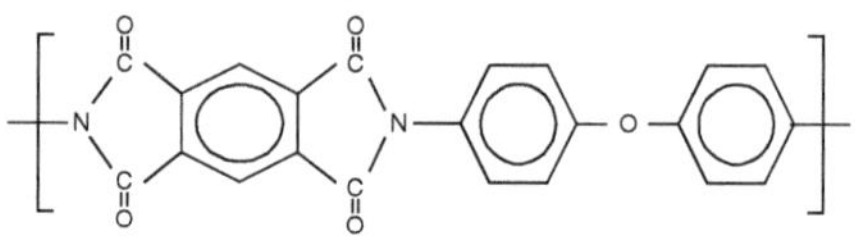	Excellent thermal stability; good chemical resistance
Polyacrylonitrile (PAN)	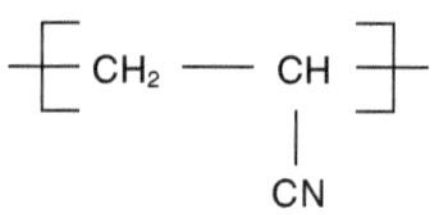	Highly crystalline, relatively hydrophilic, high resistance to hydrolysis and oxidation

polyvinylpyrrolidone or polyethyleneglycol, surface modification by polymer grafting or plasma treatment etc. Because of its high thermal stability and good chemical stability (higher than cellulosic polymers, but it cannot be used in contact with chloroform or dichloromethane), polyimide (PI) is also an interesting material for the preparation of UF asymmetric membranes. Polyacrylonitrile (PAN) membranes have also been widely used as UF membranes, usually copolymerized with more hydrophilic polymers to improve processability.

3.2.2.3 Membranes for Nanofiltration and Reverse Osmosis

Polymers which have most commonly been used for manufacturing Reverse Osmosis and Nanofiltration membranes are cellulose acetate (CA) (historically, the first one successfully tested, and still today dominating the market because of its low cost) and (aromatic) polyamide (PA), as shown in Table 3.3.

Widely used are also Thin Film Composite membranes (TFC), consisting of an ultra-thin layer (usually of polyamide) which is polymerized (interfacial polycondensation) or cross-linked onto an asymmetric porous support (usually polysulphone cast on a polyester support) [7]. Membranes based on this concept are commercialised by Dow-Filmtec [8].

Similar materials are used for NF membranes applied as water softening membranes. Examples of commercialised membranes are: Toray membranes based on modified cellulose acetate, UOP membranes made with a cellulose acetate blend or TFC polyamide, Filmtec TFC membranes with a thin top layer consisting of a fully

Table 3.3 Common polymeric materials used in the preparation of NF and RO membranes

Polymer	Structure	Properties
Cellulose acetate (CA)	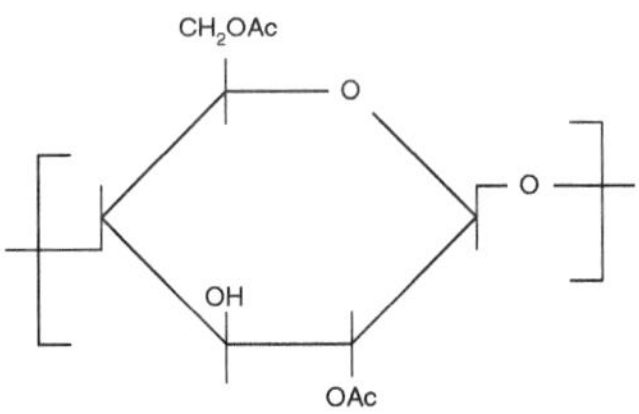	Very hydrophilic; sensitive to thermal and chemical degradation; low tensile strength
(Aromatic) Polyamide (PA)	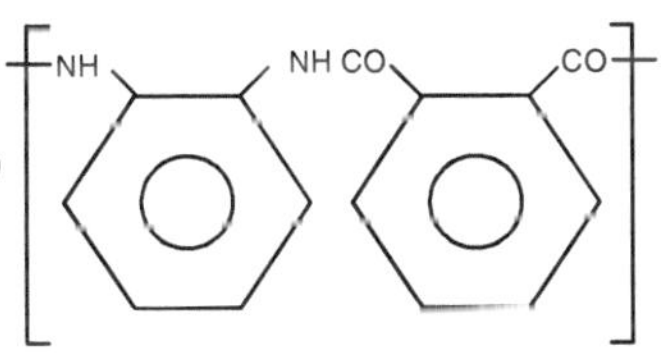	Hydrophilic; good solvent resistance

aromatic cross-linked polyamide and TFC Hydronautics and Desal/Osmonic membranes made of anionic aromatic polyamide. One of the major drawbacks of cellulose acetate membranes is the possibility of deterioration by hydrolysis, according to the generic reaction:

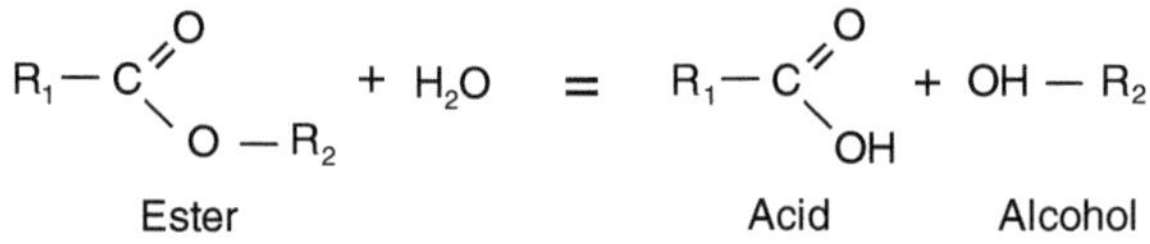

For a pH of around 4–5, membrane lifetimes can easily reach 4 years, whilst they decrease to 2.5 years when working at a pH of about 6 [9].

3.2.3 Membrane Modules

A large variety of membrane configurations, including flat sheet (plate-and-frame and spiral wound modules) and tubular (tubular, capillary and hollow fibre modules) have been tested for membrane desalination applications. The choice of a module is usually determined by economic and operative conditions. An important factor is the efficient control of concentration polarisation and membrane fouling. In plate-and-frame modules, the membranes, the porous support plates and the spacers are stacked between two endplates and placed in an appropriate housing. In this configuration, the packing density is about 100–400 m^2/m^3, depending on the number of membranes used. Various cassettes stacked together can be used for filtration purposes.

Spiral-wound modules (SWM), the most widely used for NF and RO desalination purposes, allow the efficient packing of flat-sheet membranes in a convenient cylindrical form (packing density is about 500–800 m^2/m^3). They are placed in an arrangement of two rectangular membranes back-to-back and sealed on three sides, to form an envelope. One or multiple envelopes are wound around a collector tube connected to the fourth side, which remains open. The feedwater to be treated enters one end of the cylinder, part of it (the permeate) permeates through the membrane, while the remaining brine (retentate) flows to the end of the module. The product permeate circulates between both membranes until entering the collector tube. Each spiral-wound module is contained in a pressure vessel assembly, consisting of a cylindrical housing for the modules, plumbing to connect modules together in series and plumbing to connect the feed inlet, product and brine outlet (Fig. 3.3). These modules have a good packing density, but cleaning is difficult and the glues used to seal the modules require certain precautions during use (solvents, temperature, etc.). The current industrial standard SWM element measures 20 cm in diameter [10]. Recently, Koch Membrane Systems introduced the first commercially available large diameter (MegaMagnum^{TM}) element with a nominal diameter of 45 cm [11].

A tubular module typically consists of several membrane tubes placed in a plastic cylindrical housing. The diameter of tubular membranes typically varies between

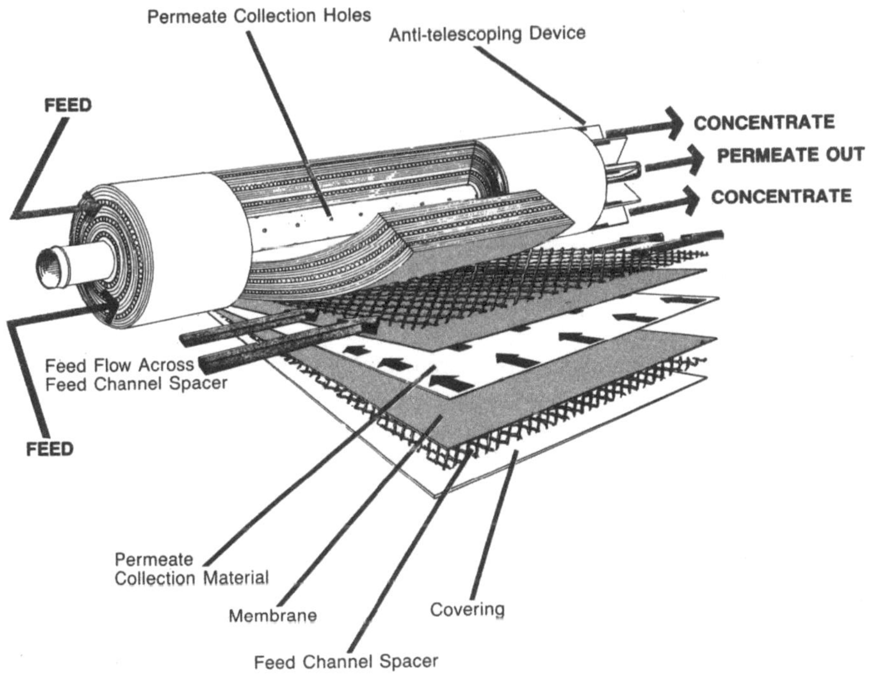

Fig. 3.3 Schematic drawing of a spiral-wound membrane module

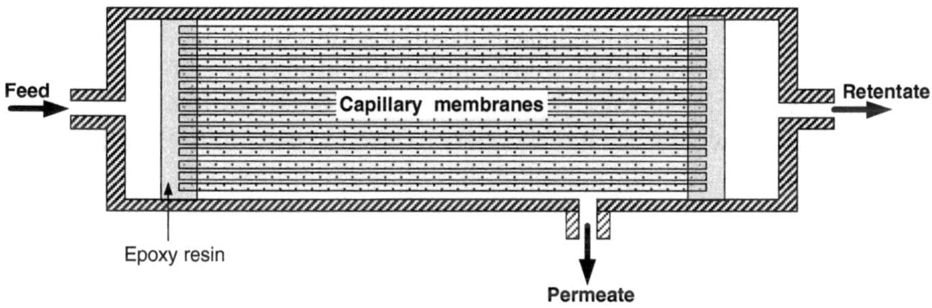

Fig. 3.4 Schematic drawing of a capillary membrane module

1.0 and 2.5 cm, with a packing density of about 300 m^2/m^3. In filtration operations, these modules allow high feed flow rates that reduce fouling tendency and polarisation phenomena.

In capillary membrane modules (Fig. 3.4), currently used for MF and UF operations, a large number of membrane capillaries (inner diameter of 0.2–3 mm) are arranged in parallel as a bundle in a shell tube. Packing density is in the order of 600–1200 m^2/m^3.

Hollow fibre membranes are very small-diameter capillaries, about the size of a human hair. The water to be treated permeates into a cylinder containing fibres grouped in bundles and cased in a pressure vessel. The fibres are asymmetric and

Table 3.4 Comparison between different typologies of membrane modules

	Tubular	Plate-and-frame	Spiral-wound	Capillary	Hollow fibre
Packing density	Low	→	→	→	High
Investment cost	High	←	←	←	Low
Fouling tendency	Low	→	→	→	High
Cleaning	Good	←	←	←	Poor

have internal and external diameters of around 42–85 μm, respectively. Thousands of fibres are combined to produce bundles of around 120 cm in length and 10–20cm in diameter. The whole assembly is contained in a cylindrical housing or shell approximately 137 cm long and 15–30 cm in diameter. Pressurised feed water enters through the central distributor tube, passes through the tube wall, and flows radially through the fibre bundle towards the outer permeator pressure shell. This type of module has the highest membrane surface area per element (>1000 m^2/m^3) but it presents the lowest flux rate. However, due to its high packaging density, this module appears to be very sensitive to the raw water quality with respect to fouling potential [12]. A qualitative comparison between different membrane modules is shown in Table 3.4.

Desalination plants based on RO membrane technology usually use multiple stage processes. The simplest plant design uses a series array configuration, with membrane elements (usually 6–8) connected in series in a single housing. The series array design is mainly limited by feed fouling potential and restrictions on pressure head loss. For higher plant throughput, multiple housings are used in parallel. Multiple passes are required when it is necessary to increase the quality of the permeate product.

3.3 Pressure-Driven Membrane Processes

Membrane technology, and in particular Reverse Osmosis, has grown rapidly in recent years. The production of asymmetric cellulose acetate membranes in the early 1960's by Loeb and Sourirajan [13] is generally recognised as a significant moment for the overall membrane technology. The discovery of an effective method for increasing the permeation flux of polymeric membranes, without significant changes in selectivity, has made their use possible in large-scale operations for desalting brackish- and sea-water by Reverse Osmosis [13].

RO is a pressure-driven membrane process able to separate particles, macromolecules, ions, etc. from a solvent (usually water). RO membranes typically reject all molecules over a molecular weight of 150 Da and a high percentage (> 99%) of those between 25 and 150 Da. The transmembrane flux is a function of the effective pressure, given by the hydrostatic pressure difference between feed and filtrate,

Table 3.5 Main characteristics of pressure-driven membrane separation processes

Separation process	Pore size or maximum MW range	Operating pressure (MPa)	Principle of separation	Substances removed	Alternative traditional treatment method
Microfiltration	0.1–10 μm	0.05–0.2	Size exclusion	Bacteria, viruses, larger colloidal particles, precipitates and coagulates	Ozonation, chlorination, sand-bed filtration, bioreactors, coagulation and sedimentation
Ultrafiltration	2–10 nm 1,000–500,000 Da	0.1–0.5	Size exclusion	High molecular weight proteins, large organic molecules and pyrogens	Sand-bed filtration, bioreactors and active carbon treatment
Nanofiltration	2–70 Å 180–10,000 Da	0.3–3	Size exclusion, diffusion, Donnan-exclusion	Large divalent and some monovalent ions, colourants and odorants	Lime/soda softening and ion exchange
Reverse osmosis	1–70 Å	1–10	Solution-diffusion mechanism	All of the above in addition to monovalent ions	Evaporation, freezing and electrodialysis

minus the difference in osmotic pressure between these solutions. Nanofiltration (NF) is a membrane softening process capable of removing a large part of bivalent ions (calcium, magnesium, etc.), dissolved organic matter, as well as compounds responsible for tastes and odours in water. Separation results from the combination of different mechanisms such as size exclusion, diffusion and Donnan-exclusion. Other pressure-driven membrane processes, such as Microfiltration (MF) and Ultrafiltration (UF) are based on sieving mechanisms and particles are separated according to their size. In general, MF is used to separate particles with diameters in the range of 0.1–10 μm, whereas UF retains macromolecules, viruses or sub-micron particles. Thus, MF and UF are destined for raw water clarification/disinfection, while RO and NF are used to remove environmental micro-pollutants, organic matter and dissolved salts. Table 3.5 provides some information on the principal characteristics of these membrane systems.

A simple and general diagram of a membrane filtration stage is depicted in Fig. 3.5. It shows the main process streams: the feed solution entering the membrane stage, the filtrate (or permeate) passing through the membrane, and the retentate (or concentrate) collecting all substances rejected by the membrane. The performance of a membrane operation is generally determined by the transmembrane (or permeate flux) and by the membrane separation properties. A parameter of practical interest in water treatment processes is the recovery ratio (RR), expressed as:

$$RR(\%) = \frac{Q_p}{Q_f} \times 100 \tag{3.1}$$

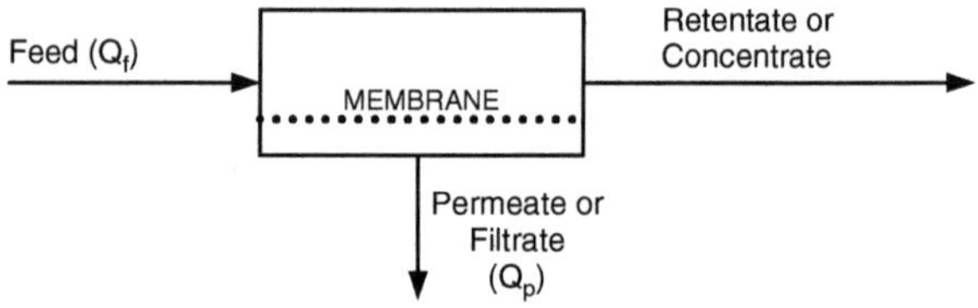

Fig. 3.5 General representation of a membrane separation process

Here, $Q_p$and Q_f are the permeate and feed flow rates, respectively. Typically, single RO elements are operated with a recovery of 10–15% [14].

The separation performance is generally expressed in terms of rejection (or retention of salts in brine) R of a certain class of particles or substances, which is given by:

$$R_i = \frac{c_i^c}{c_i^f} = \frac{c_i^f - c_i^p}{c_i^f} = 1 - \frac{c_i^p}{c_i^f} \tag{3.2}$$

where c_i is the concentration of the *i* th substance, whilst superscripts ^{p}, ^{c} and ^{f} refer to permeate, concentrate and feed solutions, respectively. The separation capabilities of the membrane filtration processes are summarized in Fig. 3.6.

3.3.1 Microfiltration and Ultrafiltration

Microfiltration and Ultrafiltration, typical pressure-driven membrane operations, are basically identical processes and differ only in the sizes of the particles to be separated and the membrane typology. In both cases, a mixture of different components is brought to the surface of a membrane; under the driving force gradient, some components permeate the membrane while others are retained. Thus, a feed solu-

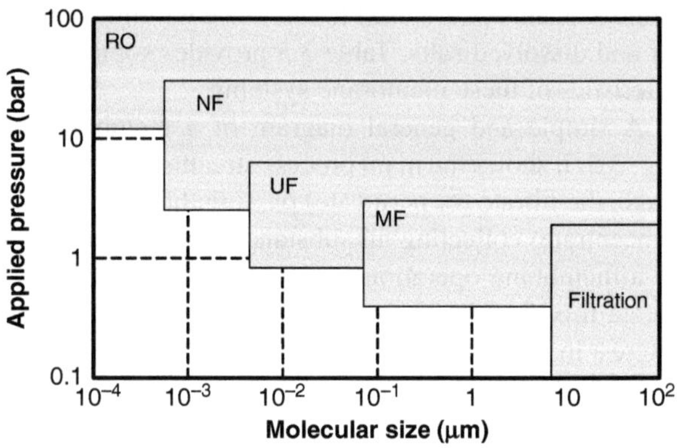

Fig. 3.6 Separation spectrum of pressure-driven membrane processes

tion is separated into a filtrate which is deplete of particles or molecules, and a retentate in which these components are concentrated.

The term Microfiltration is used when particles with diameters in the order of a few microns are separated from a solvent, or other components, of low molecular weight [15]. The separation mechanism is based on a sieving effect and particles are separated according to their dimensions. The hydrostatic pressure difference used as driving force is in the range 0.05–0.2 MPa.

The mass transport in Microfiltration membranes takes place by viscous flow through the pores, mathematically illustrated by the Hagen-Poiseuille equation:

$$J_v = L_v \frac{\Delta P}{\delta} \tag{3.3a}$$

with:

$$L_v = \frac{\varepsilon\, r^2}{8\,\eta\,\tau} \tag{3.3b}$$

J_v is the total volumetric flux through the membrane, ϵ the membrane porosity, r the pore radius, η the viscosity of the solution, τ the tortuosity factor (defined as the ratio of the actual pore length to the thickness of the membrane), ΔP the pressure difference across the membrane (not affected by the osmotic pressure), and δ the membrane thickness. L_v is known as the hydrodynamic permeability coefficient.

In UF, where generally the osmotic pressure of the feed solution is not insignificant, as some relatively low molecular weight solutes are retained, hydrostatic pressures of 0.1–0.5 MPa are used [16].

The flux of the solvent, as an approximation of total volumetric flux, can be calculated as the sum of the diffusive flow of water through the membrane and the viscous flow (Eq. (3.4) below):

$$J_v = \bar{V}_w^2 L_w \frac{\Delta P - \Delta \Pi}{\delta} + L_v \frac{\Delta P}{\delta} \tag{3.4}$$

where:

$$L_w = \frac{D_w c_w}{RT} \tag{3.5}$$

$\overline{V}$ is the partial molar volume, L_w is the diffusive permeability coefficient, $\Delta\Pi$ the osmotic pressure difference, D the diffusion coefficient, c the concentration, R the gas constant and T temperature. Subscript w refers to water. However, in various practical applications, $L_w << L_v$ and Eq. (3.4) reduces to Eq. (3.3a).

3.3.2 Nanofiltration

In Nanofiltration, a hydrostatic pressure is also applied to push a molecular mixture through the membrane surface. In general, NF is considered to be a process half way between UF and RO [17]. From a practical point of view, the applied pressure is generally one order of magnitude higher than in UF, but lower than in RO.

NF differs from UF and RO also in the separation mechanism, basically determined by two distinct properties:

1. the pore size of the membrane, ranging from 2 to 70 Å;
2. the charge of the membrane surface, that can be positive or negative, and which affects the rejection properties of the membrane due to electric interactions between ions and fixed charges.

To describe the volumetric flux as a function of hydrostatic pressure, the same basic Eq. (3.4) can be used. In this case L_w and L_v are practically of the same order of magnitude and no simplifications can be made. Complications arise in the description of the flux of each individual component (J_i) if the membrane holds positive or negative electric charges at its surface. In this case the partition coefficient for ionic components is determined both by size exclusion and Donnan exclusion (resulting from electric Donnan potential φ_{Don} established between electrolyte solution and charged membrane) [18].

For a dilute solution, with appropriate simplifications, the equation is:

$$\varphi_{Don} = \sum_i \frac{1}{z_i F}\left[RT \ln \frac{c_i^b}{c_i^m}\right] \tag{3.6}$$

Here, z_i is the valence of the i th ion, F the Faraday constant, and superscripts b and m relate to concentration values in the bulk of the solution and at the membrane surface, respectively. The flux of individual components through a charged NF membrane is given by:

$$\begin{aligned} J_i = D_i^m \left(\frac{\overline{V}_i k_{size} k_{Don} c_i^b}{RT}\frac{dP}{dz} + k_{size} k_{Don} \frac{dc_i^b}{dz} + \frac{z_i F k_{size} k_{Don} c_i^b}{RT}\frac{d\varphi_{Don}}{dz} \right) \\ + L_v k_{size} k_{Don} c_i^b \frac{dP}{dz} \end{aligned} \tag{3.7}$$

where k is the partition coefficient between the membrane and the adjacent solution ($k_i = c_i^m/c_i^b$), z is the spatial coordinate through the membrane, and all other symbols are as previously described. As a consequence of the additional driving force of the Donnan potential, cations and anions can be partially separated by charged NF membranes.

3.3.3 Reverse Osmosis

A Reverse Osmosis (RO) process allows the separation of dissolved ions, and larger dissolved species, from feed water. RO rejection levels for ions are typically around 99.5%. RO performance is essentially characterized in terms of water flux, salt rejection and recovery rates.

From a phenomenological point of view, RO separation is based on countering the natural osmotic process by artificially applying hydrostatic pressure to the side of the more concentrated salt solution (for example seawater), so driving the solvent flux in an opposite direction to that dictated by natural osmosis. This causes water to move through the RO membrane from the more concentrated salt solution (seawater) to the other side, which holds essentially desalted water.

The value of osmotic pressure Π depends on the solution concentration according to the following equation:

$$\Pi = RT \sum_i \nu_i c_i \tag{3.8}$$

where ν_i is the van't Hoff coefficient of the i th ion. The osmotic pressure of sea water is typically around 2.5 MPa. As long as the applied hydrostatic pressure ΔP is higher than the osmotic pressure difference $\Delta\Pi$ between the feed and permeate solutions, the solvent (water) will flow from the more concentrated solution (feed) to the dilute solution (permeate). The transmembrane flux is reversed from the direction of natural osmosis, as illustrated in Fig. 3.7. A solution-diffusion model is a common tool used to describe the transport in RO [19]. It is assumed that the flux of the i th component through a membrane occurs by diffusion, and can be derived from the product of its concentration and mobility in the membrane and the driving force to the transport (effective pressure difference for water, and concentration gradient for salt):

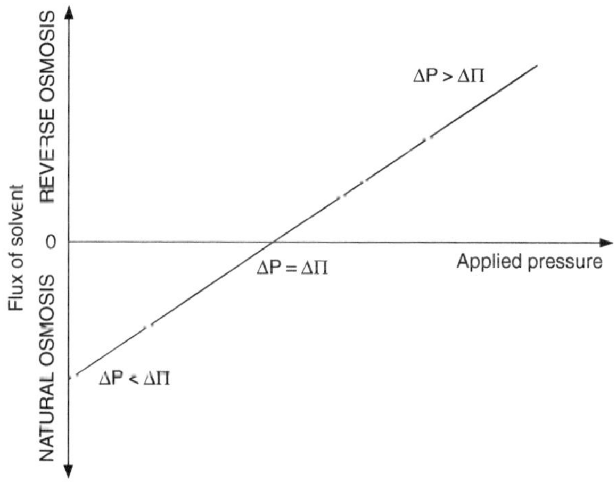

Fig. 3.7 Transmembrane flux of solvent through a semi-permeable membrane as a function of the applied pressure

$$J_v = -\frac{k_w D_w \overline{V}_w}{RT}\left(\frac{\Delta P - \Delta \Pi}{\delta}\right) \tag{3.9}$$

and:

$$J_s = k_s \frac{D_s}{\delta}\left(c_s^f - c_s^p\right) \tag{3.10}$$

Here J_v is the volumetric flux through the membrane and J_s is the solute molar flux, k the partition coefficient, D the diffusion coefficient, $\bar{V}$ the partial molar volume, δ the membrane thickness, R the gas constant, T the temperature, and c the molar concentration. Subscripts w and s refer to water and solute, while superscripts f and p refer to feed and permeate, respectively.

3.3.4 Concentration Polarisation

When a solution is brought to a membrane surface, some components will permeate the membrane under a given driving force whilst others are rejected. This leads to an accumulation of retained material and to a depletion of permeating components in the boundary layers adjacent to the membrane surface. This phenomenon is known as concentration polarisation. Assuming that, in steady state, the convective transport of solutes to the membrane surface is counterbalanced by a diffusive flux of retained solutes back into the bulk solution (Fig. 3.8), the film model gives [20]:

$$\frac{c_s^m}{c_s^b} = \frac{\exp \frac{J_v}{\kappa}}{R + (1 - R)\exp \frac{J_v}{\kappa}} \tag{3.11}$$

Here, subscript s refers to solute, superscripts m and b refer to concentrations at the membrane surface and in the bulk solution, respectively. J_v is the transmembrane volumetric flux, c the concentration, R the rejection coefficient ($R = 1 - c_s^p / c_s^m$, with superscript p referring to permeate) and κ the mass transfer coefficient, calculated using empirical correlations expressed in their general forms using Sherwood (Sh), Reynolds (Re) and Schmidt (Sc) numbers:

$$Sh = \alpha \, \mathrm{Re}^{\beta} Sc^{\gamma} \tag{3.12}$$

where α, β and γ are numerical coefficients.

Concentration polarisation has significant negative effects on NF and RO process performance, and in particular:

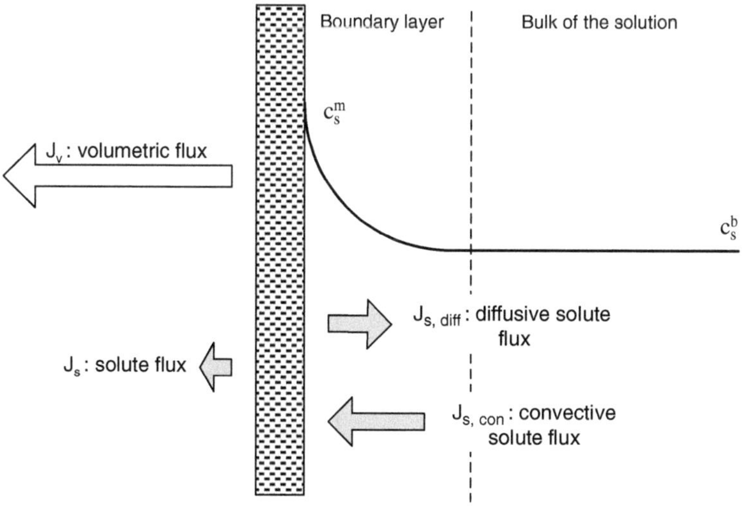

Fig. 3.8 Concentration polarisation: solute concentration profile and fluxes in a membrane system under steady state conditions

- it leads to an increase in the osmotic pressure of the solution at the membrane surface, with a consequent reduction in the transmembrane flux at constant applied hydrostatic pressure;
- for some salts based on divalent ions (Ca^{2+}, Mg^{2+}), solubility limits can be exceeded ($CaCO_3$, $CaSO_4$, $MgCO_3$, etc.), leading to their precipitation on the membrane surface, which limits the mass transfer;
- rejection decreases due to higher salt flux as a consequence of the increased salt concentrations at the membrane surface.

3.4 Electrodialyis

Electodialysis (ED) has been in commercial use since 1952 for desalting brackish well water in the Arabian Desert, over 10 years prior to RO application. As illustrated in Fig. 3.9, an electrodialyser comprises a stack of anion and cation exchange membrane pairs, with an anode at one end of the stack and a cathode at the other. When electrodes, in a solution of saline water, are connected to an external source of direct current, e.g. a battery, electrical current is carried through the solution, with the ions forced to migrate to the electrode with the opposite charge (positively charged ions migrate to the cathode and negatively charged ions migrate to the anode). The anions can move freely through the nearest anion exchange membrane, but their further progress toward the anode is blocked by the adjacent cation exchange membrane. In the same manner, the cations move in the opposite direction, through the nearest cation exchange membrane, but are then blocked from further progress by the adjacent anion exchange membrane.

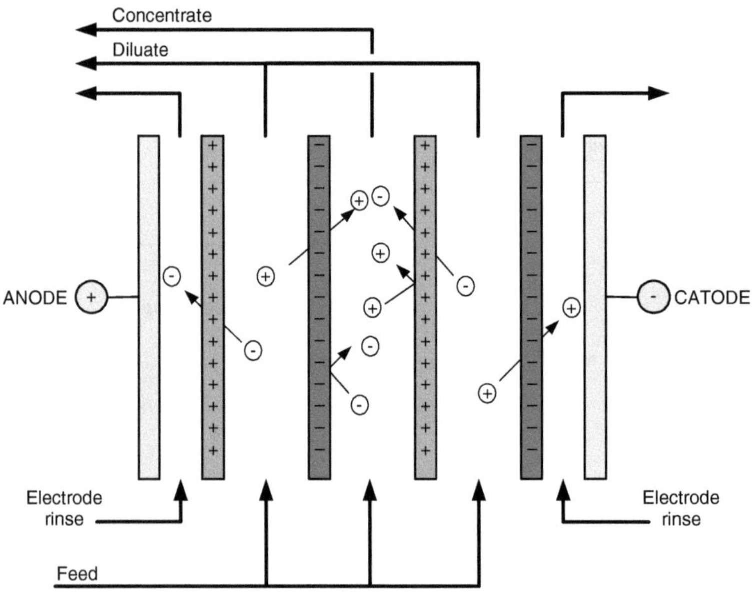

Fig. 3.9 Flow diagram of a typical electrodialysis operation

Via this arrangement, concentrated and dilute solutions are created in the spaces between the alternating membranes. Using this process, it is possible to feed brackish water into the dilution compartment inlet and obtain potable water at its outlet.

Mass transport in electrodialysis is usually described by phenomenological equations [21]. Under the ideal solution assumption, assuming neither pressure gradients nor kinetic coupling of fluxes, and expressing the phenomenological coefficient by the Maxwell-Stefan diffusion coefficient, the flux $J_i$of the i th ion is:

$$J_i = -D_i^m \left(\frac{dc_i}{dz} - c_i \sum_i \frac{T_i}{c_i} \frac{dc_i}{dz} \right) + T_i \frac{I}{z_i F} \tag{3.13}$$

where D_i^m is the diffusion coefficient, c_i the concentration, T_i the transport number, z_i the valence of the i th ion, z the coordinate through the membrane, I current and F the Faraday constant. The transport number T_i determines the fraction of current carried by ith ion:

$$T_i = \frac{z_i J_i}{\sum_i z_i J_i} \tag{3.14}$$

In a strictly permselective anion exchange membrane the current can be transported only by anions, and the anionic transport number is 1; in a strictly permselective cation exchange membrane the current can be transported only by cations, and the cationic transport number is 1.

Hydrolysis and scaling represent the major obstacles to stable and efficient electrodialyser operation. Hydrolysis occurs in the electrolyte solution near the membrane surface, only if the current density exceeds a limit, which is determined by the ion depletion layer that forms near the surface of membranes in the dilution compartments.

Hydrolysis begins when Na^+ and Cl^- ions become unavailable at the membrane surface, and the resulting H^+ and OH^- ions are then transported through the cation and anion exchange membranes, respectively, into the concentration compartments: here $Mg(OH)_2$ may be deposited as scale, thus lowering the efficiency of ED.

The appropriate operating current density is generally lower than the limiting current density. Moreover, water recovery is increased by increasing the NaCl concentration. It must be noted, however, that maintaining a high NaCl content in the concentrate may lead to scaling, if the multivalent ion content of the raw water is high, and care is therefore required. Membrane selection is generally based on, inter alia, membrane resistance and physical strength. The composition of the raw water, and particularly its organic acid content, is also an important determining factor. In fact, organic acids tend to adsorb onto the anion exchange membrane, thus creating high membrane resistance.

Electrodialysis works well with brackish water; the recovery rate of water is usually over 90%, installation and running costs are fairly economical at moderate TDS, and membranes are durable under strong acid conditions and in alkaline conditions.

3.5 Membrane Contactors

Membrane contactors are an innovative technology in which porous membranes are used as tools for inter-phase mass transfer. The membrane does not act as a selective barrier and separation is based on the principles of phase equilibria [22].

In Membrane Distillation (MD), a microporous hydrophobic membrane is in contact with an aqueous heated solution on one side (feed or retentate). The hydrophobic nature of the membrane prevents mass transfer of the liquid phase and creates a vapour-liquid interface at the pore entrance. Here, volatile compounds evaporate, diffuse and/or move via convection through the membrane pores, and are condensed and/or removed on the opposite (permeate or distillate) side of the system. Lower temperatures and pressures, compared to those usually used in conventional distillation columns, are generally sufficient to create an acceptable transmembrane flux (1–20 kg/m^{2}h^2), with consequent reduction of energy costs and mechanical requirements of the membrane. Typical feed temperatures are in the range 30–60°C, thus permitting the efficient recycling of low-grade or waste heat streams, as well as the use of alternative energy sources (solar, wind or geothermal). In addition, the possibility of using plastic equipment reduces or avoids erosion problems. When compared with the RO process, MD does not suffer limitations

Fig. 3.10 Crystals of: (**a**) calcite; (**b**) sodium chloride; (**c**) epsomite, produced by membrane crystallisation

due to concentration polarisation phenomenon and can be preferentially employed whenever high permeate recovery factors or retentate concentrations are required. In addition, the high flexibility and capability of MD operations offer attractive possibilities of integrating them into various industrial production cycles, with consequent synergic benefits.

Membrane Crystallisation (MCr) has been recently proposed as an interesting and promising extension of the Membrane Distillation concept [23]. This innovative technology uses the evaporative mass transfer of volatile solvents through microporous hydrophobic membranes to concentrate solutions above saturation limit, thus attaining a supersaturated environment where crystals may nucleate and grow (Fig. 3.10).

Seawater is the most abundant aqueous solution on earth: 3.3% of its composition is represented by dissolved salts, and seven elements (Na, Mg, Ca, K, Cl, S and Br) account for 93.5% of the ionic species. Membrane crystallisation can be used to recover valuable salts (such as calcium sulphate, sodium chloride, magnesium sulphate) from RO and/or NF reject brine, with consequent benefits in terms of overall costs reduction and reduced environmental impact.

3.6 Fouling

The composition and characteristics of the feed water used in desalination plants directly affect the RO process [24]. The physical-chemical nature of the water itself, as well as colloids, organic compounds or particles present, can decrease the efficiency of the process, or even damage the membrane and decrease its life (see Fig. 3.11 for an overview).

Damage is primarily caused by oxidation and hydrolysis of the membrane material due to compounds present in the feed water. Most membranes cannot tolerate residual chlorine concentrations, introduced in most desalination processes to prevent biological growth on membranes. Often sodium bisulphite is added to remove this excess chlorine. Sometimes it is necessary to adjust the pH of the feed water to

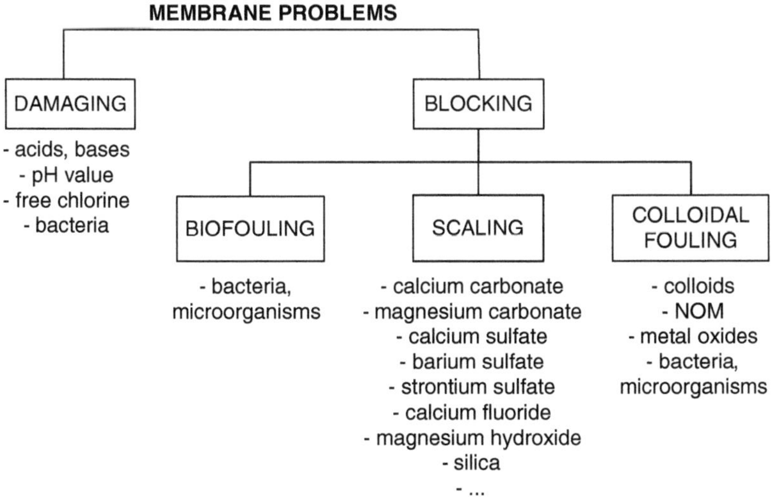

Fig. 3.11 List of common agents which create problems in membrane processes

recommended values for optimal operation by adding mineral acid (e.g. sulphuric acid). Moreover, to prevent membrane damage, oxidising agents and dissolved oxygen must be removed by adding a reducing compound such as sodium thiosulphate or sodium bisulphite. These procedures are described in Sect. 3.8.

Biological fouling is caused by the growth of colonies of bacteria or algae on the surface of the membrane [25]. Since chlorine can only be used on cellulose, a biomass can develop if care is not taken, especially when installations are shut off. Washing with biocides may be necessary. For membranes resistant to chlorine, the shock chlorination process is the simplest solution. This involves washing the membrane with heavily chlorinated water for a few seconds.

Scaling refers to the precipitation and deposition of barely soluble salts on the membranes; in fact, under particular operative conditions, the solubility limits of some of the components in the concentrate stream may be exceeded. These components include: calcium carbonate, magnesium carbonate, calcium sulphate, silica, barium sulphate, strontium sulphate and calcium fluoride [26].

In an RO unit, the final stage is subjected to the highest concentration of dissolved salts, and it is here that the first symptoms of scaling appear. Fouling by precipitation is amplified by the concentration gradient phenomenon on the surface of the membrane. The scaling tendency of a given feed water is often evaluated using the Langelier Saturation Index (LSI) for brackish water (TDS < 10,000 mg/l) [27] and the Stiff and Davis Stability Index (S&DSI) for seawater (TDS > 10,000 mg/l) [28]. LSI and S&DSI are defined as:

$$LSI = pH - pH_s \tag{3.15a}$$

$$pH_s = pCa + pAlk + pK_2 - pK_s \tag{3.15b}$$

and

$$S\ \&\ DSI = pH - pH_s \tag{3.16a}$$

$$pH_s = pCa + pAlk + K \tag{3.16b}$$

where pH_S is the pH level at which the water is in equilibrium with $CaCO_3$, pCa $= -\log_{10}[Ca^{2+}$ as $CaCO_3]$, pAlk $= -\log_{10}$[alkalinity of $CaCO_3$], $pK_2 = -\log_{10}$ (ionisation constant of HCO_3^-), $pK_s = -\log_{10}$ (solubility product of $CaCO_3$), K is the ionic strength constant at 25°C.

LSI also provides the criterion for technical water classification:

- LSI <-2.0, non scale-forming, highly corrosive;
- $-2.0 <$ LSI <-0.5, non scale-forming, moderately corrosive;
- $-0.5 <$ LSI < 0, balanced, pitting corrosion possible;
- $0 <$ LSI < 0.5, moderately scale-forming, non corrosive;
- $0.5 <$ LSI < 2.0, strong scale-forming, non corrosive.

Several solutions exist to limit scaling problems:

- operate at low water recoveries so that the solubility limits are not exceeded and no precipitation occurs;
- control pH by adding acids to regulate the calcium carbonate balance of the water, transforming bicarbonates into carbon dioxide and so preventing precipitation of $CaCO_3$;
- use precipitation inhibitors or antiscalants that, when added to the feed water, prevent compound precipitation by reacting with the surface of the ions;
- eliminate calcium using a pH base to reduce the risk of precipitation of calcium compounds (softening);
- use the *ion exchange on resin* technique for elimination of undesirable ions, such as fluorides;
- ensure that silica (SiO_2) is not precipitated on the membrane. In this case, although antiscalants are ineffective, the risk of fouling is low because the reaction is very slow.

Particulate clogging occurs when the feed water contains a great deal of particles and suspended colloidal matter (mechanical fouling), thus requiring washing. This can be avoided by limiting the amount of particles in the inlet water using coagulation-flocculation (a critical requirement is an appropriate dosing of coagulant-flocculants, because they can damage the membrane, often irreparably), Microfiltration or, even better, Ultrafiltration processes [29]. The concentration of particles in water can be measured and expressed in several ways. For relatively high concentrations, a gravimetric procedure is normally used to determine the Total Suspended Solids (TSS). Turbidity measurements, expressed in Nephelometric

Turbidity Units (NTU), provide quick and easy measurements. Turbidity measurements, however, are not valid at extremely low solid concentrations.

It is generally accepted that accumulation of foulants on the surface of RO and NF membranes can adversely affect both the permeate flux and rejection properties of a membrane. Colloidal fouling generally results in a marked deterioration in the observed salt rejection (the cake-enhanced concentration polarisation increases the diffusive transport of salt through the membrane).

Colloidal fouling is caused by the accumulation of colloidal particles on the membrane surface and formation of a cake layer. The permeate flux decline is only in part due to cake layer resistance, whereas the hindered back-diffusion of salt ions within the colloid deposit layer results in elevated salt concentration at the membrane surface, so causing an enhanced osmotic pressure that decreases the net driving force.

The most common tools for prediction of colloidal fouling are the Silt Density Index (SDI) [30] and the Modified Fouling Index (MFI) [31].

The SDI measures the time required to filter a fixed volume of water (usually 500 ml) through a standard 0.45 μm pore size Microfiltration membrane with a constant given pressure of 0.207 MPa (30 psi). The SDI is determined by the following steps:

1. measuring the time necessary to filter a fixed amount of water (usually 500 ml), t_{in} (initial time), starting with a dry filter;
2. measuring the time necessary to filter a second fixed amount of water (usually 500 ml), t_{fin} (final time), with a filter that has been operating for a time t^* (5, 10 or 15 min);
3. calculating the SDI by the following formula:

$$SDI = \frac{100 \cdot \left(1 - t_{in}/t_{fin}\right)}{t^*} \tag{3.17}$$

Membrane manufacturers often demand stringent SDI values of 2–4 (percent decay per minute).

The MFI index is evaluated using the same equipment as the SDI measurement, but with a different procedure based on the filtered volume (V) measured during time t (every 30 s over a maximum period of 20 min), according to the formula:

$$MFI = \frac{\eta\,(20^\circ C)}{\eta\,(T_{water})} \frac{\Delta P}{\Delta P_0} \tan\alpha \tag{3.18}$$

where η is the viscosity, ΔP is transmembrane pressure, ΔP_0 is transmembrane reference pressure at 20°C; and $\tan\alpha$ is the slope from the linear part of the plot of t/V plotted against filtered volume V. The MFI index recommendations, for an acceptable operation, range from 0–2 s/l^2 for RO and 0–10 s/l^2 for NF.

Natural organic matter (NOM) is considered a major foulant of all the dissolved species and suspended matter present in natural water.

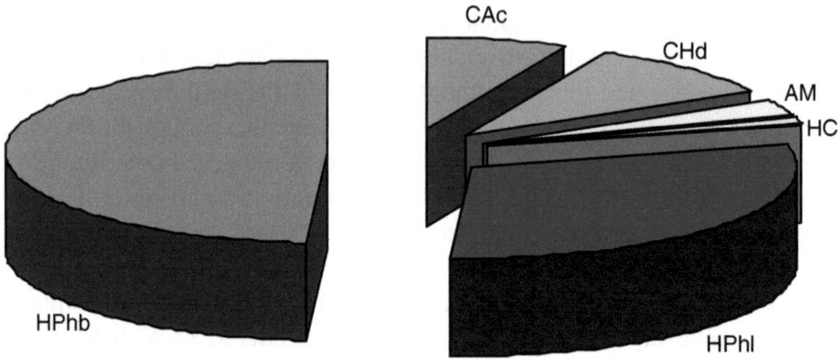

Fig. 3.12 Distribution of Natural Organic Matter in natural water (based on Dissolved Organic Carbon). CAc: carboxylic acid; CHd: carbohydrates; AM: amminoacids; HC: hydrocarbon; HPhl: hydrophilic; HPhb: hydrophobic

NOM is a complex mixture of particulate and soluble components of both inorganic and organic origin, with a wide range of molecular weight and functional groups (phenolic, hydroxyl, carbonyl groups and carboxylic acid), formed by allochthonous input, such as terrestrial and vegetative debris, and autochthonous input, such as algae. Among these components, the dissolved organic matter (DOM) has the most detrimental effect. DOM is comprised of humic substances, polysaccharides, amino acids, proteins, fatty acids, phenols, carboxylic acids, quinines, lignins, carbohydrates, alcohols, resins, and inorganic compounds such as silica, alumino-silicates, iron, aluminium, suspended solids and microorganisms (bacteria and fungi). In general, NOM can be fractionated into three segments: hydrophobic (humic substances), hydrophilic and transphilic fractions [32].

Referring to Fig. 3.12, the hydrophobic fraction represents almost 50% of dissolved organic carbon (DOC) with a larger molecular weight, the hydrophilic fraction is composed of 25–40% DOC with a lower molecular weight (polysaccharides, amino acids, protein, etc.), and the transphilic fraction is comprised of approximately 25% DOC in natural water but with a molecular weight in between hydrophobic and hydrophilic fractions. The humic fraction is the major foulant in membrane water filtration and determines the rate and extent of fouling.

Mechanisms by which NOM typically fouls MF and UF membranes include: (1) pore narrowing associated with NOM adsorption on the pore walls; (2) plugging of pore openings by colloidal NOM; and (3) formation of a continuous gel layer covering the membrane surface. The structure of the NOM layer is affected by both chemical conditions (pH, ionic strength and divalent cations) and physical conditions (initial permeate flux and cross-flow velocity). Inorganic particles can affect the fouling behaviours of organic substances, because their presence creates significant competition between NOM and inorganic particles adsorbing onto the membrane surface or into the pores. In particular, NOM-calcium complexation exacerbates the fouling problem, resulting in the formation of a thick, dense and highly resistant fouling layer [33, 34].

3.7 Cleaning

Cleaning extends the life and efficiency of a membrane process. As a rule, membrane elements should be cleaned whenever an operating parameter changes by 10–15%. If the permeate flow rate drops, the quality of desalted water changes, and applied pressure and other process parameters may need to be adjusted to maintain productivity and quality levels. An examination of feed water analyses and records, the membrane pressure differential, water flux, salt rejection and applied pressure is essential in order to establish the type and extent of fouling. If one or more parameters reach unacceptable values, it is necessary to clean the membrane, in part or completely, to restore the flux.

Membrane cleaning can be divided into four groups:

1. Mechanical cleaning;
2. Hydrodynamic cleaning;
3. Air/water cleaning; and
4. Chemical cleaning.

Mechanical cleaning is accomplished by introducing high shear forces at the membrane surface by mechanical action (for example, use of sponge-balls for the cleaning of tubular membranes, use of ultrasound waves that result in the cavitation of fluids and the consequent increase of turbulence and shear forces at the membrane surface).

Hydrodynamic cleaning methods use water flow to achieve high shear forces at the membrane surface (temporary increase in the cross-flow velocity, pulsed feed flow, temporary reversal of the flow through the membrane elements, etc.). In particular, periodic backwashing of membranes is a very effective method to remove accumulated layers of fouling, although it results in an increase of energy consumption and the loss of permeate. A backwash can only be used for ultra- and microfiltration membranes, since backwashing of spiral wound membranes damages the glue-layers of the membrane elements.

By using mixtures of water and air during membrane cleaning (periodic injection of compressed air), increased turbulence is achieved that results in high shear forces at the membrane surface [35, 36].

Chemical cleaning methods are used mainly for the removal of membrane foulants by dissolving, complexating, oxidising, inactivating, solubilizing, hydrolysing and denaturalising the membrane fouling [37, 38]. Common cleaning agents are:

- *Acid cleaning agents* , used to dissolve inorganic precipitates causing scaling (compounds like $CaCO_3$, $FeSO_4$, FeO, $FeOH$, $Al_2(SO_4)_3$, $BaSO_4$, $SrSO_4$, $CaSO_4$) and to dissolve the inorganic matrix in a biofilm. Acids largely used are citric acid, sulphuric acid and phosphoric acid;
- *Alkaline cleaning agents* (typically $NaOH$ and/or Na_2CO_3), used for dissolving organic deposits;

- *Complexing or anti-precipitating agents* , used to remove metals and other precipitating ions from the solution (for example, EDTA removes Ca^{2+} and Mg^{2+} ions, which determine the hardness of the water);
- *Biocides* (for example: chlorine, chloramines, organic peroxides, glutaraldehyde, sodium bisulphite), that inactivate microorganisms via toxicity;
- *Detergents or surfactants* , that reduce the surface tension of the water, so resulting in a better hydration of the fouling layer and therefore leading to an improvement of the solubility of the fouling layer;
- *Enzymatic cleaning agents* , based on enzymes hydrolysing the extra polymeric substances formed by microorganisms;
- *Chaotrophic cleaning agents* , that accomplish denaturation of proteins, resulting in better solvability of organic compounds.

3.8 Pretreatment of Membrane Desalination Plants

Pretreatment is necessary to preserve the performance and lifetime of RO membranes. A proper selection of pretreatment methods for feed water extends the life span of the system by preventing or minimizing particulate and colloidal fouling, biological fouling, scaling and membrane plugging, as well as reducing the need for chemical membrane cleaning. In the past, most RO plants used conventional pretreatment, which consists of an appropriate cascade of chemical and physical treatment without the use of membrane technologies. In recent desalination installations, the use of membrane-based pretreatment, prior to the RO stage, is being considered as a reliable alternative to conventional methods.

3.8.1 Conventional Pretreatment

Conventional pretreatment methods for an SWRO plant usually include the following steps, summarised in Fig. 3.13:

- screens for coarse pre-filtration;
- chlorination/acid addition for pH-adjustment/ flocculation agents;
- coagulation/flocculation;
- single- or double- media filtration;
- dechlorination/antiscalants;
- cartridge filtration.

Common disadvantages of conventional pretreatment systems for the operation of RO membranes include:

- fluctuations of feed water quality;
- difficulties in supplying a constant SDI < 3.0;

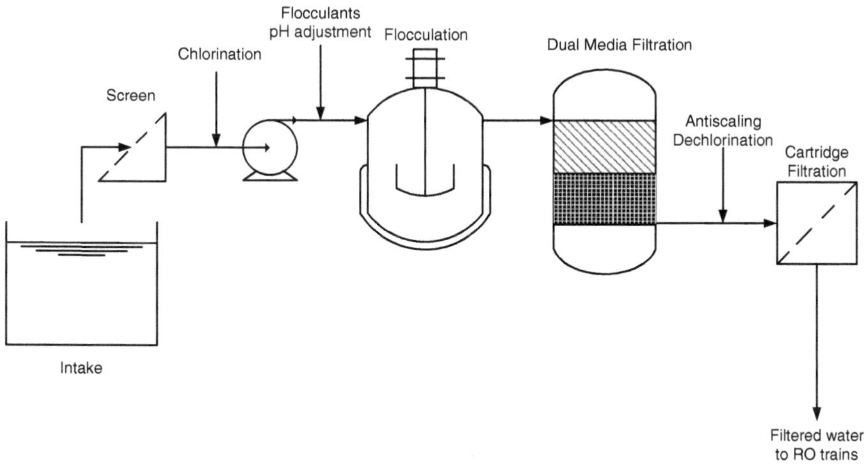

Fig. 3.13 Typical process diagram of a conventional pretreatment process

- difficulties in removing particles smaller than 10–15 μm;
- large space occupation due to slow filtration velocities.

3.8.1.1 Screens

Seawater desalination facilities require an intake system capable of providing a reliable quantity of clean seawater with a minimum ecological impact [39].

Screens are used to protect water pumps from clogging and to remove coarse floating solids from feed water. But a proper water intake is needed in order to limit the impingement phenomenon occurring when marine organisms are trapped against intake screens by the velocity and force of water flowing through them. Some species may be able to survive impingement and be returned to the sea, but the 24-hour survival rate of less robust species and/or juvenile fish may be less than 15%. Entrainment occurs when smaller organisms pass through an intake screen and into the process equipment.

Traditional travelling water screens are equipped with revolving wire mesh panels with 6–9.5 mm openings. As the wire mesh panels revolve out of the stream of flow, a high-pressure water spray removes accumulated debris, washing it into a trough for further disposal. The screens can be located onshore, or at the end of a channel that extends out beyond the surf zone, or at the end of a pipe that extends out into the sea, terminating in a vertical "velocity cap" inlet, that converts vertical flow into horizontal flow at the intake entrance to reduce fish entrainment.

3.8.1.2 Chlorination

Chlorination is a basic disinfection method aimed at destroying or inactivating pathogenic microorganisms including bacteria, amoebic cysts, algae, spores and viruses. The feed rate may be manually or automatically regulated; chlorinators are usually available with maximum feed rates ranging from 50 to 3,600 kg/d.

Sodium hypochlorite is the most commonly utilised direct solution for disinfection. The reaction occurring in solution is:

$$NaOCl + H_2O \rightarrow HOCl + NaOH$$

Chlorine gas can also be injected directly into the main flow, obviating the need for a solution phase. It dissolves readily in water to a maximum concentration of 3,500 mg/l and reacts as:

$$Cl_2 + H_2O \rightarrow HOCl + HCl$$

This type of system may be much more efficient for killing bacteria or viruses, especially when other soluble constituents in the water (such as ammonia) are competing for the chlorine. However, direct injection is not a common procedure.

In water, hypochlorous acid dissociates to hydrogen and hypochlorite ions:

$$HOCl \leftrightarrow H^{+} + OCl^{-}$$

The sum of Cl_2, NaOCl, HOCl and OCl^- is referred to as free residual chlorine. For continuous chlorination at the intake point, a free residual chlorine concentration of 0.5–1.0 mg/l should be maintained along the pretreatment line to prevent biofouling. Chlorine contact basins allow sufficient contact time for disinfection. Contact or detention times of 10–120 min may be required, before the distribution or discharge of the water, depending upon the level of residual chlorine [40].

3.8.1.3 Coagulation and Flocculation

Coagulation is applied to remove the fine suspended solids that fail to settle during the sedimentation process, as well as soluble organic and toxic substances, and trace metals (size range of 1 nm to 1.0 μm). Types of coagulants include: aluminium sulfate (or alum, $Al_2(SO_4)_3 \cdot 14H_2O$), ferric sulfate $Fe_2(SO_4)_3$ or ferric chloride $FeCl_3$, quicklime (CaO), and synthetic cationic polymers, anionic polymers and nonionic polymers. The molecular weight range of the synthetic polymers is 10^4–10^7 [41].

Initial, or flash, mixing of these chemicals is necessary to disperse the coagulant quickly and evenly. Several systems are available for this purpose including diffuser grid, chemical jet and in-line blender systems. Flocculation is achieved through slow mixing using mechanical devices such as paddles, turbines or propellers; this process facilitates the agglomeration of particles and minute flocs into large flocs, by gradually bringing the particles together. Best results are achieved if the mixing speed is gradually decreased during the flocculation process. Often several flocculation chambers are built in series with successively decreasing mixing velocities.

The most commonly used coagulant is alum. The addition of Al^{3+} in the form of alum, at concentrations less than the solubility limit for the metal hydroxide, results in the formation of the metal hydroxide that adsorbs onto particles thus causing destabilisation by charge neutralisation. Addition of aluminium and iron salts, at concentrations greater than the solubility limit of the metal hydroxide, will result in formation of the hydroxide precipitate:

$$Al_2(SO_4)_3 \cdot 14H_2O + 3Ca(HCO_3)_2 = 2Al(OH)_3(s) + 3CaSO_4 + 14H_2O + 6CO_2$$

In this situation, charge neutralisation and enmeshment in the precipitate both contribute to coagulation.

Theoretically, using the stoichiometry of the above reaction, 1 mg/l of alum will consume approximately 0.50 mg/l of alkalinity (as $CaCO_3$) and produce 0.44 mg/l of carbon dioxide. If the natural alkalinity of the water is not sufficient to react with the alum and buffer the pH, it may be necessary to increase alkalinity by adding lime or soda ash (Na_2CO_3) to the water.

3.8.1.4 Slow Sand Filtration

Slow sand filtration includes biological activity in addition to physical and chemical mechanisms to remove impurities from water. It can be successfully applied if the water turbidity is less than 50 NTU. Basic operation involves passing water through a bed of fine sand at a low velocity, which causes the retention of suspended matter in the upper 0.5–2 cm of the filter bed. The principal purification mechanism is a biological process. Scraping off the top layer cleans the filter and restores it to its original capacity. The interval between two successive cleanings ranges from a few weeks to a few months, depending on the feed water characteristics. The advantages of slow sand filtration derive from the simplicity of its design and operation and the fact that it requires little in the way of power and expensive chemicals. Slow sand filters are appropriate for the removal of organic and inorganic suspended matter as well as pathogenic organisms present in the surface waters of rural areas in developing countries. Sludge handling problems are also minimal. Close control by an operator is not necessary. However, slow sand filters require a large area, large quantities of the filter medium and labour for the required manual cleaning.

3.8.1.5 Dual Media Filtration

Dual Media Filters (DMFs) remove remaining suspended solids and lower the turbidity of the RO feed water to about 0.4 NTU with an SDI of 2.5 or less. The DMF consists of coarse anthracite coal over a bed of fine sand, which provides the final clarification step [42]. When influent water passes through the DMF, unsettled flocculated material and impurities come in contact with the media grains and adhere. This reduces the size of the water channel resulting in a straining action.

Cleaning is usually required when water head loss through the filter exceeds 1.5–2.5 m. A process called backwashing is used to break up surface scum and dirt in the filter medium. Gravity flow is more economical than pressure flow as a means of cleaning filters for large storage tanks.

3.8.1.6 pH Adjustment

Before entering the cartridge filters and the subsequent RO stages, the pH value has to be reduced (at moderate pH, RO shows better performance and cellulose acetate membranes suffer from less hydrolysis). Sulphuric acid is typically injected to adjust the pH to around 7.5 thus preventing calcium carbonate caking of the filters.

3.8.1.7 Anti-scaling Agents

The addition of anti-scaling agents is advisable for seawater RO systems that work with recoveries greater than 35% [43]. Sodiumhexametaphosphate (SHMP) was commonly used as an anti-scalant, but has been widely replaced by polymeric compounds (such as polyphosphates and polyacrylates) due to the eutrophicating properties of SHMP and associated disposal problems.

3.8.1.8 Dechlorination

Dechlorination has to be performed prior to the RO stage in order to avoid oxidation damage by residual chlorine in the feedwater [44]. Commonly, sodium metabisulphite is used for dechlorination due its high cost effectiveness. In water it reacts forming sodium bisulphite:

$$Na_2S_2O_5{+}H_2O \rightarrow 2NaHSO_3$$

The sodium bisulphite then reduces the hypochlorous acid:

$$2NaHSO_3{+}2HOCl \rightarrow H_2SO_4{+}2HCl + Na_2SO_4$$

In practice, 3.0 mg of sodium metabisulphite is typically used to remove 1.0 mg of free chlorine.

3.8.1.9 Cartridge Filters

Cartridge filters are preferred systems for low water contamination. Cartridge filters can be surface or depth-filters; depth-filters capture contaminants and particles through the whole thickness of the medium, whilst in surface filters particles are blocked at the surface. Usual pore size ranges from 5 to 10 μm.

3.8.1.10 MF/UF Pretreatment

Microfiltration is a low energy-consuming technique extensively used to remove suspended solids and to lower COD/BOD and SDI to values below 5. Ultrafiltration retains suspended solids, bacteria, macromolecules and colloids. Despite the larger pressure gradient compared to MF, the UF membrane separation method remains economically competitive compared to conventional pretreatments. Turbidity is also greatly decreased, and SDI is always below 2 in the permeate stream. The main benefits of MF/UF pretreatment technologies are [45]:

- potential for a higher RO design flux and water recovery factor (for feeds below 35,000 ppm TDS, the RO flux can be increased by 25% or more);
- low space saving with UF;
- reduced requirement for RO disinfection and cleaning;
- improved treated water quality.

Using membranes in pretreatment, the raw water is usually roughly prefiltered with a mechanical screen before it is fed onto the membrane. Chemical dosing in membrane pretreatment is significantly reduced compared to conventional pretreatment.

Continuous cost reduction of MF and UF in large applications makes these options attractive for SWRO pretreatment [46, 47]. Although the capital cost of membrane pretreatment exceeds that of conventional processes by 20–50% (for instance, membranes are about twice as expensive as dual media filters), this is compensated by cost reduction in the subsequent RO trains. A typical annual replacement rate for a conventional pretreatment system is 15–20%, which reduces to 10–15% if MF/UF technology is used. RO cleaning frequencies are also reduced by UF/MF pretreatment, thus resulting in considerable lengthening of the interval between cleans and a decrease of cleaning chemicals and chemical waste. Membranes have a moderate running cost (replacement after 5–10 years) and power cost compared to conventional units.

Notation

J_i	molar flux of the i th ion (mol/m^{2}s)
ΔP	pressure difference across the membrane (Pa)
c	concentration (mol/m^3)
c_i	concentration of the i th substance (mol/m^3)
D	diffusion coefficient (m^2/s)
F	Faraday constant (C/mol)
I	current (A)
J_i	flux of the i th component (mol/m^{2}s)
J_s	solute molar flux (mol/m^{2}s)
J_v	volumetric flux through the membrane (m/s)
k	partition coefficient (–)
k_{Don}	Donnan partition coefficient (–)
k_{size}	Steric partition coefficient (–)
L_v	hydrodynamic permeability coefficient (m^2/Pa s)
L_w	diffusive permeability coefficient (mol^2/m^4 Pa s)
Q	flow rate (m^3/s)

R	gas constant (m^3 Pa/mol K)
r	pore radius (m)
R_i	rejection coefficient of the *i* th ion (–)
Re	Reynolds number (–)
RR	recovery ratio (%)
Sc	Schmidt number (–)
Sh	Sherwood number (–)
t	time (s)
T	temperature (K)
T_i	transport number (–)
$\bar{V}$	partial molar volume (m^3/mol)
z	spatial coordinate through the membrane (m)
z_i	valence of the *i* th ion (–)

Subscripts and Superscripts

b	bulk of the solution
c	concentrate solution
f	feed solution
fin	final
in	initial
m	membrane surface
p	permeate solution
s	solute
w	water

Greek Symbols

α	numerical coefficient (–)
β	numerical coefficient (–)
δ	membrane thickness (m)
ϵ	membrane porosity (–)
γ	numerical coefficient (–)
η	viscosity of the solution (Pa s)
κ	mass transfer coefficient (m/s)
ν_i	van't Hoff coefficient of the *i*th ion (–)
Π	osmotic pressure (Pa)
$\Delta\Pi$	osmotic pressure difference (Pa)
φ_{Don}	electric Donnan potential (V)
τ	tortuosity factor (–)

Abbreviations

BOD	biochemical oxygen demand
CA	cellulose acetate
COD	chemical oxygen demand
DIPS	diffusion-induced phase separation
DMF	dual media filtration
DOC	dissolved organic carbon
DOM	dissolved organic matter
ED	electrodialysis
LSI	Langelier saturation index
MCr	membrane crystallisation
MD	membrane distillation
MF	microfiltration
MFI	modified fouling index
MW	molecular weight
NF	nanofiltration
NOM	natural organic matter
NTU	nephelometric turbidity units
PA	polyamide
PAN	polyacrylonitrile
PCA	polycarbonate
PES	polyethersulfone
PET	polyethyleneterephtalate
PI	polyimide
PP	polypropylene
PSU	polysulfone
PTFE	polytetrafluoroethylene
PVDF	polyvinylidenefluoride
RO	reverse osmosis
S&DSI	Stiff and Davis stability index
SDI	Silt Density Index
SHMP	sodiumhexametaphosphate
SWM	spiral-wound modules
SWRO	seawater reverse osmosis
TDS	total dissolved solids
TFC	thin film composite
TIPS	temperature-induced phase separation technique
TSS	total suspended solids
UF	ultrafiltration

References

1. E. Drioli, E. Curcio, G. Di Profio, State of the art and recent progresses in membrane contactors, Chem. Eng. Res. Des. 83 (2005) 223–233
2. N. A. Zubir, A. F. Ismail, Effect of sintering temperature on the morphology and mechanical properties of PTFE membranes as a base substrate for proton exchange membrane, Songklanakarin J. Sci. Technol. 24 (2002) 823–831
3. R.W. Gore, Very Higly stretched polytetrafluoroethylene and process therefor, US Patent 3, 962, 153, (1976)
4. R.L. Fleischer, P.B. Price, R.M. Walker, Nuclear tracks in solids, Sci. Am. 220 (1969) 30–39
5. P. van de Witte, P.J. Dijkstra, J.W.A. van der Berg, J. Feijen, Phase separation processes in polymer solutions in relation to membrane formation, J. Memb. Sci. 117 (1996) 1–31
6. J.C. Jansen, M. Macchione, C. Oliviero, R. Mendichi, G.A. Ranieri, E. Drioli, Rheological evaluation of the influence of polymer concentration and molar mass distribution on the formation and performance of asymmetric gas separation membranes prepared by dry phase inversion, Polymer 46 (2005) 11366–11379
7. R.A. Prakash, N.V. Desai, R. Rangarajan, Interfacially synthesized thin film composite RO membranes for seawater desalination, J. Memb. Sci. 124 (1997) 263–272
8. J.E. Cadotte, Reverse Osmosis Membranes, US Patent 4277, 344, (1981)
9. C. Fritzmann, J. Löwenberg, T. Wintgens, T. Melin, State-of-the-art of reverse osmosis desalination, Desalination 216 (2007) 1–76
10. J.Schwinge, P.R. Neal, D.E. Wiley, D.F. Fletcher, A.G. Fane, Spiral wound modules and spacers: Review and analysis, J. Memb. Sci. 242 (2004) 129–153
11. B. Antrim, B. Liu, A. von Gottberg, World's largest spiral element – history and development, Desalination, 178 (2005) 313–324
12. A. Gabelman, S.-T. Hwang, Hollow fiber membrane contactors, J. Memb. Sci. 159 (1999) 61–106
13. S. Loeb, S. Sourirajan, High flow porous membranes for separation of water from saline solutions, US Patent 3, 133, 132, (1964)
14. MWH, Water Treatment – Principles and Design, 2nd ed., Wiley (2005)
15. G. Foley, A review of factors affecting filter cake properties in dead-end microfiltration of microbial suspensions, J. Memb. Sci. 274 (2006) 38–46
16. W.R. Bowen, F. Jenner, Theoretical descriptions of membrane filtration of colloids and fine particles: An assessment and review, Adv. Colloid Interface Sci. 56 (1995) 141–200
17. N. Hilal, H. Al-Zoubi, N.A. Darwish, A.W. Mohamma, M. Abu Arabi, A comprehensive review of nanofiltration membranes: Treatment, pretreatment, modelling, and atomic force microscopy, Desalination 170 (2004) 281–308
18. A.E. Yaroshchuk, Non-steric mechanisms of nanofiltration: Superposition of Donnan and dielectric exclusion, Sep. Purif. Technol. 22–23 (2001) 143–158
19. U. Merten, Transport properties of osmotic membranes, in Desalination by Reverse Osmosis (Ed. U. Merten), Cambridge MA: The M.I.T. Press (1966)
20. S.S. Sablani, M.F.A. Goosen, R. Al-Belushi, M. Wilf, Concentration polarization in ultrafiltration and reverse osmosis: A critical review, Desalination 141 (2001) 269–289
21. H. Strathmann, Ion exchange membrane separation processes, Amsterdam: Elsevier (2004)
22. E. Drioli, A. Criscuoli, E. Curcio, Membrane contactors: Fundamentals, applications and potentialities, Amsterdam: Elsevier (2006)
23. E. Drioli, E. Curcio, A. Criscuoli, G. Di Profio, Integrated system for recovery of $CaCO_3$, NaCl and $MgSO_4{\cdot}7H_2O$ from nanofiltration ritentate, J. Memb. Sci. 239 (2004) 27–38
24. S. El-Manharawy, A. Hafez, Water type and guidelines for RO system design, Desalination 139 (2001) 97–113
25. J.S. Baker, L.Y. Dudley, Biofouling in membrane systems – A review, Desalination 118 (1998) 81–89

26. C.A.C. van de Lisdonk, B.M. Rietman, S.G.J. Heijman, G.R. Sterk, J.C.Schippers, Prediction of supersaturation and monitoring of scaling in reverse osmosis and nanofiltration membrane systems, Desalination 138 (2001) 259–270
27. G. Pátzay, G. Stáhl, F.H. Kármán, E. Kálmán, Modeling of scale formation and corrosion from geothermal water, Electrochim. Acta 43 (1998) 137–147
28. M. Al-Shammiri, M. Al-Dawas, Maximum recovery from seawater reverse osmosis plants in Kuwait, Desalination 110 (1997) 37–48
29. R.Y. Ning, T.L. Troyer, Colloidal fouling of RO membranes following MF/UF in the reclamation of municipal wastewater, Desalination 208 (2007) 232–237
30. A. Mosset, V. Bonnelye, M. Petry, M.A. Sanz, The sensitivity of SDI analysis: From RO feed water to raw water, Desalination 222 (2008) 17–23
31. J.C. Schippers, J. Verdouw, The modified fouling index, a method of determining the fouling characteristics of water, Desalination, 32 (1980) 137–148
32. A.W. Zularisam, A.F. Ismail, R. Salim, Behaviours of natural organic matter in membrane filtration for surface water treatment – a review, Desalination 194 (2006) 211–231
33. S. Lee, J. Cho, M. Elimelech, Combined influence of natural organic matter (NOM) and colloidal particles on nanofiltration membrane fouling, J. Memb. Sci. 262 (2005) 27–41
34. E. Aoustin, A.I. Schafer, A.G. Fane, T.D. Waite, Ultrafiltration of natural organic matter, Sep. Purif. Technol. 22–23 (2001) 63–78
35. E. Drioli, M. Romano, Progress and new perspectives on integrated membrane operations for sustainable industrial growth. Ind. Eng. Chem. Res. 40 (2001) 1277–1300
36. E.R. Cornelissen, J.S. Vrouwenvelder, S.G.J. Heijman, X.D. Viallefont,D. Van Der Kooij, L.P. Wessels, Periodic air/water cleaning for control of biofouling in spiral wound membrane elements, J. Memb. Sci. 287 (2007) 94–101
37. E. Zondervan, B. Roffel, Evaluation of different cleaning agents used for cleaning ultra filtration membranes fouled by surface water, J. Memb. Sci. 304 (2007) 40–49
38. S.S. Madaenl, T. Mohamamdi, M.K. Moghadam, Chemical cleaning of reverse osmosis membranes, Desalination 134 (2001) 77–82
39. D. Gille, Seawater intakes for desalination plants, Desalination 156 (2003) 249–256
40. A.M. Shams El Din, R.A. Arain, A.A. Hammoud, On the chlorination of seawater, Desalination 129 (2000) 53–62
41. B.Q. Donald, J.K. Edzwald, J. Dahlquist, L. Gillberg, Pretreatment considerations for dissolved air flotation: Water type, coagulants and flocculation, Water Sci. Technol. 31 (1995) 63–71
42. A. Zouboulis, G. Traskas, P. Samaras, Comparison of single and dual media filtration in a full-scale drinking water treatment plant, Desalination 213 (2007) 334–342
43. D. Hasson, A. Drak, R. Semiat, Induction times induced in an RO system by antiscalants delaying CaSO4 precipitation, Desalination 157 (2003) 193–207
44. M.O. Saeed, Effect of dechlorination point location and residual chlorine on the biofouling in a seawater reverse osmosis plant. Desalination, 143 (2002) 229–235
45. C.V. Vedavyasan, Pretreatment trends- an overview, Desalination 203 (2007) 296–299
46. D. Vial, G. Doussau, The use of microfiltration membranes for seawater pre-treatment prior to reverse osmosis membranes, Desalination, 153 (2002) 141–147
47. D.F. Halpern, J. McArdle, B. Antrim, UF pretreatment for SWRO. Pilot studies, Desalination 182 (2005) 323–332

Chapter 4
Commercial Desalination Technologies

An Overview of the Current Status of Applications of Commercial Seawater Desalination Processes

Hisham Ettouney and Mark Wilf

Abstract The vast majority of commercial desalination systems utilise one of four desalination processes: reverse osmosis (RO), multi-stage flash (MSF), multiple-effect distillation (MED) and mechanical vapour compression (MVC). A small fraction of desalination systems utilise electrodialysis (ED) technology to treat low salinity brackish water. Worldwide, RO desalination systems account for close to 50% of overall capacity. However, in the arid countries of the Middle East, the majority of desalination systems utilise evaporation processes: MSF, MED and MVC. Although, the energy requirement of evaporation processes is higher than that for membrane processes, distillation desalination systems will continue to dominate Middle East markets for some time to come, due to the large base of thermal desalination units, with proven high operational reliability and the convenience of their integration with power plants (dual purpose systems). With the growing trend to privatise the desalination market in the Middle East, the proportion of desalination capacity supplied by RO will increase, due to the better economics of the RO process. In this chapter, an up-to-date review of industrial units has been performed in order to present the most common features of industrial operating units. This review includes typical design parameters, operating conditions and process performances. Moreover the developments that have taken place over the years are presented, along with examples of recent installations and their production capacities, performance parameters and locations.

H. Ettouney (✉)
Department of Chemical Engineering, College of Engineering and Petroleum, Kuwait University, P.O. Box 5969 – Safat 13060, Kuwait
e-mail: ettouney@hotmail.com

M. Wilf (✉)
Tetra Tech-Infrastructure Services Group, 10815 Rancho Bernardo Road, Suite 200 San Diego, CA 92127, USA
e-mail: mark.wilf@tetratech.com

A. Cipollina et al. (eds.), *Seawater Desalination*, Green Energy and Technology,
DOI 10.1007/978-3-642-01150-4_4,

4.1 Introduction

In Chaps. 2 and 3, the most used desalination processes have been presented, discussing their theoretical aspects, fundamental phenomena and giving some high level information on the design and control of industrial units.

Although there are four adopted desalination technologies, namely, reverse osmosis (RO), multi-stage flash (MSF), multiple-effect distillation (MED) and mechanical vapour compression (MVC), the RO and MSF processes account for about 50% and 40% of the entire desalination market, respectively. The use of the RO process continues to grow at a fast rate due to the development of more efficient and less expensive membranes, and the reduction of the energy requirement. In addition, the accumulation of field experience in RO design, construction, operation and maintenance has helped its growth and expansion throughout the world. The RO process is also used for desalination of sea, brackish and river waters. The RO process is the only desalination technology used in some countries, including the USA, Spain, Cyprus and Malta.

On the other hand, over the years the output capacity of MSF units has increased significantly. Presently MSF desalination systems are being built with unit product water capacities up to 50,000–75,000 m^3/day. This development has helped in the reduction of the capital cost component of water costs, and has made the MSF process more competitive. Most MSF plants are located in the Gulf States, which adopted this process in the 1950s. The accumulated field experience and infrastructure of MSF processes in the Gulf States, as well as the low cost of local energy and the increase in MSF unit capacity, continues to favour MSF in most of the Gulf States.

The application of the other two thermal technologies, MED and MVC, has been limited in the Middle East region due to the smaller capacity achieved by each single unit, and by some strategic choices made in the Gulf countries. Nowadays, the acceptance of the MED process has increased its use somewhat, due to its lower energy consumption compared to MSF, and due to the increase in energy costs which has characterised the last decade. Finally, the application of MVC technology seems to be limited to small capacity requirements and the strong preference of end users for a thermal desalination technology.

The application of reverse osmosis to water desalination initially started with the treatment of low salinity, brackish water sources. The early membrane elements, based on cellulose acetate membrane material, did not have sufficient salt rejection performance to produce potable water, from seawater, at an acceptable cost. With the introduction of aromatic polyamide membranes in the 1980s, membrane performance improved dramatically. Current RO membranes enable the operation of desalination units at a lower pressure and reduce seawater salinity to potable standards in a single-pass process. Large scale utilisation of reverse osmosis desalination plants, as a source of potable water, started in US, mainly in Florida. Seawater RO plants were initially built at arid locations, in countries without significant oil resources and with a high influx of tourists. Examples of such locations are the island of Malta, the Canary Islands and the Caribbean. Over time, due to its low energy requirement and improved economics, the RO process has grown at a higher rate than thermal desalination processes, and is currently applied worldwide to augment water supply. The present worldwide capacity of RO desalination systems is over 15 million m^3/day of product water, and growing at an annual rate of over 10%.

In this chapter, an up-to-date review of industrial units has been performed, in order to present the most common features of commercial desalination units. In the first paragraphs, thermal units are considered, presenting typical design parameters, operating conditions and process performances of worldwide operating plants. A historical overview follows, which highlights the growth trends of the thermal desalination industry, indicating how each technology has been differently adopted and likely future trends. Finally, several examples of operating plants are provided, together with some fundamental information on start-up year, manufacturer, unit configuration and plant capacity.

A similar approach is used for membrane processes. The industrial examples are limited to Reverse Osmosis, which is the predominant membrane desalination process, applied commercially, to produce fresh water from seawater.

4.2 Design, Operating and Performance Parameters of Thermal Units

With reference to thermal desalination technologies (see Chap. 2), the main design and operating parameters for these units are shown in Table 4.1. The largest production capacity is for the MSF system and is followed by MED and MVC. All systems, except for once-through MSF, can operate over a wide range of intake seawater temperatures. All systems can also handle a wide range of intake seawater salinity. One of the most important performance parameters is the gain output ratio (GOR), defined as the amount of distillate produced per unit mass of heating steam. Another fundamental parameter is the specific power consumption, defined as the amount of power consumed per unit mass of product. For MSF and MED, the GOR

Table 4.1 Most common design and operating parameters of commercial thermal desalination systems

Common design parameter	
Intake seawater temperature (T_{cw})	5 35°C
Intake seawater salinity (X_{cw})	36,000–45,000 ppm
Brine blow down temperature (T_{bn})	5–10°C higher than intake seawater temperature
MVC design parameters	
Number of effects	1–4
Top brine temperature	70°C
Production capacity per unit	Less than 5,000 m^3/day
Brine blow down salinity	65,000 ppm
Wall thickness of evaporator tubes	0.4–1.0 mm (the lower limit is for titanium tubes)

Table 4.1 (continued)

Common design parameter	
Diameter of evaporator tubes	0.01–0.05 m
Length of evaporator tubes	5 m
MSF design parameters	
Number of flashing stages (n)	20–24 (some old units up to 52 stages [1])
Production capacity (M_d)	5,000–75,000 m^3/day
Top brine temperature (T_{bo})	90–110°C
Gain Output Ratio (GOR)	8–10 $kg_{distillate}/kg_{motive\ steam}$
Electrical power consumption	3–4 $kWh/m^3{}_{distillate}$
Brine blow down salinity for MSF/BR (X_{bn})	70,000 ppm
Brine blow down salinity for MSF/OT (X_{bn})	40,000–60,000 ppm
Wall thickness of condenser tubes	0.5–1.3 mm, the lower limit is for titanium tubes
Diameter of condenser tubes	0.01–0.05 m
Length of condenser tubes	Depends on unit production capacity and may reach 25 m for 75,000 m^3/day units.
MED design parameters	
Number of evaporation effects (n)	2–12
Production capacity (M_d)	up to 30,000 m^3/day
Gain output ratio (GOR)	8–16 $kg_{distillate}/kg_{motive\ steam}$ (the higher values for MED–TVC)
Electrical power consumption	1.2–2 $kWh/m^3{}_{distillate}$
Wall thickness of evaporator tubes	0.4–1.0 mm, the lower limit is for titanium tubes
Diameter of evaporator tubes	0.01–0.05 m
Length of evaporator tubes	5–8 m
Top brine temperature (T_{bo})	70°C
Brine blow down salinity (X_{bn})	65,000 ppm

ranges between 8 and 10. For MED-TVC the GOR increases to values up to 16. For this reason, most existing MED-TVC systems can be operated in standalone mode. Conversely, MSF and MED largely benefit from being coupled with thermo-electric power stations, which easily provide low cost low pressure steam and electricity.

The source of thermal energy for MSF and MED desalination processes is low pressure steam. In addition, electric energy is required for internal pumping. Mechanical Vapour Compression and RO processes utilise only electric energy

Table 4.2 Typical features of energy input in seawater desalination processes

Process/energy type	MED	MSF	MVC	RO
Motive steam pressure, ata	0.2–0.5	1.0–2.5	–	–
Average electrical energy equivalent, kWh/m^3	4.5	14.0		
Electrical consumption, kWh/m^3	1.2–2.0	3.0–4.0	8.5	3.5–5
Total electrical energy equivalent, kWh/m^3	5.7–6.5	17–18	8.5	3.5–5

to drive the compressors (MVC) and process pumps (RO). A summary of energy requirements is provided in Table 4.2, adapted from the chapter by Leon Awerbuch in the book by Wilf [2].

In the following paragraphs some more detailed information will be given for each of the above mentioned technologies.

4.3 Multiple-Effect Distillation (MED)

The MED process is a well established technology in the desalination industry. The process dates back to 400–500 years ago. The single stage evaporation process originated on ship decks in the sixteenth century, where wood or coal stoves were used to evaporate seawater to generate fresh water. At that early stage, product quality was difficult to control due to a large entrainment of brine in the vapour product. Further developments in the evaporation process took place in the sugar, pulp and paper industries [3]. The use of evaporation processes for water desalination was practised by the oil industry during the first and second world wars. The early years of the desalination industry, during the 1950s, involved the use of evaporation as well as flash desalination processes. These processes had small production capacities of less than 500 m^3/day. Further developments in the MED process took place during the 1960s and 1970s. This led to the present form of the MED process, with capacities of around 5,000 m^3/day. In 2006, the MED unit capacity increased to a value of 36,000 m^3/day [4]. Most MED processes operate at low temperatures (lower than 70°C). This is to limit the rate of scale formation on the outside surface of the evaporator tubes. Also, it allows efficient use of thermal or mechanical vapour compression, where the vapour from the last effect is re-compressed over a temperature range of 30–40°C.

The MED process remains limited to smaller system capacities (compared with MSF). The MED market share is about 12.5% of the entire thermal desalination processes market and less than 6% of the entire desalination market, which is dominated by RO and MSF. The market share of MED processes is presented in terms of cumulative production capacity in the Gulf countries and citations for the most recent installed large MED units across the world. Figure 4.1 shows the cumulative production capacity of MED in the Gulf countries. The production capacity for Saudi Arabia, Oman, and Qatar is less than 100,000 m^3/day, which is close to the production capacity of a single large scale MSF unit. The largest production capacity of MED is found in the UAE, with a value of 600,000 m^3/day. However this production capacity accounts for less than 10% of the MSF production capacity of the UAE. It should be noted that the figure does not include the state of Kuwait, where no MED units are used for seawater desalination.

Table 4.3 shows recent MED installations around the world. The majority of these units are installed in the UAE. Several large units were also installed in the Jamnagar oil refinery (India), with a total production capacity of more than 150,000 m^3/day. The most important feature of the newly installed units is the increase in unit capacity to values above 20,000 m^3/day, whereas previously, the capacity for most MEDs was below 10,000 m^3/day. In addition, most of the installed units

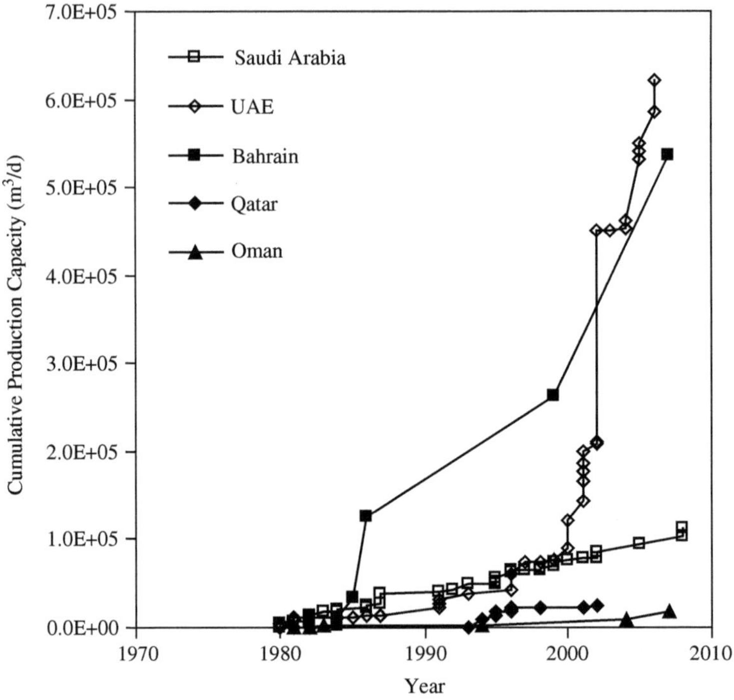

Fig. 4.1 Cumulative production capacity of MED plants in Gulf countries

Table 4.3 Examples of recent MED installations around the world

Country	Location	Year	Number units	Unit capacity (m^3/day)	notes	Manufacturer
Italy	Trapani	1993	4	9,000	TVC	VWS Sidem
Netherlands Antilles	Curaçao	1996	1	12,000	TVC	VWS Sidem
United Arab Emirates	Jebel Dhana	1996	2	9,090	TVC	VWS Sidem
United Arab Emirates	Ras Al Khaimah	1997	2	6,820	TVC	VWS Sidem
India	Jamnagar	1998	2	12,000	TVC	IDE
Italy	Priolo Gargallo	1998	2	7,200	Co-generation	VWS Sidem
Netherlands	Rotterdam	1999	2	12,000		VA TECH
Bahrain	Askar	2000	4	10,750	TVC	VWS Entropie (ex Weir)
Spain	Las Palmas	2000	2	17,500	TVC	IDE

Table 4.3 (continued)

Country	Location	Year	Number units	Unit capacity (m^3/day)	notes	Manufacturer
United Arab Emirates	Umm Al Nar	2000	2	15,911	TVC	VWS Sidem
United Arab Emirates	Layyah	2001	2	22,848	TVC	VWS Sidem
United Arab Emirates	Ajman	2001	2	6,819		Sasakura
Libya	Tobruk	2002	3	13,333	TVC	VWS Sidem
United Arab Emirates	Al Taweelah	2002	14	17,143	TVC	VWS Sidem
Iran	Band Azzaluyeh	2004	5	7,500	TVC	VWS Sidem
United Arab Emirates	Kalba	2004	1	9,090	TVC	VWS Entropie (ex Weir)
China	Huanghua	2005	2	10,000	TVC	VWS Sidem
India	Jamnagar	2005	1	15,000	TVC	IDE
Libya	Zuara	2005	3	13,333		VWS Sidem
United Arab Emirates	Ras Al Khaimah	2005	3	22,730	TVC	VWS Sidem
United Arab Emirates	Layyah	2006	4	9,092		VWS Sidem
United Arab Emirates	Layyah	2006	1	36,368		VWS Sidem
India	Jamnagar	2007	4	24,000	TVC	IDE
Bahrain	Hidah	2007	10	27,300	TVC	VWS Sidem

have horizontal tube arrangements and use thermal vapour compression. The Al-Taweelah MED plant in the UAE is one of the largest and has 14 units, a total production capacity of 240,000 m^3/day and a unit production rate of 17,143 m^3/day. Another milestone of MED technology is to be found in the new MED plants in Bahrain and UAE, which were commissioned in 2007 and 2006 respectively. The main feature of these plants is the increase in unit capacity to 27,300 and 36,000 m^3/day. In particular, the former plant has 10 units and a total product water capacity of 273,000 m^3/day.

4.4 Mechanical Vapour Compression (MVC)

The MVC process for seawater desalination was developed in the early 1980s [5]. System development was motivated by the need to have a thermal desalination system that utilised solely electric power. Originally, the MVC process was pursued to

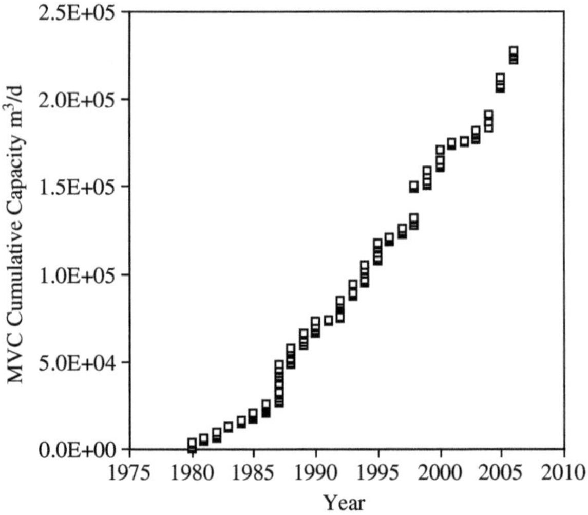

Fig. 4.2 Cumulative production capacity of MVC plants across the world

compete against the RO process. Although during the 1980s the desalination industry was mainly dominated by the MSF process, it was the start of rapid growth of the RO process. As a result, the MVC process remained limited to a very small portion of the desalination market. Review of the market data shows that the entire production capacity of the MVC process is less than 250,000 m^3/day; which is less than a single large scale production plant for any of the other desalination process (RO, MSF or MED).

Figure 4.2 shows the cumulative production capacity of the MVC process. It is evident that there is a continuous, but small, increase in the MVC market size. Review of the data in Table 4.4, shows that most of the MVC units are small in capacity, with values less than 3,000 m^3/day. The entire MVC plant capacity is also small with values less than 20,000 m^3/day. Over the past three decades, the cumulative production capacity of the MVC process increased at an approximate rate of 9,000 m^3/day each year. Considering the dominant status of the RO and MSF processes, the growth rate of the MVC process is expected to remain constant during the current and following decade. In this regard, MVC would remain to provide small amounts of desalinated water to industrial sites or small communities.

4.5 Multi-Stage Flash Units (MSF)

The MSF process accounts for about 90% of all thermal desalination processes and about 40% of seawater desalination [4]. Over the years, the MSF process has proved to be highly reliable. This has been reflected in continuous development and growth of the process. The MSF process, as well as the MED process, consumes a larger amount of energy than the RO process (about 17–18 kWh_{equiv}/m^3 for MSF, 5.7–6.5 kWh_{equiv}/m^3 for MED and 3.5–5 kWh_{equiv} /$m^3$for RO). However the reliable performance of the MSF and MED thermal desalination processes has made

Table 4.4 Examples of recent MVC[a] installations around the world

Country	Location	Year	Number of units	Unit capacity (m^3/day)	Manufacturer
Cape Verde	Praia	1995	2	1,250	IDE
Italy	Agrigento	1995	2	1,500	IDE
Spain	Bahía de Palma	1995	1	1,500	IDE
Chile	Antofagasta	1997	1	1,344	IDE
Saudi Arabia	Dammam	1997	1	1,242	Hadwaco
Australia	Freemantle	1998	1	1,800	IDE
Egypt	Abu Soma	1998	1	1,500	IDE
Greece	Lavrion	1998	2	1,200	IDE
Italy	Sardinia	1998	6	2,880	IDE
Greece	Athens	1999	1	1,920	IDE
Guinea	Guinea	1999	2	1,363	Alfa Laval
Turkmenistan	Turkmenistan	1999	2	1,500	VWS Sidem
Chile	Antofagasta	2000	1	1,560	IDE
Turkmenistan	Turkmenbashi	2000	2	3,000	IDE
USA	California	2003	1	1,450	IDE
Algeria	Arzew	2004	1	1,440	VWS Sidem
Australia	Burrup Penin	2004	3	1,200	VWS Sidem
China	Qingdao City	2004	1	3,000	
Kazakhstan	Kazakhstan	2005	1	1,100	GE Ionics
Peru	Moquegua	2005	2	1,320	IDE
India	Kudankulam	2006	4	2,560	IDE
Peru	Moquegua	2006	1	1,500	IDE

[a]All the units reported in the table are single effect.

both processes highly competitive against the RO process. Recent reports by Borsani and Rebagliati [6] and AlBahou et al. [7] show that the unit product cost for the three processes is almost the same, with a value of $0.5/m^3. In addition, field reports by Thirumeni [8], Helal [9], and Schaefer [10] show that MSF plant life approaches 40 years. Several old units installed in the 1970s and 1980s remain in operation and have been rehabilitated to continue to be operational for the subsequent 10–20 years. This fact would reduce further the unit product cost, taking into consideration that plant capital may account for 30–40% of the unit product cost. At present very large MSF units, with production capacities that range between 50,000 and 75,000 m^3/day, are being installed in several countries, including Kuwait, Saudi Arabia, and the United Arab Emirates. The large increase in unit capacity contributes further to the reduction in unit product cost.

The first MSF designs were developed in the late 1950s by Silver [11]. In the 1960s the first MSF units were installed with a capacity of 500 m^3/day. The unit capacity then increased to the current capacity of 27,000–32,000 m^3/day, which was introduced in the late 1970s. In later years, the unit capacity increased to a range of 50,000–75,000 m^3/day [6]. Although MSF is found in several countries across the world, large scale MSF plants are only found in the Gulf countries. Cumulative increase in MSF production capacity in the Gulf countries is shown in Fig. 4.3. The most striking feature of this data is the rapid increase in production capacity in the

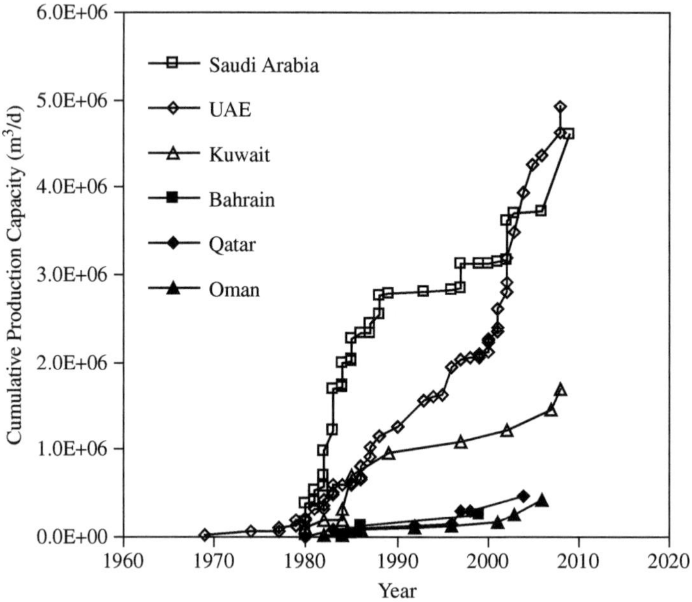

Fig. 4.3 Cumulative production capacity of MSF plants in the Gulf countries

UAE. The MSF production capacity in Saudi Arabia approached 4.6 million m^3/day in 2006, while in the UAE the MSF production capacity approached 4.9 million m^3/day in 2008. The desalination markets in Kuwait, Qatar, and Oman are fairly similar, where the MSF process is the dominant technology. However, the MSF desalination capacity in Kuwait is almost three times that of either Oman or Qatar, where the total production capacity approached 1.7 million m^3/day in 2008. The desalination industry in Bahrain is unique among the Gulf States. This is because of the stronger role of the RO process in production of desalinated water mainly from brackish wells. The RO share in the Bahrain desalination market is 27% and the remainder is accounted for by the MSF process [4].

Examples of recent MSF desalination units constructed in the Gulf countries are shown in Table 4.5. The main features of these units are the increase in unit production capacity to a range of 50,000–75,000 m^3/day, (which is becoming the industrial standard) and the increase in operating temperature to a range of 105–112°C. However, the number of stages and the performance ratio remain well within those of older units, with 20–24 stages and performance ratios of 8–10. It is noted that all the cited units are constructed by either Italian or Korean companies and all of them have been installed in Gulf Countries. Another important feature is the size of the entire plants, which have approached 1 million m^3/day.

Table 4.6 shows the main design characteristics for a number of new MSF installations. It is worth noting how system design allows for an increase in the stage width and weir loading, which is necessary in large capacity units.

Table 4.5 Examples of recent MSF installations in the Gulf countries

Plant	Year	Number of units	Unit capacity (m^3/d)	No of stages	Top brine temperature (°C)	PR
Al Taweelah "B" (Abu Dhabi-UAE)	1995	6	57,600	20	112	8
Al Hidd (Bahrain)	1999	4	37,000	21	107–112	9
Ruwais (UAE)	2001	2	15,000	15	105–112	6
Jebel Ali "K" (Dubai UAE)	2001	2	45,480	21	105	9
Jebel Ali "K" 2 (Dubai UAE)	2003	3	60,530	19	105	8
Mirfa (Abu Dhabi-UAE)	2002	3	34,000	21	110	8.9
Umm Al Nar Station "B" (UAE)	2002	5	56,825	22	110	9
Fujairah (UAE)	2003	5	56,750	22	110	9
Az Zour South (Kuwait)	1999	12	32,731	24	110	8.8
Shuweihat (Abu Dhabi-UAE)	2004	6	75,670	21	111	9
Subyia (Kuwait)	2007	12	56,825	23	110	9.5
Ras Laffan (Qatar)	2007	4	68,190	22	110	9.5
Sohar (Oman)	2008	4	37,504	24	110	9.5
Shoaiba (Saudi Arabia)	2009	12	73,645	22	110	9.5

4.6 Trends in Thermal Desalination

Over the past two decades several new trends have been demonstrated by the thermal desalination industry. These trends are dictated by the competitiveness of the market and the need to optimise and reduce unit product cost [6, 7, 13].

- Continuous increase in unit production capacity has resulted in a decrease in the unit product cost. MSF unit capacity increased from an average of 27,000–32,000 m^3/day to 50,000–75,000 m^3/day. Similarly, MED unit capacity increased from 12,000 to 20,000 m^3/day to more than 35,000 m^3/day.
- Improved-performance construction materials are used in place of conventional materials. For example, Duplex stainless steel has replaced 316L stainless steel. Duplex stainless steel provides higher corrosion resistance and has a longer service life. Similarly, titanium tubing provides superior performance compared to copper-nickel tubing, in low temperature sections of MSF and MED.
- The massive field experience in construction and operation of thermal desalination plants has resulted in streamlining of the process flow diagram. This has eliminated expensive redundant stand-by units (controls, valves, by-passes, pumps, etc.). Also, plant owners are currently using less stringent design specifications, which reduce plant capital and operating costs. Examples include use of smaller fouling factors and heat transfer areas and reduction of tube wall thickness.

Table 4.6 Design characteristics of recently installed MSF units [12]

Plant	Stage width (m)	Average stage length (m)	Average stage height (m)	Average weir loading (kg/(m s))	Average demister length (m)	Average specific heat transfer area ($m^2/(kg/s)$)	Manufacturer
Al Taweelah "B" (Abu Dhabi, UAE)	19.0	5.58	6.2	293	2.3	174	Italimpianti
Al Hidd (Bahrain)	14.2	4.66	5.6	260	2.1	198	Italimpianti
Ruwais (UAE)	8.0	3.64	4.7	193	2.3	141	Italimpianti
Jebel Ali "G" (Dubai, UAE)	14.0	4.25	5.4	226	2.0	184	Italimpianti
Jebel Ali "K" (Dubai, UAE)	17.8	4.7	5.8	237	2.0	189	Italimpianti
Jebel Ali "K" 2 (Dubai, UAE)	23.0	5.7	6.4	282	2.5	194	Italimpianti
Mirfa (Abu Dhabi, UAE)	14.0	4.4	5.5	242	2.0	197	Italimpianti
Ras Laffan (Qatar)	18.0	5.0	5.9	270	2.2	199	Doosan
Shuweihat (Abu Dhabi, UAE)	23.8	6.0	6.6	311	2.3	199	Italimpianti
Subyia (Kuwait)	19.0	4.5	4.5	288	1.7	253	Doosan

– Use of the Build-Own-Operate-Transfer (BOOT) contracts have reduced considerably construction and operating costs of desalination plants. This is directly reflected in the unit product cost.

4.7 Reverse Osmosis (RO)

4.7.1 Commercial RO Membrane Products

4.7.1.1 Configuration and Performance of Commercial Membranes

While thermal processes are complex industrial units characterised by typical features of large process apparatus, reverse osmosis plants are relatively simple and their core is undoubtedly the separation membrane.

The majority of RO membranes are made almost exclusively from two polymers: cellulose acetate blends and aromatic polyamides. Chemical structures of both membrane materials have been discussed in Chap. 3. Cellulose acetate was the first polymer used for manufacturing reverse osmosis membranes. Loeb and Sourirajan [14] developed the cellulose acetate membrane for RO application in the late 1950s. It is derived from a chemical process using cellulose, a material naturally present in plant tissue. Current cellulose acetate membranes are made from a blend of cellulose diacetate and triacetate polymers. Peterson et al. [15] introduced the second membrane material, aromatic polyamide, in the early 1980s.

Cellulose Acetate (CA) material forms membranes with asymmetric structures. This means that both the membrane barrier, responsible for separation of water and dissolved ions, and the supporting layer, are made of the same material but with different structures. The membrane barrier is dense, with high ion rejection and low water permeability. The supporting layer is porous, providing mechanical support to the thin membrane barrier. Cellulose acetate can be produced with a relatively wide range of salt rejection and water permeability properties. It is tolerant to low levels of chlorine in feed water and has a low fouling tendency, attributable to its smooth surface and neutral surface charge. Cellulose acetate membranes are susceptible to hydrolytic degradation, and so cellulose acetate membranes cannot be cleaned using chemicals with very low or very high pH. Performance and longevity of cellulose acetate membranes is not as good as for membranes made of aromatic polyamide. Therefore today, CA membranes are mainly used in applications that require the presence of free chlorine in feed water i.e. some pharmaceutical processes and food processing applications.

Composite membranes are mainly based on aromatic polyamide membrane barrier material. The composite structure means that the membrane barrier and the supporting porous layer are made of different materials. This configuration enables the independent optimisation of the properties of each component. Composite membranes are characterised by high water permeability and high salt rejection. However, the aromatic polyamide membrane barrier material is susceptible to degradation in the presence of free chlorine. The membrane has a high fouling tendency in the presence of dissolved organics. These membranes have a high tolerance to both high and low pH, therefore they can be cleaned using harsh chemicals. Due to their superior performance and longevity, composite polyamide membranes are today's "membrane of choice" for all types of applications.

The composite polyamide membrane barrier material is formed during an interfacial polymerisation reaction of two aromatic compounds: trimesoyl chloride (TMC) and meta-phenyl diamine (MPD). The resulting polymeric compound is very stable under all conditions encountered in the reverse osmosis process, with exception of the presence of strong oxidants. Strong oxidants, mainly those which contain free chlorine, would damage the salt rejection properties of this membrane material.

Figure 4.4 shows an ESM (electron scanning microscope) picture of a cross section of a composite polyamide membrane. The picture shows the distinct features of the membrane: on the bottom are fibres of polyester fabric, above is a polysulfone support which penetrates into the fabric. The polysulfone support shows a uniform porous structure. The dense surface on top of the polysulfone layer is the aromatic polyamide barrier. This very thin barrier (~ 2,000 Ä) is responsible for the separation property of the membrane. The membrane can withstand continuous operation at high feed pressure of about 7 MPa (1,000 psi) for a number of years. Aromatic polyamide membranes can be exposed to a wide range of salinity and pH values and will maintain stable performance with respect to water and salt transport.

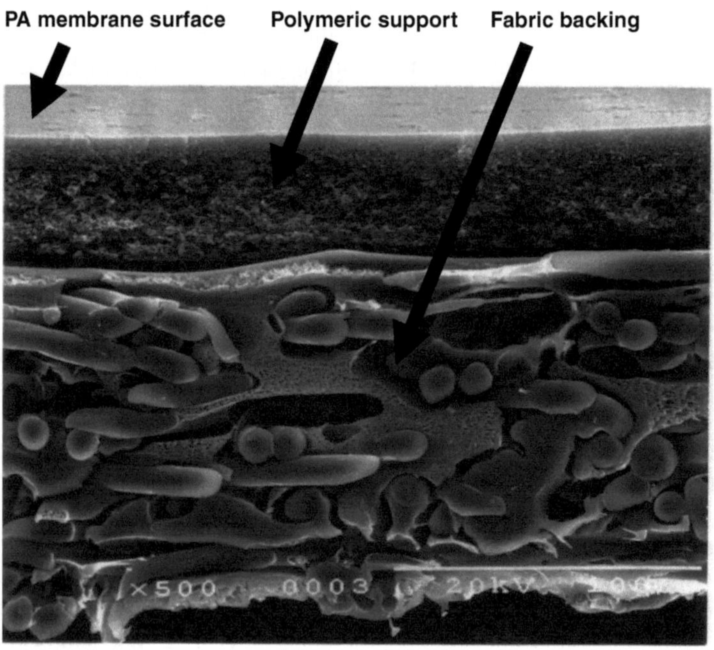

Fig. 4.4 Electron Scanning Microscope picture of a cross section of composite polyamide membrane. Magnification × 500
Source: [2], with kind permission.

4.7.1.2 Commercial Modules Configuration: Spiral Wound Elements

The application of the reverse osmosis process requires packaging of membranes into modules that provide high packing density and facilitate convenient separation of feed, permeate and concentrate streams.

The two major membrane module configurations, which have been used for reverse osmosis applications, are hollow fibre and spiral wound. At present, the spiral wound configuration is most commonly used in commercial desalination systems.

Two other configurations, tubular and "plate and frame", have found good acceptance in the food and dairy industry and some specialised applications.

In most common spiral wound configurations, two sheets of flat membrane, about 1 m long and 1 m wide, are joined together, separated by permeate spacer fabric material. This assembly is glued on three sides forming an envelope, with one side open. The open side of the membrane envelope is attached to a perforated permeate tube. After the required number of membrane envelopes are attached to the permeate tube, the membrane pack is wound around the permeate tube forming a cylindrical shape. On the feed side of the membrane envelopes, a polypropylene net maintains the separation of adjacent membrane envelopes, forming a feed channel. Commercial elements have between 20 and 40 membrane envelopes attached to the permeate tube, forming an element with a 20 cm (8") diameter and a length of 1 m. Such an element would contain 37–41 $m^2$of active membrane area. It is worth noting that other constructions are possible. For example, small under-sink elements have just one leaf, while 16" diameter elements could have up to about 150 leaves.

The schematic configuration of a spiral membrane element is presented in Fig. 4.5. As is shown in this figure, the feed water stream enters the feed channels

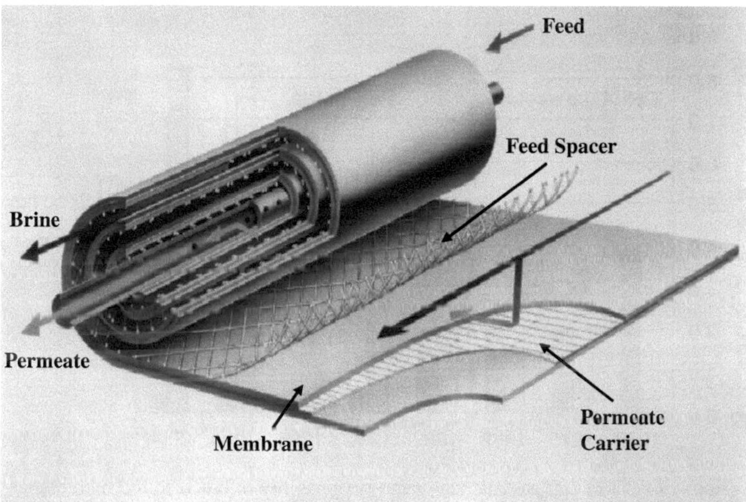

Fig. 4.5 Schematic configuration of a spiral wound element

of the membrane element. This stream splits into two: the permeate and the concentrate. Part of the feed water stream permeates through the membrane and enters the permeate channel. Inside the permeate channel there is a porous fabric called the permeate carrier. This fabric acts as a spacer, maintaining separation between the two adjacent membrane layers, thus forming the permeate channel. The permeate flows along the permeate channel to the central permeate tube. The remaining part of the feed water flows to the other end of the feed channel and exits the element. The feed channel is about 0.7 mm deep. Based on this configuration and the dimensions of the spiral wound element feed channel, it is evident that feed water to the RO device has to be very clean, in order for the membrane element to function properly. Any suspended solids present in the feed water could easily be trapped inside the feed spacer mesh, and block the flow of the feed water.

4.7.2 Performance of RO Membrane Elements

Reverse osmosis process development started with cellulose acetate (CA), the first membrane material to demonstrate that the reverse osmosis process could produce low salinity water from high salinity feedwater. The performance of CA membranes has gradually improved. CA membranes were mainly used to desalt low salinity brackish water. The introduction of aromatic polyamide composite membranes, in the early eighties, brought a dramatic improvement in performance. Nominal salt passage decreased to below 1% (better than 99% salt rejection). At the same time, water permeability has increased, thus reducing the feed pressure required for the RO process. The time line of improvement of brackish membranes is shown in Fig. 4.6.

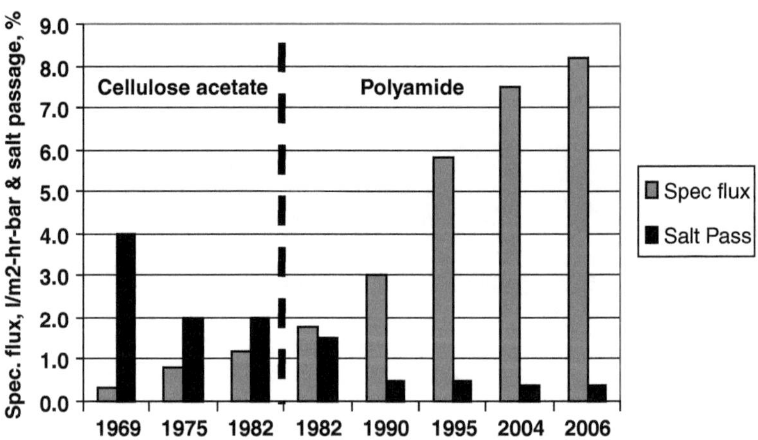

Fig. 4.6 Improvement of salt passage and water permeability of brackish RO membranes against time
Source: [2], with kind permission.

Utilising CA membranes, with relatively low salt rejection, required repeated membrane treatment steps to produce water of potable quality. Introduction of polyamide membranes made commercial desalination of seawater a reality. Polyamide composite membranes demonstrated significant improvement in salt rejection, combined with a much higher water permeability. It was then possible to produce potable water from seawater in a single desalination step, at a feed pressure lower than with CA membranes. As shown in Fig. 4.7, initial development efforts were concentrated mainly on producing seawater membranes with a higher salt rejection. More recently, seawater membranes with higher water permeabilities have been developed.

Commercial membrane elements for the removal of solutes from water solutions are manufactured for three major applications:

- Nanofiltration
- Brackish water Reverse Osmosis
- Seawater Reverse Osmosis

Nanofiltration (NF) membranes are characterised by high water permeability and high salt passage. NF membranes are used for the treatment of low salinity water, usually below 10,000 ppm TDS (total dissolved solids). Usual NF applications involve partial reduction of hardness, dissolved organics and/or colour reduction of source ground water. Illustrative commercial offerings of NF membranes are provided in Table 4.7.

The water permeability of polymer surface layers in brackish membranes is usually lower than the permeability of membrane barriers in nanofiltration membranes. However, the major performance differentiator between these two membrane types is salt rejection. The salt rejection of brackish membranes is significantly higher than

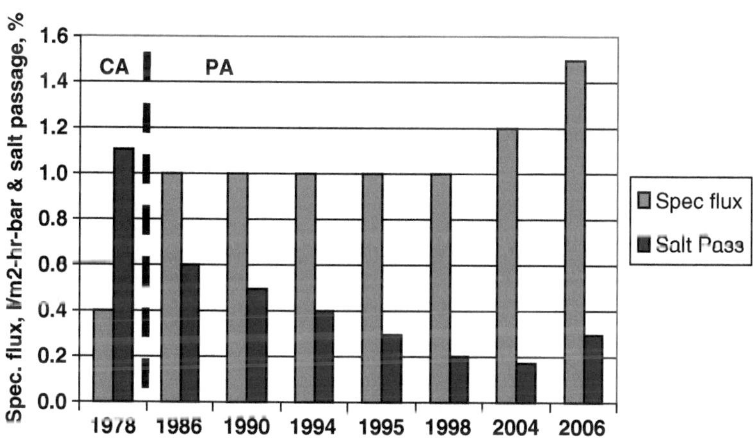

Fig. 4.7 Improvement in salt passage and water permeability of seawater RO membranes against time
Source: [2], with kind permission.

Table 4.7 Illustrative commercial offerings of NF membranes[a]

Element model	Hydracore	ESNA-LF	SU620F	NF-90	NF-270
Membrane area, m^2	37.1	37.1	37.1	37.1	37.1
Permeate flow, m^3/day	31.0	29.5	21.9	37.9	47.3
Salt rejection, %	50.0	80.0	55.0	97.0	97.0
Salt passage, %	50.0	20.0	45.0	3.0	3.0
Test flux rate l/m^2h	34.8	33.2	24.7	42.5	55.9
Permeability, l/m^2-h-bar	7.7	7.2	8.7	11.9	15.7
Relative salt transport: salt passage × flux rate	17.4	6.6	11.1	1.3	1.7

[a]Performance under nominal test conditions.

Table 4.8 Illustrative commercial offerings of brackish water membranes[a]

Element model	ESPA2+	ESPA4+	TMG20-430	BW30-XLE440	BW30 LE-440
Membrane area, m^2	40.0	40.0	40.0	40.9	40.9
Permeate flow, m^3/day	41.6	49.2	41.6	48.1	48.1
Salt rejection, %	99.60	99.60	99.50	99.0	99.30
Salt passage, %	0.40	0.40	0.50	1.00	0.70
Test flux rate, l/m^2-h	43.5	51.3	43.5	49.1	49.1
Permeability, l/m^2-h-bar	5.0	8.2	6.2	7.7	6.0
Relative salt transport: salt passage × flux rate	0.261	0.308	0.218	0.491	0.344

[a]Performance under nominal test conditions.

for NF membranes, as shown in Tables 4.7 and 4.8, which list membrane element performance values for the corresponding membrane categories.

Table 4.9 lists performance values of membrane elements for seawater applications. Water permeability is significantly lower than water permeability of brackish water membrane elements. The values for salt rejection of seawater membranes are only marginally higher than the salt rejection values of brackish membranes. However, when comparing salt passage or the values of relative salt transport (last row

Table 4.9 Illustrative commercial offerings of seawater RO membranes[a]

Element model	SWC4+	SWC5	TM820-400	SW30HR-LE	SW30HR-XLE
Membrane area, m^2	37.1	37.1	37.1	37.1	37.1
Permeate flow, m^3/d	24.6	34.1	24.6	26.5	34.1
Salt rejection,	99.80	99.80	99.75	99.75	99.70
Salt passage, %	0.20	0.20	0.25	0.25	0.30
Test flux rate, l/m^2-h	27.6	38.2	27.6	31.3	38.2
Permeability, l/m^2-h-bar	1.0	1.5	1.0	1.2	1.5
Relative salt transport: salt passage × flux rate	0.055	0.076	0.069	0.078	0.114

[a]Performance under nominal test conditions.

of Tables 4.7 and 4.8) between these two membrane categories, it is evident that the transport of dissolved ions through seawater membranes is significantly lower than the corresponding property of brackish membranes.

4.7.3 Configuration of RO Membrane Modules

In Reverse Osmosis systems, membrane elements are installed in pressure vessels to enable the application of feed water pressure and the convenient separation of feed, permeate and concentrate streams. In pressure vessels, a number of elements are placed in series, with permeate tubes of subsequent elements connected together, as shown in Fig. 4.8. Feed water enters the pressure vessel through a feed port and flows in the direction of the concentrate port, at the other end of the pressure vessel. During the passage of feed through an element, part of the feed permeates through the membrane. The volume of the feed stream is reduced thus increasing the concentration of dissolved constituents. As the feed water cascades from one element to the next, each subsequent element in series operates at a higher feed salinity. Due to increasing feed salinity and higher osmotic pressure, the flow of product water reduces and the permeate salinity increases in the direction of the feed flow.

In RO units the pressure vessels may also operate as a group, with feed, concentrate and permeate ports connected in parallel to the corresponding unit manifolds. Each group of parallel connected pressure vessels forms a "stage". A number of stages connected in series form an "array" (see Fig. 4.9).

Figure 4.9 shows a schematic configuration of a two-stage membrane unit. The pressure vessel array in the unit consists of two stages, the first stage with 4 pressure vessels and the second stage with 2 pressure vessels. The configuration ratio of 2:1 (i.e. the ratio of the number of pressure vessels from one stage to the next) is a common engineering practice in the RO industry. As the feed water flows through the first stage, approximately 50% of the feed water is converted into permeate. The configuration of a 2:1 pressure vessel ratio results in a similar average feed flow velocity within all pressure vessels in the system, therefore preventing excessive concentration polarisation (see Chap. 3).

Feed water flows through a pump, a feed control valve, the pressure vessels array and a concentrate valve. The diagram also shows feed and concentrate pressure

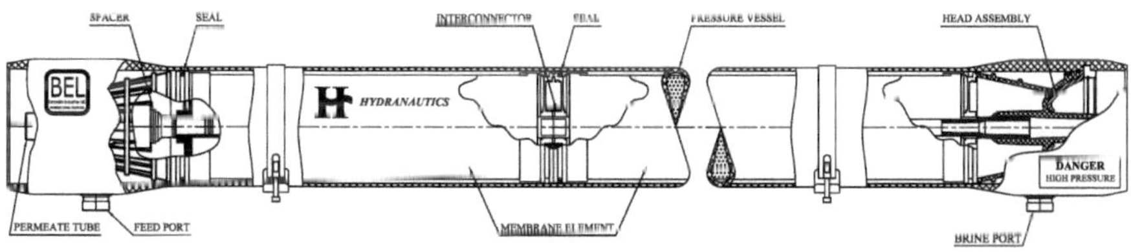

Fig. 4.8 Configuration of pressure vessels for RO applications
Courtesy of Bell Industries.

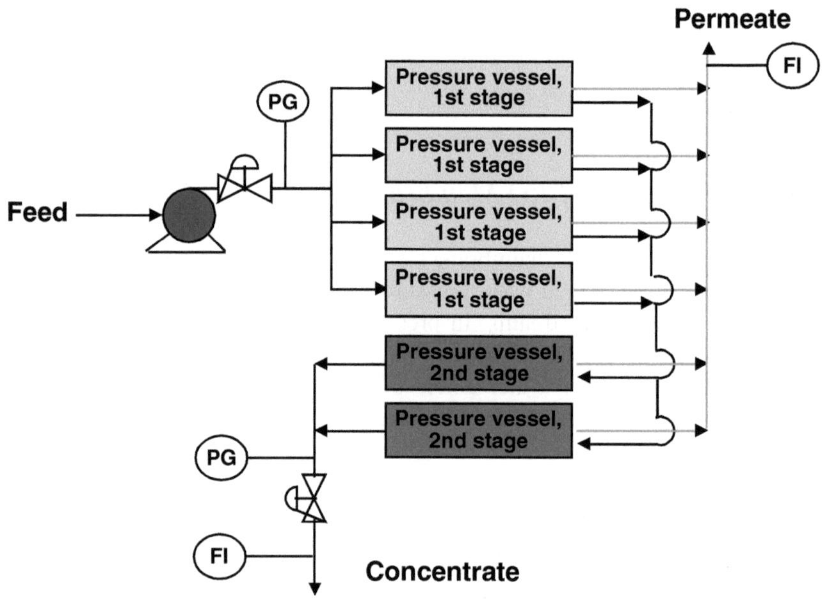

Fig. 4.9 Schematic configuration of a two-stage RO unit
Source: Handbook, EDS Course: RO, NF, Membrane Filtration and MBR Technology, L'Aquila, with kind permission.

gauges (PG) and flow indicators (FI) on the permeate and concentrate manifolds. The concentrate from the first stage is collected and becomes feed for the second stage. Permeate from both stages is collected in a common manifold.

Seawater RO (SWRO) desalination units are designed to operate almost exclusively as single-stage units. All pressure vessels are connected in parallel via feed and concentrate ports. Recovery rates of seawater systems vary with feed salinity and range between 40 and 50% for a seawater feed with average salinity (35,000 ppm). Seawater systems that treat lower salinity feeds are designed and operated at higher recovery rates, of up to 60%.

Figure 4.10 is a photograph of a brackish water RO train. It shows a two-stage unit with a pressure vessel array configured as 32 pressure vessels in the first stage and 14 pressure vessels in the second. Starting from the right, there are two feed manifolds, each connected to 16 pressure vessels. The concentrate manifold (on top of the membrane unit) is connected to 14 pressure vessels. The permeate manifold is shown on the top and on the left side of the membrane unit. The feed and concentrate manifolds are made of stainless steel because of the high pressure operation. The permeate manifold, operating at very low pressure, is made of PVC. A panel for local instruments display and a panel for sampling permeate flow from individual pressure vessels, are shown also. Each pressure vessel holds 7 spiral elements, a total of 322 elements for the entire train. Nominal permeate capacity of such a unit is 8,000 m^3/day. The dimensions of the unit are 4m (height) $\times$ 2.9m (width) $\times$ 8m (length).

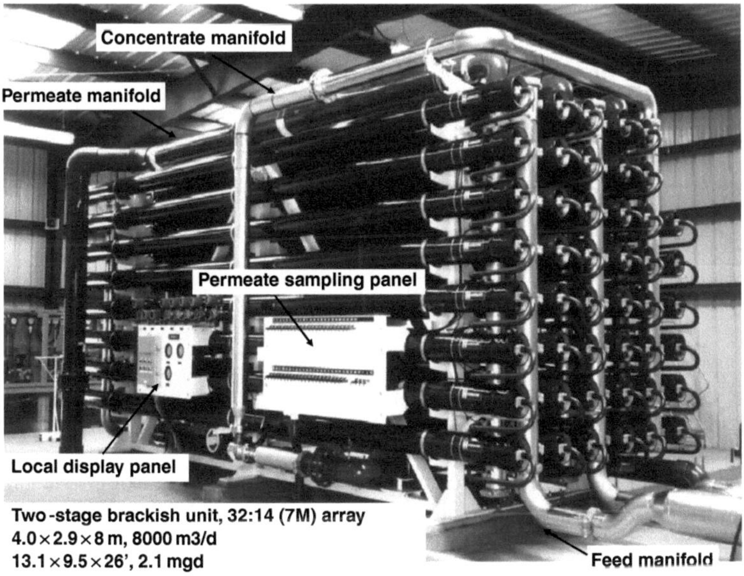

Fig. 4.10 Brackish water RO train
Source: [2], with kind permission.

4.7.4 Configuration of RO Desalination Plants

The configuration of membrane desalination systems depends to a great extent on the quality and source of the feed water. Systems treating brackish well water are relatively simple. In the majority of cases, well water is very clean, due to slow infiltration of water through a porous layer of underground aquifers. As shown schematically in Fig. 4.11, feed water pretreatment is limited to acid dosing and/or the addition of a scale inhibitor. Cartridge filtration, shown before the high pressure pump, serves mainly as protection for the pumping equipment against potential release of sand particles from the well.

Brackish waters usually show high concentrations of alkalinity and corresponding high concentrations of CO_2 . Removal of CO_2 in a degasifier is the most cost effective option to increase water pH. An alternative, the addition of NaOH to convert CO_2 to an alkali, is more expensive.

A seawater RO unit, treating surface seawater from an open intake, will require significantly more and extensive pretreatment. Surface seawater treatment steps usually include initial screening of large objects, followed by coagulation and flocculation. As a coagulant, ferric salts are used at the dosing rate of 1–15 ppm. In some cases, seawater pH is adjusted with sulphuric acid to improve the flocculation process. Suspended solids are removed in a single or two stage media filtration. The filtrated effluent could be further conditioned by pH adjustment with acid and/or addition of a scale inhibitor. If chlorine has been added in the process of pretreatment, it has to be reduced to chloride using sodium bisulfite, prior entering the RO

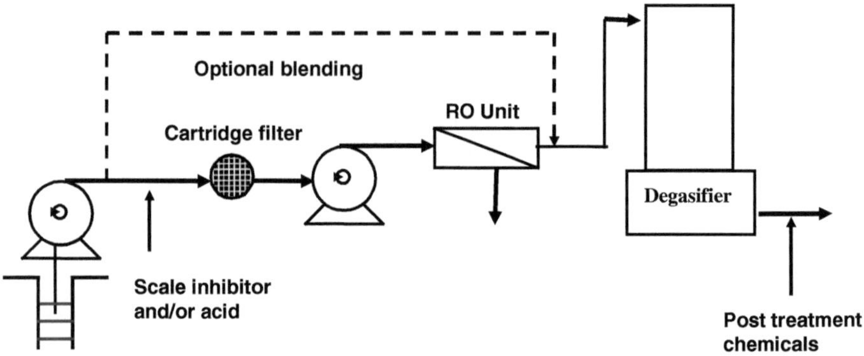

Fig. 4.11 Brackish RO unit treating well water

membranes. After the addition of chemicals, feed water flows through the cartridge filters under the suction of high pressure pumps. Currently, increased efforts are being made to replace media filtration with membrane filtration. Configuration of such a system is shown in Fig. 4.12. This diagram represents an immersed, vacuum driven, membrane filtration system, but pressurized systems are also used for this application [16]. The separation of feed water into low salinity product and high salinity concentrate is accomplished in the membrane unit. The configuration of the membrane unit has been discussed in the previous section.

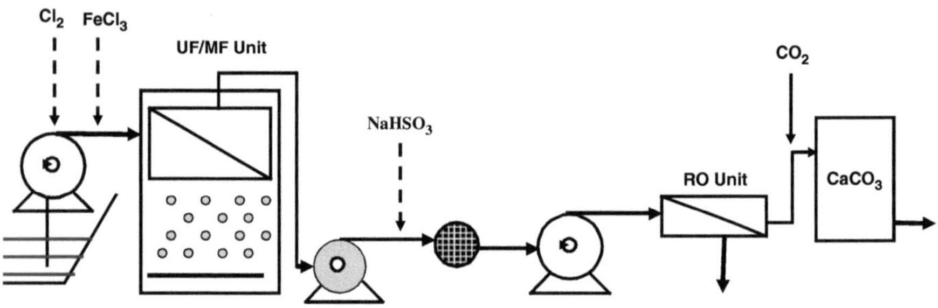

Fig. 4.12 Seawater RO unit with membrane filtration (UF/MF) pretreatment

Seawater exhibits low alkalinity therefore the permeate in seawater RO systems has a low concentration of CO_2. Thus, the stabilization of seawater RO permeate, involving dissolution of $CaCO_3$, requires the addition of CO_2, ahead of the limestone vessel, to achieve the required level of hardness and alkalinity.

4.7.5 Operating Parameters and Performance

The operating parameters for commercial RO systems are derived from the composition and quality of the feed water, the type of the RO membrane elements used, the average permeate flux rate and the recovery rate. The above information forms a set of design parameters which are interrelated. During the design process, the system operating parameters are determined based on the experience of the design engineer, recommendations from membrane manufacturers and calculations of projected system performance. Two major operating parameters, that uniquely define the economics of the desalination process, are the recovery rate and the feed pressure. For given conditions of feed water composition, temperature and average permeate flux, the feed pressure is a function of the recovery rate and the water permeability of the membrane. Higher water permeability and lower recovery rate will result in a lower feed pressure, at which the system has to operate, in order to produce a given capacity of product water. Water permeability is a property of the membrane material and the condition of the surface. It usually declines with time. System design determines the recovery rate. In brackish and nanofiltration systems, the tendency is to design systems for the highest recovery rate possible. The usual recovery rate range is 75–85% for brackish systems, and 80–90% for nanofiltration systems. Feed pressure in nanofiltration systems is in the range 0.7–1.0 MPa. In brackish systems, the feed pressure is in the range 1.0–2.0 MPa. High recovery results in a lower volume of concentrate, which is usually associated with a lower disposal cost. For low and medium salinity brackish water, higher recovery rates usually result in lower energy requirements.

The recovery rate in brackish systems is limited by the scaling tendency of the concentrate, mainly due to the presence of sulfate and calcium ions, and silica compounds. Even with the application of effective scale inhibitors, there is a limit to how much these scale-forming compounds can be concentrated without precipitation.

Scaling concerns in seawater RO systems are negligible and an optimum recovery rate is determined by economic considerations. With increasing recovery rates, average salinity and osmotic pressure increases and increasingly higher feed pressures are required. On the positive side, with a higher recovery, a lower feed flow rate and a smaller pretreatment system, is required. For the prevailing economic parameters, the optimum recovery rate in seawater RO applications is in the range 45–55%. This recovery rate results in feed pressures below 7.0 MPa and product water salinities below 500 ppm TDS, in a single-pass process [2, 17].

4.7.6 Optimisation of Energy Requirements

The energy usage in RO systems is a composite of the energy for the feed water intake, operation of pretreatment process, operation of the high pressure pumping system, treatment of product water, operation of the control system and auxiliary needs. The energy usage of the high pressure pumping systems is the largest one, being close to 80% of the total in seawater RO systems. The energy requirement of the high pressure pumping unit is a direct function of the required feed pressure and the recovery rate. In nanofiltration systems, energy use of RO processes is about 0.5 kWh/m^3. In nanofiltration systems, the concentrate pressure is low and the flow rate is small (due to a high recovery rate). Therefore, the application of concentrate energy recovery devices would not be cost effective. In RO systems treating medium salinity brackish water, the corresponding energy use is about 0.8 kWh/m^3. Depending on the feed pressure applied and the recovery rate, in some cases the use of energy recovery devices may be justified. In seawater RO systems, the energy use of the high pressure pump is much higher, due to feed pressures being in the range 6.0–7.0 MPa. However, part of the energy is usually recovered via operation of energy recovery devices. In the past, the most frequently used energy recovery device was the Pelton Wheel. The Pelton Wheel consists of a drum with buckets, mounted on a common shaft, with an electric motor and a high pressure centrifugal pump (Fig. 4.13).

The concentrate stream, which leaves the RO unit under pressure, impinges on the bucket, turns the drum and provides additional torque to the pumping device. The efficiency of current Pelton Wheel energy recovery devices are in the range 84–88%.

Recently even more efficient energy recovery devices have been introduced commercially. They are called, "pressure exchangers" or "isobaric devices". These devices are positive displacement type, energy transfer units. The energy exchange efficiency is in the range 94–96%. Representative examples of energy requirements of pumping systems, in an RO seawater system, are given in Table 4.10.

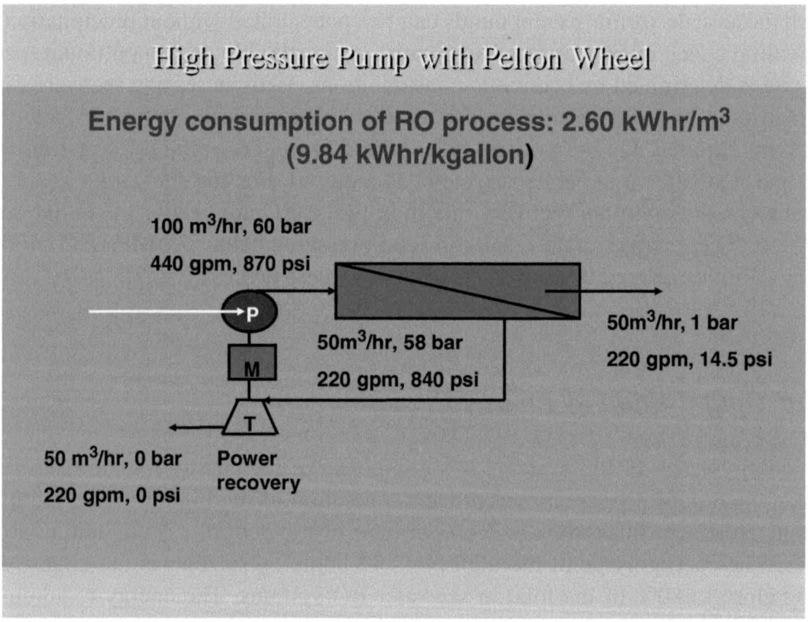

Fig. 4.13 Configuration of a seawater RO unit with a Pelton Wheel energy recovery device Source: [2], with kind permission.

Table 4.10 Typical values of energy requirements for high pressure pumping equipment in RO seawater desalination processes. Feed pressure 6.0 MPa, recovery rate 50%

Configuration	High pressure pump only	High pressure pump with Pelton wheel	High pressure pump with isobaric device
Energy use, kWh/m^3	3.9	2.6	2.1
Energy usage reduction, %	–	33	46

4.7.7 Operating RO Plants Worldwide

As in the case of thermal plants, the number and production capacity of Reverse Osmosis plants have also been rising in the last few decades. In particular, while MSF and MED technologies have experienced a dramatic increase in Middle East countries, RO technology has led the growth of the desalination market in western countries (Europe, America, Australia).

A complete overview of contracted and operating RO desalination units is outside the scope of this book, however Table 4.11 shows some of the most representative RO operating plants. Such data clearly indicates that the production capacity of an SWRO plant is now approaching capacities up to now typical of only thermal processes.

Table 4.11 Examples of recent RO installations

Plant	Year	Unit capacity (m^3/d)	Energy consumption kWh/m^3	Manufacturer
Ashkelon (Israel)	2005	325,000	< 4	IDE technologies
Tuas (Singapore)	2005	136,000	4.1	Hyflux
Fujairah (UAE)	2003	170,000	3.8	Doosan
Carboneras (Spain)	2005	120,000	4	Hydranautics
Rabigh (Saudi Arabia)	2008	200,000	4	Mitsubishi
Larnaka (Cyprus)	2001	54,000	4.5	IDE technologies
Florida (USA)	1999	95,000	4	Stone and Webster – Poseidon

Source: [18]

4.8 Post-treatment in Desalination Plants

The different uses of desalinated water often require further post-treatments which produce a final product of the needed quality standard. Water for the process industry is usually required at very low salt concentrations. Fresh water produced by thermal processes is therefore almost always suitable for immediate use, while permeate from RO units usually requires a further de-ionizing step using ionic resins or a final high rejection RO step.

Water produced for civil usage requires post-treatment mainly focused on the adjustment of water pH and hardness to reduce corrosion of the water distribution network and to comply with local health regulations [19, 20]. The product water is also disinfected to prevent biological growth and to guarantee its biological sterility.

Post-treatments can be performed according to several different procedures which depends on the capacity and type of the plant, the quality of the produced water, the standards to be achieved and the availability of specific chemicals.

For a typical large scale post-treatment following a thermal process, the required steps are: CO_2 absorption; limestone dissolution by reaction with the absorbed CO_2; degasification of the unreacted CO_2 gas; and final pH adjustment by addition of NaOH. The treated product is chlorinated to prevent biological growth and any possible reduction of ferrous compounds (red water phenomena) by iron-reducing bacteria [21]. A brief description of the above steps is given below.

Part of the distillate enters into a packed absorption column and absorbs the CO_2 gas in counter-current flow. The absorption column operates at pressures close to 0.5 MPa in order to maximise the rate of CO_2 absorption. The outlet gassed water, which contains a large amount of CO_2, is mixed with another part of the by-passed distillate and is sent on to the limestone dissolution filters. The reaction between the CO_2 gas and the limestone produces calcium bicarbonate ($Ca(HCO_3)_2$) and increases the pH of the carbonated water (acidified in the absorption column).

The treated water still contains some un-reacted CO_2 which is degasified into air, via a packed stripping column in counter-current mode. The pH of the remaining by-passed distillate product is adjusted by adding NaOH. The entire product stream is then blended and disinfected, prior to final blending with brackish water, to adjust its salt content. Approximately 10 parts distillate water is mixed with 1 part brackish water (or 100 parts distillate with about 1 part seawater). In addition, a final dose of anti-corrosion agent may be added to the blending chamber [22].

In RO units, post-treatment is mainly achieved by the addition of chemicals which stabilise the final product. The stabilisation process involves the control of CO_2 concentration (usually by removing the excess in BWRO or adding CO_2 in SWRO) and increasing the alkalinity and hardness, with a final pH adjustment. Common chemicals used for the stabilisation process are shown in Table 4.12, along with their effect on alkalinity and hardness.

Table 4.12 Usage of reagents in RO product post-treatment and corresponding hardness and alkalinity increases per 1 mg/l of CO_2 present in the product water

Reagent	Usage, mg/l	Compound formed	Hardness increase, mg/l	Alkalinity increase, mg/l	TDS increase, mg/l
$Ca(OH)_2$	0.84	$Ca(HCO_3)_2$	1.33	1.33	1.84
$CaCO_3$	2.27	Ca(HCO3)2	2.27	2.27	3.27
NaOH	0.91	$NaHCO_3$	0.0	1.13	1.91
Na_2CO_3	2.40	$NaHCO_3$	0.0	2.27	3.40

RO permeate usually has a higher salinity than distillate produced in thermal units. Thus the final step of mixing post-treated water with brackish or seawater, to adjust the final salinity, is not usually necessary.

Small desalination units, often follow slightly different procedures, which present economic advantages compared to the above-described post-treatment processes. For example, in cases of small thermal plants it may be more convenient to avoid the use of absorption columns and limestone filters, and to simply use commercial chemicals to achieve the desired concentration of Calcium Carbonate (hardness), pH and salinity.

4.9 Economics of Industrial Desalination Plants

The economics of desalination processes are affected by several general factors including:

- Plant capacity: larger plant capacities reduce the cost per unit product, despite a higher initial capital investment (i.e. "economies of scale").
- Site conditions: installation of new units as an addition to existing sites would eliminate cost associated with feed water intake, brine disposal and feed water pretreatment facilities.
- Qualified manpower: the availability of qualified operators, engineers, and management personnel would result in higher plant availability and production capacity, and shorter downtimes.
- Energy cost: the availability of inexpensive sources of low cost electrical power and heating steam (for thermal units) has a significant impact on product unit cost.
- Plant life and amortisation: increase in plant life reduces capital product cost.

Given the above and other specific factors, estimating the cost of desalinated water cannot be generalised, and depends heavily on plant design, type and location. Some general information is provided below to give an impression of what is possible in estimating production water costs.

For example, the costing of thermal desalination processes can be obtained by using simple relationships based on specific cost values. This approach requires careful selection of model data and the need for recent estimates of this data, which reflects the current industry status [23]. A more detailed and accurate approach requires the solution of system material and energy balance equations to determine the energy requirements (heating steam and electrical power), flow rates, stage dimensions and heat transfer areas. This data, together with cost values for electrical power, heating steam, condensers and heaters, pumping units and other related construction parameters, are used to determine unit product cost [24].

Table 4.13 illustrates a simple costing procedure for thermal plants, which is based on specific costing values. The most critical of these is the specific capital cost, which averages \$1,000/(m^3/day). The values used for the MSF and MED

Table 4.13 Costing parameters and unit product cost of thermal desalination processes

Process	MED	MSF	MVC
Capacity (m^3/day)	30,000	50,000	5,000
Plant factor	0.9	0.9	0.9
Interest rate	0.05	0.05	0.05
Plant life (year)	40	40	40
Specific capital cost (\$/(m^3/day))	951	878	1,100
Amortisation factor (per year)	0.05827	0.05827	0.05827
Electrical power cost (\$/kWh)	0.05	0.05	0.05
Electrical power consumption (kWh/m^3)	1.5	3	14
Labour specific cost (\$/m^3)	0.01	0.01	0.01
Chemical specific cost (\$/m^3)	0.04	0.06	0.04
Spare parts (% of capital cost)	0.01	0.01	0.01
Loss of electrical power (kWh/m^3)	10	10	
Unit cost of power loss (\$/kWh)	0.03	0.03	
Capital (\$'000)	28,530	43,900	5,500
Annual amortisation (\$'000/year)	1,662	2,558	320
Annual spare parts cost (\$'000/year)	285	439	55
Specific power cost (\$/m^3)	0.075	0.15	0.7
Specific cost of power loss (\$/m^3)	0.3	0.3	
Unit product cost (\$/m^3)	0.622	0.702	0.978

processes were taken from a recent study by Borsani and Rebagliati [6]. The unit product costs shown in the Table are consistent with current published data. The higher cost value for the MVC process is due to the large difference in its production capacity compared with MSF and MED.

A similar procedure may also be applied to an RO unit, provided that information on costs values can be found in literature or from reliable sources. For example investment costs of a BWRO plant can average between 300 and 600 \$/m^3/day, while for an SWRO plant it can rise up to 800–1,100 \$/m^3/day depending heavily on the quality of feed water and the need for extensive pretreatment.

Reported costing of already installed and proven desalination units varies widely and depends on the characteristics of the desalination site. For example, the average desalinated water cost in a small RO unit, operating on underground brackish water and with a capacity of less 1,000 m^3/day, might range between 3 and 5\$/m^3. This is mainly caused by the low production capacity and the high annual depreciation charges. On the other hand, an RO plant with a capacity range of 50,000–100,000 m^3/day, such as the Florida RO plant [25], produces desalinated water at an approximate cost of \$0.5/m^3. For this reason, most of the new RO plants are constructed at capacities exceeding 40,000 m^3/day and approaching 200,000 m^3/day in some cases, in order to secure water production cost close to \$0.5/m^3[26, 27].

In the Gulf countries, where thermal desalination is the dominant process, desalination takes place in very large co-generation plants with total capacities largely exceeding 100,000 m^3/day, with unit capacities varying between 20,000 m^3/day and 75,000 m^3/day. In these large plants the water cost is expected to be close to \$0.5/m^3, irrespective of the type of the desalination process [6].

Table 4.14 Reported and estimated production costs for the main industrial desalination processes

References	Process	Capacity (m^3/day)	Unit product cost (\$/m^3)
Frioui and Oumeddour [28]	MVC	1,000	1.02
	RO	1,000	1.8
	MSF	1,000	1.2–1.34
	MED	1,000	1.38–1.45
Karagiannis and Soldatos [29]	MVC	1,000–1,200	
	RO	12,000–60,000	0.44–1.62
	MSF	23,000–528,000	0.52–1.75
	MED	12,000–55,000	0.95–1.95
Díaz-Caneja and Farinas [30]	RO	65,000–170,000	0.7
Ophir and Lokiec [31]	MED	100,000	0.54
Borsani and Rebagliati [6]	RO	205,000	0.45
	MSF	205,000	0.52
	MED	205,000	0.52

Table 4.14 shows a summary of desalination costs for commercial desalination processes, which include MSF, RO, MED and MVC. Costs vary widely as stated above, however a clear indication is given of the influence of plant capacity on product cost.

Finally some estimates of water costs relevant to certain operating plants are reported in Table 4.15. These costs are "total water costs", i.e. they include both operational and investment (amortization) costs. However, due to the significant differences in plant location, technology, and operations, this information must be considered only as an indication of the main trends of desalination costs.

Table 4.15 Total water costs of some industrial desalination plants

Plant	Technology	Water cost [\$/m^3]	Plant capacity [m^3/d]	Date of estimate
Shuweihat (UAE)	MSF	1.13	454,610	2008
Ras Laffan (UAE)	MSF	0.80	272,520	2008
Hidd (UAE)	MSF	0.69	400,000	2008
Tenes (Algeria)	SWRO	0.59	200,000	2008
Taunton (Massachusets)	SWRO	1.53	18,925	2008
Palmachim (Israel)	SWRO	0.86	83,270	2008
Oued Sebt (Algeria)	SWRO	0.68	100,000	2008
Hadera (Israel)	SWRO	0.86	330,000	2008
Ashkelon (Israel)	SWRO	0.78	326,144	2008
Tianjin (China)	SWRO	0.95	150,000	2007
Dhekelia (Cyprus)	SWRO	0.88	40,000	2007
Carlsbad (California)	SWRO	0.77	189,250	2007
Pert (Australia)	SWRO	0.75	143,700	2006
Marafiq (Saudi Arabia)	MED	0.83	758,516	2006
Shoaiba 3 (Saudi Arabia)	MSF	0.57	881,150	2005
Reliance refinery (India)	MED	1.53	14,400	2005

Source: [32].

Abbreviations

BOOT	Bulid-Own-Operate-Transfer
BWRO	Brackish Water Reverse Osmosis
MED	Multiple Effect Distillation
MED-TVC	Multiple Effect Distillation – Thermal Vapour Compression
MSF	Multi Stage Flash
MSF-BR	Multi Stage Flash – Brine Circulation
MSF-OT	Multi Stage Flash – Once Through
MVC	Mechanical Vapour Compression
RO	Reverse Osmosis
SWRO	Sea Water Reverse Osmosis

References

1. Cipollina, A., Micale, G., and Rizzuti, L., A critical assessment of desalination operations in Sicily, Desalination 182 (2005) 1–12.
2. Wilf, M., with chapters by C. Bartels, L. Awerbuch, M. Mikley, G. Pearce, and N. Voutchkov, "The Guidebook to Membrane Desalination Technology. Reverse Osmosis, Nanofiltration and Hybrid Systems. Process, Design and Applications" by Balaban Desalination Publications (2006)
3. Birkett, J.D., Desalination Development during the Industrial Revolution. Proceedings of the IDA World Congress on Desalination and Water Sciences, Abu Dhabi, November, 1995, Vol. II.
4. IDA, The 19th worldwide desalting plant inventory, International desalination association, Topsfield, MA, USA, 2006.
5. Matz, R., and Fisher, U., A comparison of the relative economics of sea water desalination by vapor compression and reverse osmosis for small to medium capacity plants, Desalination, 36 (1981)137–151.
6. Borsani, R., and Rebagliati, S., Fundamentals and costing of MSF desalination plants and comparison with other technologies, Desalination 182 (2005) 29–37.
7. AlBahou, M., Al-Rakkaf, Z., Zaki, H., and Ettouney, H., Desalination experience in Kuwait, Desalination 204 (2007) 403–415.
8. Thirumeni, C., Deutsche Babcock Rehabilitation and uprating of Ras Abu Fontas MSF, desalination units: process optimisation and life extension, Desalination 182 (2005) 63–67.
9. Helal, A.M., Uprating of Umm Al Nar East 4-6 MSF desalination plants, Desalination 159 (2003) 43–60.
10. Schaefer, K.P., Computer simulations aid Sharjah's upgrade of its old MSF plants, Intern. Desalination and Water Reuse 13(4) (2004) 12–15.
11. Silver, R.S., Multi-stage flash distillation – The first 10 years, 3rd Int. Sym. On Fresh Water from the Sea, Athens, Greece, 1 (1970) 191–206.
12. Ettouney, H., Abdel-Jabbar, N., Mjalli, F.S., and Qiblawy, H., Development of Web Based Computer Package for Simulation of Thermal and Membrane Desalination Processes, MEDRC, Project ID:MEDRC-04-AS-001, final report, 2008.
13. Kronenberg, G., and Lokiec, F., Low-temperature distillation processes in single- and dual-purpose plants, Desalination 136 (2001) 189–197.
14. Loeb, S., and Sourirajan, S., Adv. Chem Ser. 38 (1962) 117.
15. Peterson, R., Cadotte, J., and Buttner, J., Final report of development of FT-30 membrane in spiral wound modules, Contract No 14-34-0001-8547 for UD Department of Interior, October 1982.

16. Bartels, C., Pearce, G., and Wilf, M., "Improving Total Cost of Desalination by Membrane Pre-Treatment", Proceedings of IDA Desalination Conference, Las Palmas, October 2007.
17. Wilf, M., and Bartels, C., Optimization of seawater RO system design, Desalination 173 (2005) 1–12.
18. Fritzmann, C., Lowenberg, J., Wintgens, T., and Melin, T., State-of-the-art of reverse osmosis desalination, Desalination 216 (2007) 1–76.
19. World Health Organization Guidelines for Drinking Water Quality. 3rd ed., Vol. 1, Geneva 2004.
20. Water Treatment Plant Design, American Society of Civil Engineers and American Water Works Association, 4th ed., McGraw-Hill, 2005
21. Al-Deffeeri, N., The release of CO_2 in MSF distillers and its use for the recarbonation plant: a case study, Desalination 222 (2008) 590–602
22. Migliorini, G., and Meinardi, R., 40 MIGD potabilization plant at Ras Laffan: design and operating experience, Desalination 182 (2005) 269–276
23. Ettouney, H.M., El-Dessouky, H.T., Faibish, R.S., and Gowin, P., Evaluating the economics of desalination, Chem. Eng. Prog., 98 (2002) 32–39.
24. Tanvir, M.S., and Mujtaba, I.M., Optimisation of design and operation of MSF desalination, process using MINLP technique in gPROMS. Desalination 222 (2008) 419–430
25. Leitner, G., "Developer Selected for 25 MGD (94,625 m^3/d) Florida West Coast Seawater Desalting Plant," Desalination and Water Reuse 9 (1999) 11–16.
26. Khawaji, A.D., Kutubkhanah, I.K., and Wie, J-M., A 13.3 MGD seawater RO desalination plant for Yanbu Industrial City, Desalination 203 (2007) 176–188
27. Sanz, M.A., Bonnélye, V., and Cremer, G., Fujairah reverse osmosis plant. 2 years of operation, Desalination 203 (2007) 91–99
28. Frioui, S., and Oumeddour, R., Investment and production costs of desalination plants by semi-empirical method, Desalination 223 (2008) 457–463
29. Karagiannis, I.C., and Soldatos, P.G., Water desalination cost literature: review and assessment, Desalination 223 (2008) 448–456
30. Díaz-Caneja, J., and Fariñas, M., Cost Estimation Briefing for Large Seawater Reverse Osmosis Facilities in Spain, International Conference on Desalination Costing, MiddleEast Research Center, Lemesos, Cyprus, 6–8 December, 2004, pp 121–140.
31. Ophir, A., and Lokiec, F., Advanced MED process for most economical sea water desalination. Desalination 182 (2005) 187–198
32. Pankratz, T., Water Desalination Report, Vol. 44, N.33–34, September 2008.

Chapter 5
Nuclear Desalination

A Review of Desalination Plants Coupled to Nuclear Power Stations

B.M. Misra and I. Khamis

Abstract The rising cost of fossil fuel, its uncertain availability and associated environmental concerns have led to a need for future desalination plants to use renewable and other sustainable energy sources, including nuclear. The desalination of seawater using nuclear energy is an option with a proven track record, with over 200 reactor-years of operating experience worldwide. Economic feasibility studies generally indicate that water costs from nuclear seawater desalination are in the same range as costs associated with fossil-fuelled desalination, at today's prices. Different approaches utilising the waste heat from nuclear reactors have been proposed to further reduce the cost of nuclear desalination. Safety concerns have also been addressed in order to find the best solutions to guarantee high water quality levels and minimise the possibility of radioactive contamination. Nuclear desalination has thus the potential to be an important option for safe, economic and sustainable supply of large amounts of fresh water to meet the ever-increasing worldwide water demand.

5.1 Why Nuclear Desalination

There are many reasons that favour a possible revival of nuclear power production in the years to come: the development of innovative reactor concepts and fuel cycles, with enhanced safety features that are expected to improve public acceptance; the production of energy less expensive than in conventional power plants and novel RE units; the need for prudent use of fossil energy sources and increasing requirements to curtail the production of greenhouse gases (GHGs). The combined heat and power (CHP) applications of nuclear power, such as desalination, district heating and industrial process heat applications, are also likely to be seriously considered in the coming years.

B.M. Misra (✉)
501 Emerald Heights, 32 Union Park, Chembur, Mumbai 400071, India
e-mail: bmmisra@gmail.com

A. Cipollina et al. (eds.), *Seawater Desalination*, Green Energy and Technology,
DOI 10.1007/978-3-642-01150-4_5, © Springer-Verlag Berlin Heidelberg 2009

> Nuclear desalination is defined as the production of potable water from seawater, in a facility in which a nuclear reactor is used as the source of energy for the desalination process. Electrical and/or thermal energy may be used in the desalination process. The facility may be dedicated solely to the production of potable water, or may be used for both the generation of electricity and production of potable water, in which case only a portion of the total energy output of the reactor is used for water production [1].

Use of energy from nuclear reactors for seawater desalination has a proven track record; it is environmentally friendly from a gaseous pollutants emission point-of-view and can be a sustainable energy source. Feasibility studies indicate that current costs of water produced from nuclear desalination plants are similar to those of fossil fuel-based desalination plants. Thus nuclear desalination is an important option for the safe, economic and sustainable supply of large amounts of fresh water to meet the ever-increasing worldwide water demand.

The principal attraction of nuclear desalination is the potential availability of a large amount of low-grade heat from nuclear power plants. The present generation of water cooled reactors have net electrical efficiencies of power conversion typically ranging from 30 to 33% which are smaller than coal or oil/gas based thermal power plants. In fact, the enthalpy of the steam at the inlet of the high pressure turbine of a nuclear power plant is lower due to the lower pressure and temperature of the saturated steam, and thus, the specific steam consumption for nuclear power plants is higher. This leads to availability of a higher amount of "waste" steam that could be utilised for thermal desalination [2].

Being nearly free of carbon generation, a long-term sustainable solution and potentially competitive with fossil fuels, nuclear energy has the potential to be considered as a suitable choice for sustainable desalination purposes. This is particularly true in those cases when the alternative for power and heat generation for desalination is using fossil fuels, such as heavy crude oil or coal. Fossil fuels entail significant pollution control costs and are an inefficient generating solution, resulting in a significant increase in the penalties for CO_2 emission and greenhouse impact [3].

Some recent cost estimates made for fossil and nuclear based desalination plants (as shown in the following Sect. 5.6.2) clearly indicate that even with the higher investment costs for nuclear powered plants, the overall water costs are much lower for nuclear desalination, mainly due to the lower nuclear fuel costs. These costs do not take into account penalties for CO_2 emission for fossil fuel-based systems, which would further improve the prospects of nuclear desalination.

Nuclear desalination, at present, appears to be the only technically feasible, economically viable and sustainable solution to meet future water demands which will require large-scale seawater desalination [4]:

- Nuclear reactors provide heat across a large range of temperatures, thus allowing easy adaptation to any desalination process.
- Some nuclear reactors furnish waste heat (normally evacuated to a heat sink) at ideal temperatures for desalination.

- Desalination is an energy intensive process. Over the long term, desalination using fossil energy sources will not be compatible with sustainable development. Fossil fuels reserves are, in fact, finite and must be conserved for other essential uses, whereas demands for desalted water will continue to increase.

5.2 Environmental Impact of Desalination Processes

The combustion of fossil fuels produces large amounts of greenhouse gases and toxic emissions. It is estimated that a water production of 10 million m^3/day from seawater desalination using fossil fuels would release 200 million t/year of CO_2, 200,000 t/year of SO_2, 60,000 t/year of NO_x and 16,000 t/year of VOCs (Volatile Organic Compounds). Thus, for the current global desalting plant capacity of 40 million m^3/day, total emissions would be four times these values, while this could be avoided if nuclear or renewable energy sources were used for desalination. It is estimated that to produce fresh water with the present desalination capacity, using nuclear energy, the required nuclear capacity would be about forty 1,000 MWe nuclear reactors [4].

Environmental impact studies of common commercial desalination processes have been carried out in recent years, using life cycle assessment including materials used in the plant, assembly, operation and waste management. Typical data from the life cycle assessment [5] of fossil-fuelled desalination plants based on MSF (multi-stage flash), MED (multi-effect distillation) and RO (reverse osmosis) processes are shown in Table 5.1.

The values of emissions are directly related to the energy (fuel) consumed in the various desalination processes. The environmental impact of desalination plants is now a matter of concern in areas where the desalination capacities are large and ever increasing. For MSF and MED plants utilising waste heat, the relevant emission data is presented in Table 5.2.

These results show a drastic reduction in the emissions per cubic meter of desalted water produced in the thermal desalination plants utilising waste-heat sources, such as those from industry or from power stations. In the case of nuclear and renewable energy-driven desalination plants, there will always be lower emissions compared to fossil-driven plants.

As most of the desalination capacity is needed in the water-scarce areas of developing countries, there could be a greater incentive of availing carbon credits as part

Table 5.1 Relevant airborne emissions produced by desalination systems based on fossil fuels

Emission per m^3 desalted water	MSF	MED	RO
kg CO_2	23.41	18.05	1.78
g dust	2.04	1.02	2.07
g NO_x	28.29	21.41	3.87
g NMVOC	7.90	5.85	1.10
g SO_x	27.92	26.29	10.68

Table 5.2 Relevant airborne emissions produced by MSF and MED when driven by waste heat

Emission per m^3 desalted water	MSF	MED
kg CO_2	1.98	1.11
g dust	2.04	1.02
g NO_x	4.14	2.38
g NMVOC	1.22	0.59
g SO_x	14.79	16.12

of the Clean Development Mechanism (CDM) and a resulting cost reduction, if the heat for desalination is obtained from clean energy sources such as renewable or nuclear energy (the latter will also be accepted as CDM under the Kyoto protocol). A typical example of an MED plant, based on Chinese nuclear heating reactor NR-200, indicates a potential product water cost reduction of nearly 20% from CDM, assuming a cost of $20/tonne of carbon [6].

5.3 Nuclear Desalination Systems

Various types of water-cooled, as well as gas-cooled reactors, are candidates for nuclear desalination. Small and medium-sized reactors (SMRs), once they become commercially available and economically competitive, may offer the largest potential as coupling options for nuclear desalination. The development of advanced and innovative reactor concepts is expected to provide the optimal choice in the medium-term future.

The design approaches for a nuclear desalination plant are essentially derived from those of the nuclear reactor alone, with some additional aspects to be considered in the design of a desalination plant and its integration with the nuclear system. All types of nuclear reactors can provide the energy required by the various desalination processes. The development of innovative reactor concepts and fuel cycles, with enhanced safety features, as well as their attractive economics, are expected to improve public acceptance and further the prospects of nuclear desalination.

Coupling nuclear reactors with desalination plants has already been demonstrated in several countries. The coupling scheme is usually dictated by the maximum economic and practical benefits that can be achieved, in terms of water and electricity production. In general, coupling is technically feasible but imposes conditions such as avoiding radioactivity cross-contamination and minimising the impact of the thermal desalination plant on the nuclear reactor.

Two types of coupling are identified: thermal coupling with distillation desalination systems (i.e. MSF or MED) and contiguous coupling with electrically–driven desalination systems (RO and MVC).

Thermal coupling is normally fulfilled via one of two options: (1) back pressure turbine (suitable for large water to power production ratios), and (2) turbine extraction/prime steam (suitable for small water to power ratios, but with better operational flexibility compared with the first option). In both cases, direct fluid coupling between the reactor and the desalination plants introduces the risk of contamination of the product water. However, design measures to reduce such risk, using intermediate heat exchangers or pressure reversal in the coolant loops, have been demonstrated.

In a contiguous coupling of an electrically–driven RO desalination system with a nuclear power plant, any sudden cessation of the electricity demand of the RO system causes a loss of electrical load. However, the amount of electrical energy used by such a desalination process is only a small fraction of the total electrical energy generated by the nuclear plant. Therefore, the nuclear plant is able to tolerate the shutdown of several RO modules, powered directly from the nuclear plant, without the need for reactor trip.

A typical coupling of a PWR with an MED desalination plant [7] is shown in Fig. 5.1. The steam from the secondary circuit of the reactor is sent to a seawater heater (an isolation heat exchanger) and the generated steam is sent to a flash tank, and then on to the desalination plant. This ensures no transgress of any radioactivity to the desalination plant. Pressure reversal is also utilised to further ensure that no radioactivity carries over into the product water. Similar schemes are used in most cogeneration applications, such as district heating and industrial heat supply from nuclear reactors. To date, no problems have been reported of radioactivity carry over.

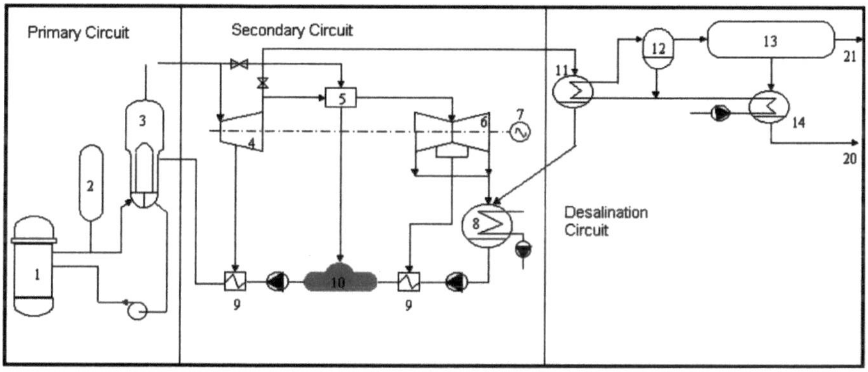

Fig. 5.1 Conventional coupling of a nuclear reactor with an MED plant. *1*: reactor core; *2*: pressuriser; *3*: steam generator; *4*: high pressure turbine; *5*: intermediate steam heater; *6*: low-pressure turbine; *7*: generator; *8*: main condenser; *9*: pre-heaters; *10*: de-aerator; *11*: seawater heater; *12*: flash tank; *13*: MED plant; *14*: MED output condenser; *20*: fresh water out; *21*: brine out-fall

5.4 Safety of Nuclear Desalination Systems

The three basic safety objectives, already existing for typical nuclear reactors, apply to nuclear desalination. These objectives cover the safety of individuals, society and the environment. They ensure that radiation exposure is kept below prescribed limits and the mitigation of radiological consequences of any accidents. In addition, they cover all reasonably practical measures to prevent such accidents and ensure that the likelihood of accidents, with serious radiological consequences, is extremely low [8].

Safety issues, concerns and considerations for nuclear desalination are identical to those for any other typical nuclear reactor plant. In some cases, safety measures of cogeneration are further enhanced via the addition of another isolation loop. Therefore, the coupling of a nuclear power system with a desalination process does not impose further specific safety-related measures. However, for an extra conservative approach, coupling should still be assessed from the point of view of the overall nuclear desalination system safety. In doing so, the effect that one system might have on the other, is examined as part of the safety analysis of the integrated system. In fact, some additional requirements may arise and be conceptualised in the design of a desalination plant and in the integration within the overall nuclear system.

An integrated desalination system consists, in general, of a nuclear power reactor coupled with a typical desalination plant through an isolation heat exchanger (see Fig. 5.2). Further considerations which may have an impact on safety are the interlinked interaction between nuclear and desalination plants, the potential impact of shared resources such as intake and outfall structures, the siting of a nuclear desalination system close to centres of population and the environmental issues arising from the discharge of concentrated brine [9].

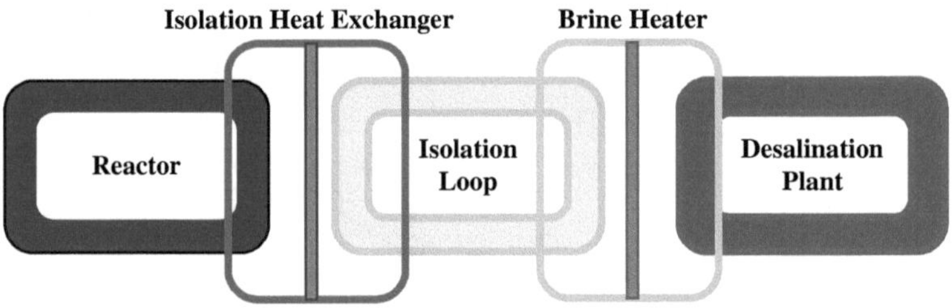

Fig. 5.2 Schematic representation of an integrated nuclear desalination system

5.4.1 Transients and Accidents Induced by Couplings

As mentioned before, specific safety issues caused by the coupling of a reactor system with a desalination plant are related to:

- The potential for transfer of radioactive materials from the nuclear plant to the desalination system, during normal operation, or as a result of an incident or accident. In general, two approaches are followed to minimise such risk (1) install an intermediate loop between the reactor and the desalination plant, and (2) use continuous monitoring of the radioactivity level of both product water and the coolant in the intermediate loop.

- The potential for more severe reactor system transients induced by transients in the desalination plant, either during normal operation or as a result of an accident. This should be overcome by design modifications and operating procedures.

As with any nuclear installation, a nuclear desalination plant should be designed to withstand a large variety of abnormal conditions. The choice of desalination technology is a major factor in determining the way in which the desalination plant is coupled to the reactor. Thermal coupling could have direct safety implications through operational transients that may exist in the nuclear desalination plant. In fact, such transients may cause the systems to have operational effects on each other. Hence, an intermediate heat transfer loop that serves as an isolation loop, e.g. condenser cooling circuit, is required in most designs. In contiguous coupling, however, the desalination system draws its electrical energy either from the grid or by direct connection to the nuclear system, with an auxiliary connection to the grid. In this case, the possibility of interaction effects between the nuclear and desalination systems are minimal. In cases where limited thermal coupling is required, such as in an RO preheat configuration, potential safety impacts should also be assessed [10].

The partial or total unavailability of the thermal plant (e.g. MED system), which provides a redundant heat sink for the nuclear facility, could result in a partial or total loss of the heat sink, with consequent possible turbine trips and reactor trips. Major causes of transients are:

- Major potential disturbances imposed by the desalination plant, such as a turbine trip due to disturbances in the desalination plant, excess load due to increased steam flow to the desalination system, loss of load due to the shutdown of the desalination plant, etc.
- Loss of condenser vacuum.
- Main condenser tube leakage.
- Loss of re-circulating cooling water flow.

Transients induced by most potential disturbances, leading to the unavailability of the desalination plant (loss of load due to the shutdown of the desalination plant), are not expected to be more severe than other analysed transients. However, the transient frequency could change as a consequence of connection to the desalination plant.

5.4.2 *Product Water Quality and Monitoring*

Monitoring of product water radioactivity is a key safety feature. In fact, detection of radioactivity in the product water is considered as the key indicator that an incident has taken place. General safety practice calls for specified limits of radioactive discharge, especially into the environment [11]. Thus, regular monitoring of product water and discharge, both of cooling water and brine, must be ensured. Such monitoring should also be evaluated with respect to possible radioactive contamination. In order to implement batch monitoring, the product water needs to be collected in storage tanks or in a reservoir, prior to its dispatch to the distribution system. The hold time must be sufficient to enable completion of full monitoring, before certifying that the product water is safe for public distribution. Unfortunately, such sampling and batch monitoring of product water cannot be reduced to much less than 1 h, resulting in massive consequences on the storage capacity required by the desalination plant.

National and international drinking water standards, as well as the ALARA (As Low As Reasonably Achievable) principle, should be applied to the product water of nuclear desalination. Radiological protection limits for drinking water are, in general, specified in national regulations and international guides (e.g. the IAEA Basic Safety Standards set a maximum annual effective dose, for an exempted practice or source, of 10 microsievert (μSv) per person).

5.5 Experiences and New Plans

The desalination of seawater using nuclear energy is a demonstrated option with over 200 reactor-years of operating experience worldwide, of which over 170 reactor-years are in Japan [12]. Kazakhstan (the Aktau fast reactor BN-350) had accumulated 26 reactor-years, producing 80,000 m^3/day of potable water, before shutting down in 1999. Recently India and Pakistan have been setting up nuclear demonstration projects at their existing PHWRs (Pressurised Heavy Water Reactors). Operating experience for all non-electric applications, including desalination, district heating and process heat, is around 1,000 reactor-years.

Figures 5.3, 5.4, and 5.5 show photographs of nuclear desalination plants in Aktau (Kazakhstan), Ohi (Japan) and Kalpakkam (India). Figure 5.6 is the

Fig. 5.3 Evaporators, Aktau, Kazakhstan

Fig. 5.4 Operating plant, Ohi, Japan

Fig. 5.5 Hybrid (MSF/RO) plant, Kalpakkam, India

KANUPP (Karachi Nuclear Power Plant) site for the proposed nuclear desalination demonstration project in Pakistan. Russia has recently launched a project to build a cogeneration facility in the city of Severodvinsk, in the Arkhangelk region, based on floating nuclear reactors (Fig. 5.7).

Table 5.3 summarises past experience, as well as current developments and future plans for nuclear-powered desalination, using different nuclear reactor types. Most

Fig. 5.6 SWRO plant, KANUPP, Pakistan

Fig. 5.7 Floating nuclear power and desalination complex, Severodvinsk, Russia

of the technologies in Table 5.3 are land-based, but the table also includes a Russian initiative for barge-mounted floating desalination plants. Floating desalination plants could be especially useful for responding to emergency demands for potable water.

Table 5.4 shows the operating nuclear desalination plants in Japan.

Table 5.3 Reactor types used, or considered for, desalination

Reactor type	Location	Desalination process	Status
LMFR	Kazakhstan (Aktau)	MED, MSF	In service up until 1999
PWRs	Japan (Ohi, Takahama, Ikata, Genkai)	MED, MSF, RO	Currently in service, with an operating experience of over 150 reactor-years
	Republic of Korea, Argentina and others	MED, RO	Integral SMRs of the PWR type; under design or to be constructed
	Russia	MED, RO	Under consideration (barge-mounted floating unit with the KLT-40)
	USA (Diablo Canyon)	RO	Operating
BWR	Japan (Kashiwazaki-Kariva)	MSF	Not brought in service following testing in 1980s, due to alternative freshwater sources; dismantled in 1999
HWR	India (Kalpakkam) India (Trombay)	MSF/RO LT-MED	Under commissioning. In service since 2004
	Pakistan (KANUPP)	MED	Existing CANDU modified to be coupled to an MED plant (under construction)
NHR-200	China	MED	Dedicated heat-only integral PWR; under design
HTRs	France, The Netherlands, South Africa	MED, RO	ANTARES, multipurpose reactor, GT-MHR and PBMR; under design and development

LMFR-Liquid Metal Fast Reactor; PWR- Pressurised Water Reactor; BWR-Boiling Water Reactor; HWR-Heavy Water Reactor; NHR-Nuclear Heating Reactor; HTR-High Temperature Reactor

5.6 Economics of Nuclear Desalination

Several factors affect desalination costs and thus determine the successful implementation of desalination systems, using either nuclear or other energies. These factors include site characteristics, plant capacity and feed water quality. The selection of power plant and desalination plant combinations for cogeneration (simultaneous production of power and water) depends on several factors, of which the most

Table 5.4 Operating nuclear desalination plants in Japan

Plant name	Location	Application	Start of operation: reactors/ desalin	Net power (MW_e)	Water capacity (m^3/day)	Remarks
Ikata-1,2	Ehime	Electricity/ desalination	1977–1982/1975	566	2,000	PWR/ MED, MSF
Ikata-3	Ehime	Electricity/ desalination	1994/1992	566	2,000	PWR/MSF (2 × 1,000 m^3/day)
Ohi-1, 2	Fukui	Electricity/ desalination	1979/1973–1976	1,175	3,900	PWR/MSF (3 × 1,300 m^3/day)
Ohi-3,4	Fukui	Electricity/ desalination	1991–1993/1990	1,180	2,600	PWR/RO (2 × 1,300 m^3/day)
Genkai-4	Fukuoka	Electricity/ desalination	1997/1988	1,180	1,000	PWR/RO
Genkai-3,4	Fukuoka	Electricity/ desalination	1995–1997/1992	1,180	1,000	PWR/ MED
Takahama	Fukui	Electricity/ desalination	1985/1983	870	1,000	PWR/RO

important is the water-to-power ratio (W/P). This is defined as the ratio of the total water production capacity installed (m^3/day) to the power installed (MW_e). Other factors include the desalination plant's energy consumption, the power plant's specific fuel consumption, the effect of seasonal loads and the specific investment costs of the water and power plants.

5.6.1 The IAEA Tool for Economic Evaluation of Desalination Systems (DEEP)

The Desalination Economic Evaluation Programme (DEEP) is derived from a desalination cost evaluation package developed in the 1980s, by General Atomics, on behalf of the International Atomic Energy Agency (IAEA). The old version, called the Co-generation and Desalination Economic Evaluation Spreadsheet (CDEE), was used for feasibility studies relating to nuclear desalination in IAEA Member States and other countries. Due to its increasing popularity, a user-friendly version was issued by the Agency, towards the end of 1998, under the name DEEP. The last few years of continuous development culminated in the development of

DEEP 3.0, released in August 2005 [13]. Following further development, the latest version, DEEP 3.2, is currently available for user download, without charge, from the IAEA website.

The DEEP package consists of several EXCEL modules. The tool separates the performance and cost calculations called "case" on one side, and the support for data input, modification and output reporting on the other side. The interface between these two modules is such that future development of the whole package may be performed by independent developers, and new cases may be incorporated into DEEP. An example of the DEEP spreadsheet is shown in Fig. 5.8.

DEEP provides a user-friendly interface when working with a single case, modifying input data and browsing in the output sheets, as well as when comparing variations from different input parameters. The DEEP main calculation sheet supports both nuclear and fossil power options. It covers heating and power plants, as well as heat-only plants, MSF and MED distillation processes and the reverse osmosis (RO) membrane process. Table 5.5 shows the options available for energy sources.

The commercially established desalination processes included in DEEP are presented in Table 5.6.

DEEP 3.1 now includes economic evaluation and hydraulic models of water transport systems, solved using Excel software calculations. Both models were

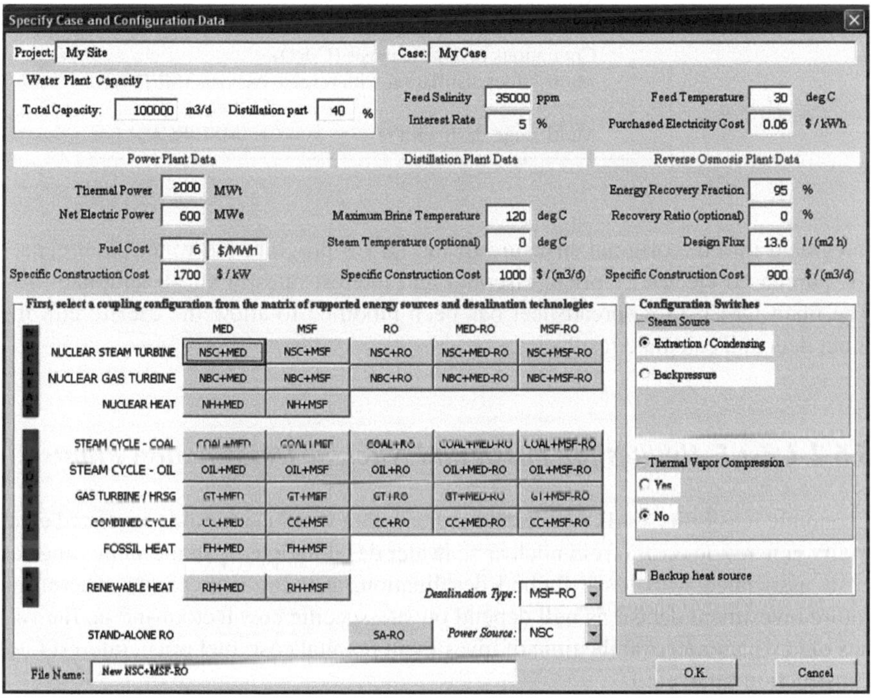

Fig. 5.8 DEEP spreadsheet

Table 5.5 The various energy options available in DEEP

Energy source	Description	Plant type
Nuclear	Pressurised light water reactor (PWR)	Cogeneration plant
Nuclear	Pressurised heavy water reactor (PHWR)	Cogeneration plant
Fossil – coal	Superheated steam boiler (SSBC)	Cogeneration plant
Fossil – oil/gas	Superheated steam boiler (SSBOG)	Cogeneration plant
Fossil	Open cycle gas turbine (GT)	Cogeneration plant
Fossil	Combined cycle (CC)	Cogeneration plant
Nuclear	Heat only reactor: steam or hot water (HR)	Heat-only plant
Fossil	Boiler: steam or hot water (B)	Heat-only plant
Nuclear	Gas turbine modular helium reactor (GT-MHR)	Power plant
Fossil	Diesel (D)	Power plant
Nuclear	Small PWR (SPWR)	Cogeneration plant

Table 5.6 The desalination processes considered in DEEP

Process	Description
Distillation	Multi-effect distillation (MED)
	Multi-stage flash (MSF)
Membrane	Standalone reverse osmosis (SA-RO)
	Contiguous reverse osmosis (C-RO)
Hybrid	Multi-effect distillation with reverse osmosis (MED/RO)
	Multi-stage flash with reverse osmosis (MSF/RO)

integrated into the original structure of the DEEP program via water plant capacity, purchased electricity price, discount and interest rates in all 38 template files. The main DEEP.XLS spreadsheet has been modified to allow the user to edit the input data and calculate results.

5.6.2 Cost Estimates for Fossil and Nuclear Desalination Plants

Economic feasibility studies generally indicate that water costs (and associated electricity generation costs) from nuclear seawater desalination are in the same range as costs associated with fossil-fuelled desalination, at today's prices [7]. Therefore, future investment decisions will depend on site-specific cost factors and on the values of key parameters at the time of investment (capital cost, fuel price, interest rate, construction time, etc.).

Recent cost estimates for large desalination plants, based on typical nuclear and fossil power units, are shown in Table 5.7.

Table 5.7 Estimated water costs from fossil and nuclear desalination plants, using IAEA's DEEP 3.1

–	PWR900 MED/RO	PWR900 MSF/RO	CC900+ MED	CC900+ MED/RO
Required capacity (range) [m^3/day]	75,000–300,000	75,000–300,000	250,000	75,000–300,000
Construction cost [$/kW]	1,763	1,763	685	685
Thermal power [MWth]	2,881	2,881	1,523	1,523
Electric power [MWe]	951	951	900	900
Fuel cost	7 $/MWh	7 $/MWh	140 $/barrel	140 $/barrel
Specific construction cost for the desalination plant [$/$m^3$]	900 for MED and 900 for RO	1,000 for MSF and 900 for RO Plant	900	900 for MED and 900 for RO
Total water cost [$/$m^3$]	0.82	0.935	2.1	1.49

It can be seen that even with higher investment costs in the nuclear case, due to lower fuel costs, the overall water costs are much lower for nuclear desalination.

5.7 Future Cost Reduction in Nuclear Desalination

Different approaches have been proposed to reduce the cost of nuclear desalination [14]. The first one is the use of waste heat from nuclear reactors for desalination. For example, the waste heat rejected by Pressurized Water Reactors (PWRs) into the heat sink, via their condensers, can be profitably used to preheat the feed water for RO systems (RO-ph process, illustrated in Sect. 5.7.1). The resulting cost reductions are 7–15% compared to traditional RO systems.

Similarly, the waste heat from the pre-cooler and intercooler exchangers of new generation High Temperature Reactors (HTRs), such as the Gas Turbine Modular Helium Rector (GT MHR) and the Pebble Bed Modular Reactor (PBMR), can lead to drastic cost reductions in MED systems coupled to such reactors.

A third approach to cost reduction would be the use of hybrid thermal/RO systems, leading to a considerably enhanced flexibility of the combined system to meet the varying water demands, and for which the overall cost of the system is significantly lower.

Another approach would be to increase the overall efficiency of the desalination systems by extracting strategic and valuable materials from the concentrated brine reject. This would also render nuclear desalination systems even more environmentally friendly.

5.7.1 Preheat Reverse Osmosis

The net electrical efficiencies of power conversion systems in most water-cooled reactors are of the order 30–33%. This means that nearly two thirds of the net thermal power produced in the reactors is evacuated to the heat sink through the condensers. The temperature of the water from the condensers is too low (30–32°C) for effective desalination using thermal processes.

However, this relatively hot water can be fed into an innovative variant of the RO process with feed preheating, called the RO-ph process. In hybrid systems, it is also possible to use the cooling seawater return stream, from the thermal desalination unit, as a feed into the RO plant.

The viscosity of the feed water is strongly dependent on its temperature. As the temperature increases, water viscosity decreases and the RO membranes become more permeable, with a consequent increase in production. However, there is also a simultaneous enhancement in the rate of salt diffusion with the rising temperature, leading to a slight increase in product water salinity. This aspect is to be kept in mind when selecting the optimum temperature of the RO-ph.

As can be seen from Fig. 5.9, there is a 2–3% increase in water production for every degree rise in seawater feed temperature. The increase in temperature can also help to reduce the applied pressure and hence the pumping power required (Fig. 5.10). Thus, an energy saving of nearly 10% is achievable in RO plants using preheated feed water.

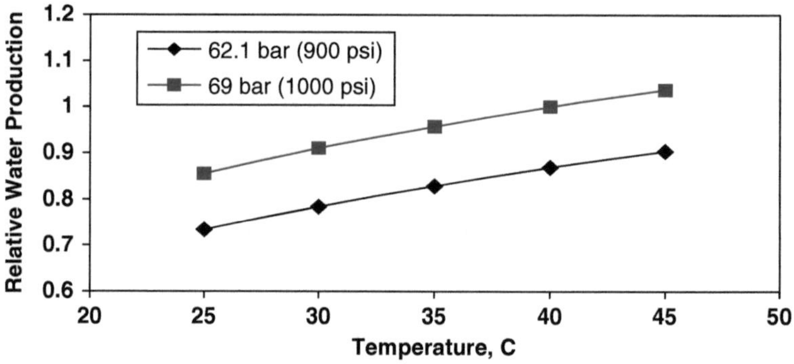

Fig. 5.9 Increase in permeate flow with rise in feed temperature

5.7.2 Waste Heat from Moderator Circuits of Pressurised Heavy Water Reactors (PHWRs)

The typical amount of waste heat available in the moderator circuit of a 500 MWe PHWR is about 100 MW_{th}. The possibility of using waste heat from PWHRs is being investigated in India, to produce fresh water using low temperature vacuum evaporation, to meet the make-up water needs of reactors, as well as for in-plant use. Figure 5.11 shows the schematics of a 1,000 m^3/day, low temperature vacuum

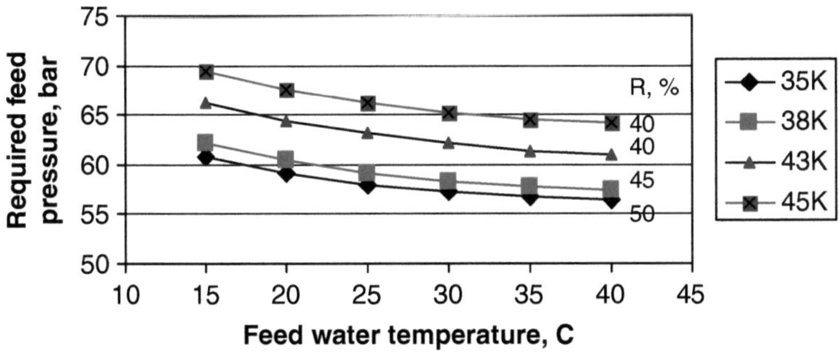

Fig. 5.10 Required feed pressure plotted against feed water temperature and salinity. R is the process recovery and K the feed water salinity, from 35 to 45 gr/lt

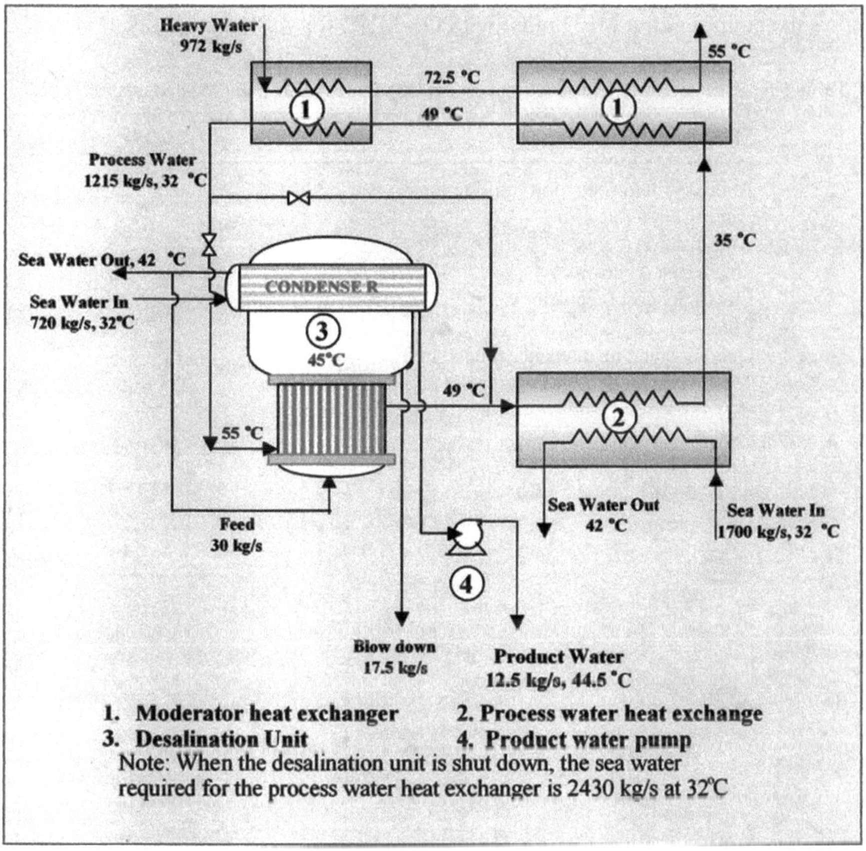

Fig. 5.11 PHWR500 coupling scheme with vacuum evaporation for the production of distilled water utilising waste heat

evaporation plant utilising the waste heat of the moderator circuit. The desalination plant will produce a high purity distillate, which will meet the entire steam generation plant make-up requirement. This will reduce the maximum load of the DM (demineralisation) plant. The water can be also utilised in the plant for other uses.

5.7.3 Waste Heat from High Temperature Gas-Cooled Reactors (HTGRs)

In the case of HTGRs, the available waste heat is suitable for evaporative desalination, thus resulting in significant energy cost savings. Assuming an MED plant life of 25 years, an availability factor of 90% and an interest rate of 5%, a total production cost of water around US$ 0.50/m^3 is estimated (Table 5.8). This is competitive compared to current desalted water prices. The introduction of an RO unit utilising the electrical power of a reactor in off-peak periods is also possible. Figure 5.12 shows the coupling of a MED plant with GT-MHR for waste-heat utilisation.

Table 5.8 Estimated water production cost (in US$) for a MED plant located at an HTGR

MED module cost ($'000)	20,000
Annual water production (million m^3)	6.935
Amortisation rate (%)	8
Amortisation (cents/ m^3)	23
Steam cost (cents/ m^3)	0
Electricity cost (cents/ m^3)	7
Chemical cost (cents/ m^3)	3
Operation and maintenance (cents/ m^3)	12
Total water production cost (cents/ m^3)	45

Source: [15]

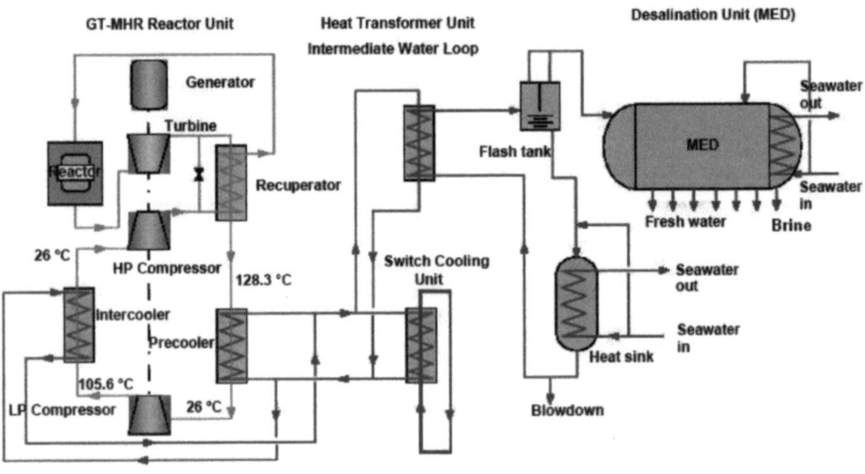

Fig. 5.12 Principle of waste heat utilisation from a HTGR

5.7.4 Hybrid Plants

The best option for future generation reactors is to have multiple product reactors, i.e. a combined production of power, hydrogen and desalinated water. These reactors are able to produce hydrogen by utilising high temperature heat of up to 950°C. The waste heat from the reactors is perfectly suitable for thermal desalination processes. However, the desalination plant could be based on either distillation or RO membrane processes. A distillation plant would utilise the available waste heat from the reactor. An RO plant would utilise the cooling water return from the distillation plant (RO-ph) and cheaper electricity from the reactor. The detailed economics of such a hybrid plant, providing multiple products, still need to be studied.

Hybrid desalination systems, having both thermal and membrane plants together at one location, appear to offer many advantages and significant potential for improvements in economics. The advantages mentioned are redundancy, savings in intake/outfall costs, production of two qualities of water for meeting the domestic, commercial and industrial demands, savings in the pretreatment and post-treatment costs and other operating costs.

5.8 Challenges and Opportunities in Nuclear Desalination

Today, 16% of the world's electricity is generated by nuclear power plants. There are 435 nuclear reactors operating in 33 countries. The existing power plants are very competitive and their load factors have remained high. Nuclear power production is now a mature technology. Its role in combined heat and power applications, such as desalination, is indeed significant.

Unfortunately, existing nuclear power plants have been branded as "cash hungry" for most worldwide utilities. In the last few years there has been remarkable change in the view of the capital markets, and nuclear is not being seen as so capital intensive. Economic competitiveness appears to be no longer an issue. A large number of reactors are now planned in many developing countries, due to their increasing energy demands and their meagre sources of fossil fuel.

However, the lack of confidence in political stability, nuclear regulatory policies and financial aspects in many countries interested in nuclear technology, has been a negative factor. To overcome this, partnership of utilities with large industrial companies will be needed. The role of governments in recognising the social benefits, and in reducing the various risks, is also desirable.

Most of the countries suffering from scarcity of water are generally not the holders of nuclear technology, do not have nuclear power plants, and do not have a nuclear power infrastructure. The utilisation of nuclear energy in these countries will require infrastructure development and institutional arrangements, e.g. for financing, legal liability, safeguards, safety and security. The acquisition of fresh fuel and the management of spent fuel will also need to be addressed.

Nuclear heat applications, including desalination, have been considered for a long time but not many have succeeded. Effective and practical measures against climate change/greenhouse gas reduction need to be taken. Nuclear technology and its related institutions need to advance and address the current climate, as do other technologies and environmental institutions. Practical application could possibly be based on exchange of experiences and further international collaboration.

Abbreviations

ALARA	As low as reasonably achievable
BWR	Boiling water reactor
CANDU	Canadian deuterium uranium reactor
CAREM	Argentine PWR
CDM	Clean development mechanism
CHP	Combined heat and power
DEEP	Desalination economic evaluation programme
DM	Demineralisation
GHG	Green house gases
GT-MHR	Gas turbine modular helium reactor
HTGR	High temperature gas cooled reactor
IAEA	International Atomic Energy Agency
KANUPP	Karachi Nuclear Power Plant
LMFR	Liquid metal fast reactor
MED	Multi-effect distillation
MSF	Multi-stage flash
MVC	Mechanical Vapour compression
MW_{th}	Thermal Mega Watt
NHR	Nuclear heating reactor
NMVOC	Non Methane Volatile organic carbon
PBMR	Pebble bed modular reactor
PHWR	Pressurised heavy water reactor
PWR	Pressurised water reactor
RO	Reverse osmosis
RO-ph	Reverse osmosis with pre-heating
SMART	System integrated modular advanced reactor
TVC	Thermal Vapour compression
VOC	Volatile Organic Compound

References

1. International Atomic Energy Agency, Introduction of Nuclear Desalination-A Guidebook, IAEA-TRS No. 400 (2000)
2. Tewari P.K. (2007) Current activities on nuclear desalination in India, Nuclear India, 41 (01–02): 22–25.

3. Awerbuch L. Proceedings of International Conference on Non-electrical Applications of Nuclear Power, 16–19 April, 2007, Oarai, Japan
4. International Atomic Energy Agency, Status of Nuclear Desalination in IAEA Member States, IAEA-TECDOC-1524 (2007)
5. Raluy RG, et al. (2005) Life cycle assessment of desalination technologies integrated with renewable energy, Desalination, 184 (1–3), 81–93
6. Tian Li. et al. (2003) A comparative economic analysis of the contribution of nuclear seawater desalination to environmental protection using clean development mechanism, Desalination, 157, 289–296
7. International Atomic Energy Agency, Economics of Nuclear Desalination-New Developments and Site-Specific Studies, Final Report of a Coordinated Research Project 2002–2006, IAEA-TECDOC-1561 (2007).
8. Carnino A, Gasparini N. Safety Aspects of the Desalination of Seawater Using Nuclear Energy, Proc. Symp. Nuclear Desalination of Seawater, Taejon, Korea (1997)
9. International Atomic Energy Agency, Safety of Nuclear Plants Coupled with Seawater Desalination Units, IAEA-TECDOC-1235 (2001)
10. International Atomic Energy Agency, Design Concepts of Nuclear Desalination Plants, IAEA-TECDOC-1326 (2002)
11. Masriera N, Doval A. et al. (2006) Water Monitoring as Safety Feature of Nuclear Desalination, Int. J. Nucl. Desalination, 2 (1), 44–45
12. Tewari P.K, Khamis I. Proceedings of International Conference on Non-electrical Applications of Nuclear Power, 16–19 April, 2007, Oarai, Japan
13. Desalination Economic Evaluation Program (DEEP) User's Manual, IAEA Computer Manual Series No. 15 (2005)
14. International Atomic Energy Agency, Advanced Application of Water Cooled Nuclear Power Plants, IAEA TECDOC 1584 (2008)
15. Methnani M. Coupling and thermodynamic aspects of seawater desalination using high temperature gas cooled reactors, Proceedings of IDA World Congress on Desalination & Water Reuse, Bahamas (2003)

Chapter 6
Solar Thermal Processes

A Review of Solar Thermal Energy Technologies for Water Desalination

M.T. Chaibi and Ali M. El-Nashar

Abstract The use of solar energy for desalination purposes was one of the first processes developed for producing fresh water from salt water. The process is based on the use of solar thermal energy to evaporate water, thus separating pure water from brine. In this chapter an overview of solar thermal desalination processes is presented. The first sections introduce the use of the simplest devices, i.e. solar stills, as stand alone systems or integrated into more complex geometries such as greenhouses. Couplings of the most recent technologies for solar thermal energy collectors with conventional thermal desalination processes are then presented. In all cases, a study of the feasibility and potential for further developments are presented. Solar thermal systems seem to be a very good option in remote areas with high insolation and where the use of conventional energy sources for desalination is not a viable option. Moreover the rising energy cost and continuous improvements of solar technologies are making such technologies more and more competitive compared with conventional ones.

6.1 Introduction

Arid lands generally have great solar energy potential. This potential may best be developed by solar desalination concepts and methods specifically suited to supply dry regions with freshwater.

Direct solar desalination systems have low operating and maintenance costs but require large installation areas and high initial investment. However, this is an appropriate solution for remote areas and small communities in arid and semi-arid regions lacking water.

M. Chaibi, (✉)
National Research Institute for Rural Engineering Water and Forestry,
B. P 10 2080 Ariana Tunisia
e-mail: chaibi.medthameur@iresa.agrinet.tn

Ali M. El-Nashar (✉)
22 Ahmed Gharbo Street, Apt. 703, Zizinia, Alexandria, Egypt
e-mail: elnashar100@hotmail.com

A. Cipollina et al. (eds.), *Seawater Desalination*, Green Energy and Technology,
DOI 10.1007/978-3-642-01150-4_6,

Most studies published in the last decade have considered small scale solar desalination systems for application in remote areas. Some of them have proposed medium and large scale systems which already are, or could be, effective in the near future.

The first part of this chapter is concerned with the analysis of direct solar desalination systems that could be applied in rural arid areas. It also describes research work carried out in the field of solar energy desalination, in conjunction with greenhouses, in arid areas. Operational and environmental limitations and features of direct solar desalination technologies are reviewed and discussed.

A new approach to improve the efficiency of solar desalination plants, incorporating three main research topics is reviewed. This concerns the basic blocks which are included within a production plant: collectors, heat storage arrangements and desalination units.

The second part of this chapter is concerned with the analysis of the various types of solar thermal collectors currently available for converting solar radiation into thermal energy suitable for desalination. Relevant examples of couplings with conventional thermal desalination technologies are also presented and critically discussed.

6.2 Direct Solar Desalination Systems

6.2.1 Historical Background

The first conventional solar still plant was built in 1872 by the Swedish engineer Charles Wilson in the mining community of Las Salinas in Northern Chile. This still was a large basin-type still used to supply fresh water from brackish feed water to a nitrate mining community. The plant had wooden basins with bottoms blackened using logwood dye and alum. The total capacity of the distillation plant was about 23 m^3/day. This first stills was in operation for 40 years until the mines were exhausted.

During World War II, efforts were increased to produce a solar still that could be utilised on life rafts, for ships and aircraft. Telkes [1] invented a small inflatable plastic unit for this purpose and hundreds of thousands of these units were produced.

Most stills built and studied since then have been based on the same concept, though with many variations in geometry, materials, methods of construction and operation.

From 1958 to 1965 the Office of Saline Water (OSW) at the Research Station in Florida tested a number of different types of solar stills and concluded that high fixed charges associated with still construction would not be offset by the savings resulting from free solar energy.

Past research work has been focused on the construction cost obstacle of the solar still. For example, to this aim, various plastic films have been used instead of the more durable but also more expensive glass coverings.

The next stage was to improve the operating efficiency of the various types of solar distillation devices. Several researchers [2–5] have attempted to enhance the vapour condensation rate by forcing air circulation in stills, and to increase the output of the stills by using latent heat of condensation/vaporisation in either multi-effect systems or for preheating the brine.

6.2.2 Different Designs of Solar Stills

Solar still designs described by Malik et al. [6] and Kudish [7] range from the simple to high-tech design.

The simplest and most practical design for a solar still is the single-basin type (Fig. 6.1).

The still consists of an air tight basin, usually made of galvanized iron sheet which contains the salt water. It has a top cover made of a transparent material, e.g. glass, and the interior surface of its base is blackened to enable absorption of solar energy to the maximum possible extent. The glass cover allows solar radiation to pass into the still. Here the radiation is mostly absorbed by the blackened base. The water begins to heat up and the moisture content of the air, trapped between the water surface and the glass cover, increases. The base also radiates energy in the infra-red region, which is reflected back into the still by the glass cover, so trapping the solar energy inside the still. Heated water vapour evaporates from the basin and condenses on the inside of the glass cover. Condensed water trickles down the inclined glass cover to an interior collection trough, placed at the lower edges of the cover to collect the distillate.

A common variant of this type is known as the inclined or tilted-tray solar still. It operates in a cascade formation and the rate of the continuous feed just compensates the rate of evaporation (Fig. 6.2). This type of construction results in higher efficiency than the simple basin type. However, the additional costs of construction are not compensated for by increased performance.

Another type of solar still that is designed to operate with a very low heat capacity is the tilted-wick still. It was proposed by Telkes [1] and consists of a tilted solar still in which the water basin is replaced by a porous wick through which water flows down by gravity. Tanaka et al. [8] proved the superiority of the tilted-wick type solar still over the basin type, and confirmed an increase in productivity of

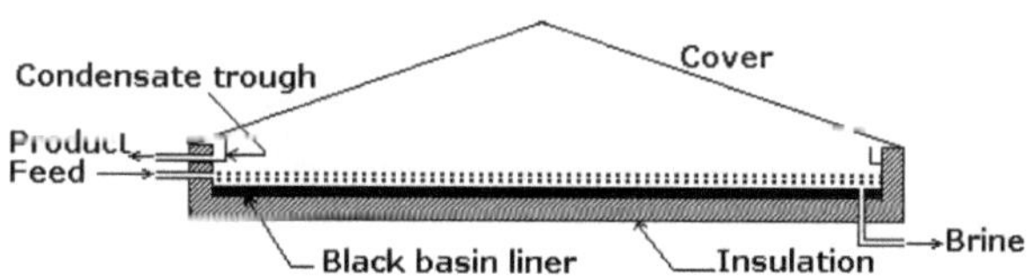

Fig. 6.1 Single basin type solar still

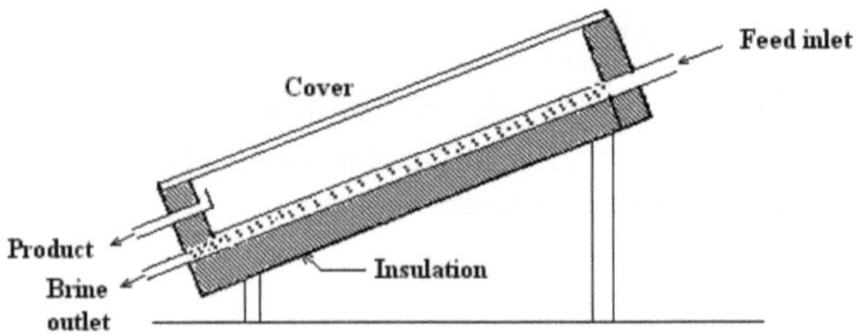

Fig. 6.2 Tilted-tray solar still

20–50%. However, the difficulties in keeping the wick wet and clean, and the cost of the construction, prevent its use on a large scale.

Sodha et al. [9] described a multiple-wick solar still in which blackened jute cloth pieces, of different lengths, are layered one upon the other and separated by polythene sheets.

The feasibility of using solar energy as a heat source for a multiple-effect diffusion solar still process was first developed by Cooper and Appleyard [10] and has been extensively studied by many researchers since, mainly because of the great potential to improve productivity. The process consists of a number of parallel plates in contact with saline soaked wicks, with narrow air gaps between the plates. The evaporation and condensation processes occur in all gaps between the plates. The hot water is applied to the first plate, in order to heat it up and to cause evaporation from the plate. The latent heat of condensation is recovered to induce evaporation from the parallel plate. The process is repeated for all plates to increase the production of distillate (Fig. 6.3). This new type of still was improved by Tanaka

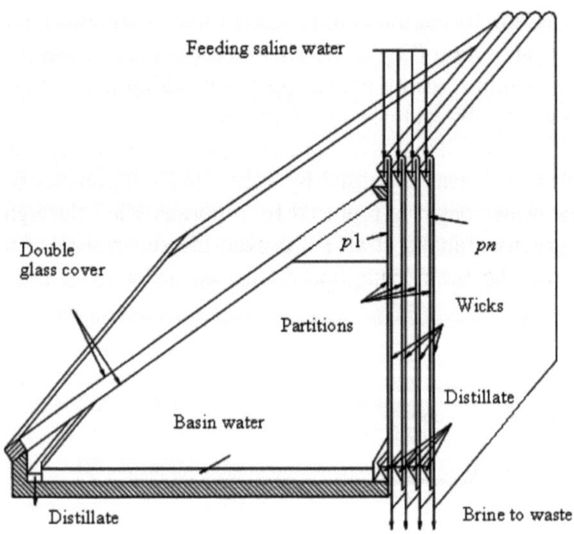

Fig. 6.3 Multiple-effect diffusion solar still

et al. [11, 12], who modified the configuration of the partitions and the diffusion gaps between them, the thermal insulation materials and the geometry of the unit. The average daily productivity of this proposed solar still was found to be 14.8–18.7 l/day distillate per unit effective area of glass, under an incident solar radiation range of 20.9–22.4 MJ/m^2/day. The process yields are reported to be doubled [12] when coupling the unit with a flat plate collector to preheat feed water, and with wheels attached to the still bottom for manual azimuth tracking. This equipment, required in addition to the vertical still, contributes to increase solar absorption at the proper angle on the first partition of the unit.

6.2.3 Solar Still Greenhouse Combination

Water production from a solar still would generally not be adequate to meet the total water requirement of a crop grown in an open irrigated field, but may be sufficient to supply freshwater for protected cultivation. Solar desalination should be used in combination with water efficient greenhouse concepts, based on a controlled environment.

The main arguments behind this combination are:

- Water requirements of crops under protected cultivation have a diurnal and seasonal fluctuation which is similar to the productivity variation of solar desalination. Both processes are primarily driven by the varying solar irradiation and therefore well correlated.
- Water use efficiency in protected cultivation is higher than in open fields, especially if the cost of the crop is dependent on it. This could result in higher income for small-scale producers, especially where there are limited water resources in the region.

However, use of solar energy desalinated water for agricultural purposes has not yet progressed beyond the experimental phase. A complete survey of a system, combining a solar still with a greenhouse, was first presented by Trombe and Foex [13] (Fig. 6.4) and later an improved version of the concept was developed by Boutiere [14] and by Bettaque [15].

The concept utilised partly transparent absorbing materials, instead of opaque ones, and consisted of a double-glassed roof. The inner layer of the glass roof was covered with a shading material. Salt water flowed down between the two layers over the top of the shading material. Part of the global irradiation was absorbed at the inner layer of the roof where the salt water evaporated. The water vapour condensed on the inner surface of the outer layer, ran down along the glass, and was collected by gutters for distribution to the crop.

According to Selsuk and Tran [16], although successful in operation, the operating conditions required for the most efficient performance of the solar still and the

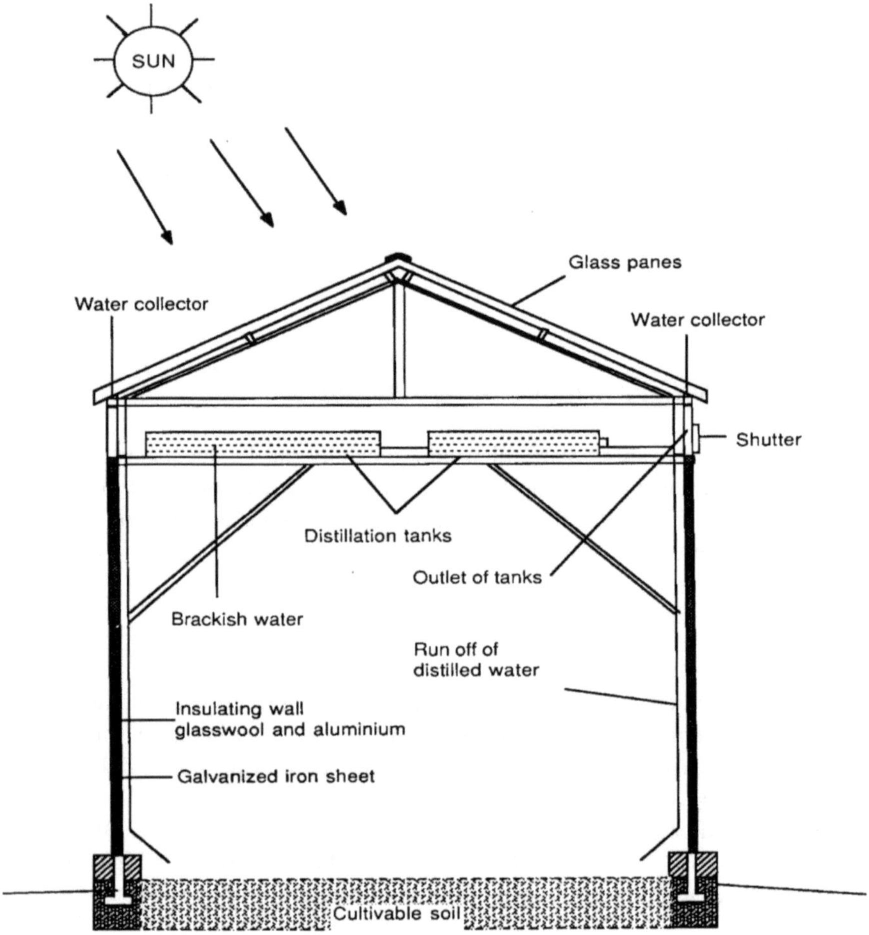

Fig. 6.4 Greenhouse solar still

greenhouse of this integrated system solution with double-glassed roof, could not have been maintained.

Consequently, they proposed a solar still concept completely separated from the greenhouse, by placing an insulation layer beneath the still, to secure a more effective control of the plant environment and to increase the solar still yield. This concept was also recommended, after experimentation and analysis, by the Brace Research Institute in Canada [16].

From 1979 to 1984, a closed greenhouse with integrated solar water desalination, called ITG-system, was developed at the University of Hanover in Germany. It was evaluated by Strauch [17], and compared with a modified Bettaque system.

The concept of the ITG-system is suitable for tropical desert conditions. It is based on the combination of collecting the water evaporated and transpirated in the

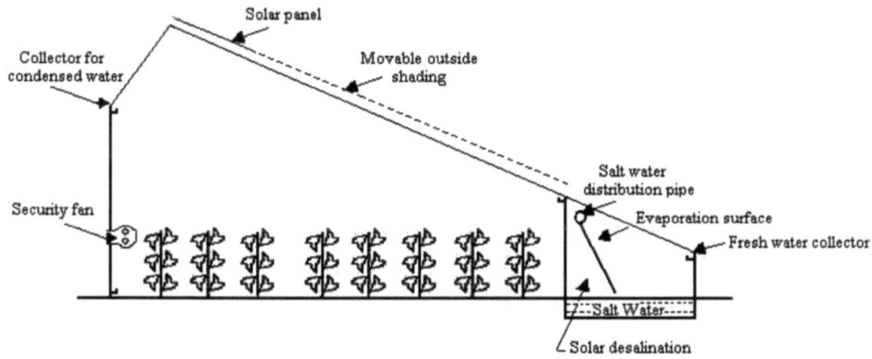

Fig. 6.5 Integrated solar water desalination green house system (ITG-system)

enclosure by condensation on its roof, and the water produced by the solar still attached to the southern side wall of the greenhouse (Fig. 6.5).

The modified double-glassed roof integrated still includes a second glass sheet below the south-oriented outer glass. This glass sheet absorbs a high amount of the near-infrared-radiation (NIR) and has a sufficient transmittance of the visible light (PAR – Photosynthetically Active Radiation) thus acting as absorber and evaporator.

Results of experiments carried out on these systems show the increase in efficiency of the ITG-system compared to the Bettaque system: 29% compared to 16%, and a productivity of 2–2.5 l/m^2/day for the ITG-system compared to 1 l/m^2/day for the Bettaque system.

However, in their study from 1978 to 1981, Dumont and de Cachart [18] proposed a greenhouse with solar distillation using a developed Bettaque concept. The flow of the water for distillation as a film is maintained by a series of sprinklers located at the top of the frame. The results of experiments carried out from 1978, show that the average daily production of fresh water was in the range of 2–3.5 l/m^2/day, under favourable conditions.

From 2000 to 2003, Chaibi and Jilar [19] developed an integrated solar system greenhouse based on the Bettaque system at the Tunisian National Research Institute of Rural Engineering, Water and Forestry. In their system, light transmission through the roof is reduced, as solar radiation is absorbed by a layer of water flowing between two layers of glass. Freshwater is evaporated, condenses on the upper glass and is then collected under the eaves of the roof (Fig. 6.6). Experimental results have shown that such a system, integrated into 50% of the roof area of a wide-span greenhouse, has the capacity to match the annual water demand for a low canopy crop. It has a water production capacity of about 1.5–2 l/m^2/day for days with high irradiation, and exceeds water demand by 2.3–1.6 times during early and late periods of the growing season.

In 2007, Zaragoza et al. [20] proposed a new concept for a single closed greenhouse for advanced horticulture use, which acts as a climatisation system and as a process for grey or saline water desalination. It consists of a greenhouse with a solar

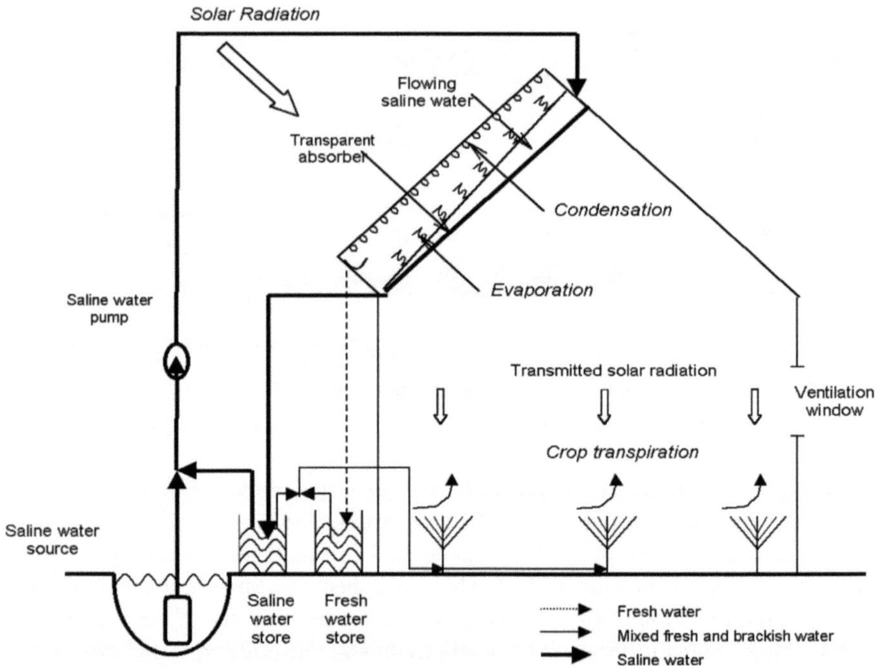

Fig. 6.6 Integrated solar system greenhouse based on the Bettaque system developed at the Tunisian National Research Institute of Rural Engineering, Water and Forestry

chimney, inside which a cooling duct contains an air-to-water heat exchanger connected to a heat accumulator. The process starts with the heating, by evapotranspiration of the plants and the soil, of the humidified air inside the greenhouse, which then rises up inside the solar tower by natural buoyancy. This rising air becomes saturated via evaporation of saline water from a pan placed on the top of the planted area. On the surface of the heat exchanger, cooling of the saturated humid air creates condensation, thus releasing additional thermal energy and distilled water. The first prototype of this concept was built in Almeria, Spain. The testing phase of the concept is on-going and technical improvements and research will present further challenges.

The different designs and concepts of an integrated greenhouse solar still presented above constitute an exciting option for the support of small scale agricultural production, in places where only saline water is available, especially when it is applied to high-value crop cultivation, such as vegetables and flowers grown in greenhouses.

6.2.4 Productivity and Efficiency of a Solar Still

Although the construction and operation of a solar still is simple, the theoretical analysis is complex and mostly depends on experimental observation. Basic concepts of solar stills were developed by several scientists. They found that the operation of a solar still is governed by various heat transfer modes and therefore a proper understanding of modes of heat transfer, mainly convection and radiation, is crucial to the study of a solar still.

Fresh water production of a solar still can be predicted based on the temperature of the feed salt water and the area of the solar still using mass diffusion relations. The net mass flux of water vapour is estimated using the analogy between heat and mass transfer.

In their analysis of water productivity, Malik et al. [6] consider the rate of mass transfer of the vapour from the water surface to the transparent cover (glass) as:

$$m_d = h_e \frac{(p_w - p_g)}{\lambda} \tag{6.1}$$

where h_e is the equivalent mass transfer coefficient, which may be expressed in terms of the heat transfer coefficient in the still ($h_{c,wg}$), p_w and p_g are the vapour pressures of the water in the basin at basin temperature (T_w), and of water at cover temperature (T_g) [6] and λ is the latent heat of water.

Based on experimental data for solar stills, Malik et al. [6] defined the relationship between the heat transfer coefficient by convection to that of evaporation as:

$$\frac{h_e}{h_{c,wg}} = 1627 \cdot 10^{-3} \tag{6.2}$$

Introducing this relationship into Eq. (6.1), the mass transferred per unit area per unit of time, by evaporation from the water surface to the glass cover is:

$$m_d = 1627 \cdot 10^{-3} h_{c,wg} \frac{(p_w - p_g)}{\lambda} \tag{6.3}$$

where $h_{c,wg}$ is the convection coefficient of the still. For simple basin types of solar stills Dunkle [21] gave the following equation, applicable for normal ranges of operation:

$$h_{c,wg} = 0.884[T_w - T_g + \frac{(p_w - p_g)}{(268900 - p_w)} \cdot T_w]^{1/3} \tag{6.4}$$

Still productivity, after a certain period of time, is the accumulated mass of freshwater collected during the period. The total time, t, is taken as 1 h, during which time dependency of $h_{c,wg}$ and Δp are approximated by average constant values. The latent heat of vaporisation, λ, is determined each hour at hourly average water temperature T_w.

From the Eq. (6.3), distilled water production per hour and per m^2 of still basin can be written as:

$$m_{dh} = \left[1627 \cdot 10^{-3} h_{c,wg} \frac{(p_w - p_g)}{\lambda}\right] \cdot 3600 \tag{6.5}$$

The overall efficiency of the solar still, η, is the ratio between the productivity multiplied by the average value of latent heat of vaporisation, and the cumulative solar irradiation over the relevant time period:

$$\eta = \int_0^t (md_d \lambda / G)\, dt \tag{6.6}$$

From the equations which govern internal and external heat transfer, and in which convection, evaporation and radiation to the inside glass surface are included, it can be shown that combined modes of heat and mass transfer constitute the main factor which limits overall efficiency and which reduces the maximum attainable efficiency. To operate the solar still at a high mean temperature, the evaporative fraction is increased and hence efficiency is greatly improved. The productivity rate, at a given water temperature, is particularly improved by water–glass temperature differences on the evaporation and condensation surfaces, as well as the difference between the saturation pressure in the basin (P_w) and the saturation pressure at the cover (P_g). This is shown in Eq. (6.1), which gives the heat transfer on evaporation from the water surface to the cover. Thus, the overall efficiency of the solar still is directly related to $h_{c,wg}$.

6.2.5 Limitations and Outlook

Apart from the economical problems associated with operating solar desalination systems, a number of technical problems can cause difficulties. These difficulties usually result in reduced efficiency compared to theoretical efficiency and any reduction is a serious matter for solar desalination technologies.

Even with a good basic design for solar desalination devices, the following points and problems need to be considered.

- Vapour leakage: in order to operate efficiently, solar stills should be reasonably airtight to avoid loosing water vapour before condensation.
- Heat absorption capacity: it is desirable to keep the basin temperature as high as possible, which increases the ratio of heat transfer by evaporation-condensation to that by convection and radiation.
- Low thermal efficiency and productivity: this can be improved by separating the heat collection and the distillation processes.
- Salt and organism accumulations: the growth of algae and other micro-flora particles on the surface of the brine and in the basin generally reduces heat transfer to the brine. Water supplies may be vulnerable to microbial contamination mainly during low temperature operating conditions.
- Availability of spare parts: lack of spare parts can affect the operation of the system and contribute to the abandoning and/or the decommissioning of solar desalination facilities.

Many solar distillation plants have been abandoned or dismantled due either to the above operational problems or to the presence of other water resources within the same area as the solar desalination facilities.

6.2.6 Environmental Analysis

Solar energy for direct solar desalination technologies is free, naturally available on sunny days and the net useful energy yield is moderate to low. The technology is well developed and can be installed easily. No carbon dioxide is added to the atmosphere and the environmental impacts from air pollution and water pollution are low. Land disturbance is also low because direct solar desalination technologies are mainly intended for remote rural areas where land is vacant, accessible and affordable. For combined solar stills and greenhouses, land disturbance is considered to be non-existent because the desalination system is incorporated into existing structures and placed on the top of the greenhouses. By using integrated solar stills in greenhouses, during sunny periods temperatures inside the greenhouses can be reduced by 2–7°C, and the daytime temperature and humidity variation has a less peaky pattern for the hours around noon, compared to the same conditions in conventional greenhouses [19]. These steady climate conditions are important in reducing the energy amount required for cooling the greenhouse and to prevent stress which is harmful and growth-reducing to crop cultivation.

It was concluded that a solar still system integrated into a greenhouse roof could reduce the energy required for greenhouse cropping, both in summer and winter seasons.

However, when using solar energy for water desalination plants, operating on a small to medium scale, there are some environmental risks, which are noted below. But these are minor issues compared to those of conventional energy desalination plants.

The disposal of the waste brine, after the desalting process, can present significant environmental and ecological degradation. There are often environmental and legal constraints from discharging from a desalting plant into surface or underground waters. For facilities located near to the sea, disposal does not generally pose a problem, but it can be a serious issue for inland facilities. Some methods that have been used include evaporation by solar means, injecting the brine into an existing zone of very saline groundwater, or transporting the brine to a saline water body [22, 23].

Site hydrology should be considered to avoid flooding of the plant. Local meteorology should also be taken into account. Gomkale [24] reported that in 1968 some Indian plants were damaged by heavy rains at Bhavnagar (1,000 l/day) and in 1982 by a cyclone at Awania village (5,000 l/day).

6.2.7 Research and Recommendations

Cooper [25] proved, using equations governing internal and external heat transfer, that a total efficiency of about 60% is the upper limit for a perfect solar still. Cooper even showed that the efficiency of solar still devices rarely exceed 50% in application, and outlined features required to attain higher efficiencies. The average value of overall efficiency of a solar still is around 30–40% with a daily production in the range of 2–4 l/m^2.

A review of factors influencing the total efficiency shows that there are three major aspects which appear to affect future solar desalination trends . These aspects concern the basic blocks which are included in a production plant: the collectors, the heat storage arrangement and the desalination unit.

For the first item, a wide variety of materials are used for solar desalination collectors. The key material is the transparent covering for the solar still, which is usually a glass pane or plastic material. Glass has proved to be an excellent and durable material but it is often very costly and subject to breakage during transport, installation and use.

Compared to other coverings, plastic films are lighter, cheaper, and easier to transport but are subject to wind and rain damage during use. Although research may continue to improve the suitability of plastic films, at present it is not appropriate to recommend them for permanent application in arid countries.

Primarily on an experimental basis, several types of rigid and semi-rigid plastic sheets have been used as covers for solar stills. These covers present the advantages of being light weight, formable into self-supporting shapes, and to have higher impact strength compared to glass. On the other hand, they do present some disadvantages, including their low thermal conductivity (of importance for condensation properties of stills), high thermal rate of expansion, high transmittance of infrared radiation, high cost, and rarity or complete absence in some arid countries.

Research and development to improve collector efficiency should concentrate on the following aspects:

- Thermal insulation materials and geometry (conduction losses, mainly through non-glassed parts, are still a major efficiency problem).
- Thermo-optical properties of surface structures and coatings, relating to manufacturing technologies especially for films or plastic sheet coverings.
- Proper sealing of the joints between cover sheets and frames.

The second main field of research is heat storage. The possibility of operating at night could double the economical advantages of a solar desalination plant compared with a conventional plant. Research should be directed towards the identification of new chemical reaction storage systems and storage system studies.

The third research field should undertake studies of the solar desalting plant as a whole with the following research objectives:

- Thermal insulation of the bottom and side walls, aimed at decreasing the heat losses by conduction to the surroundings, by using low cost, long-life insulating materials.
- Geometry and orientation of collectors aimed at maximising solar heat collected with variable solar irradiation.
- Thermo-optical properties of unit surfaces, side walls and the bottom.
- Reliability and lifetime studies to estimate, with some statistical confidence, the expected degradation in performance of the collector and solar plant materials. This is particularly important given the extremely corrosive environment caused by saline solutions combined with different treatments (degreasing, hydrazine, phosphate, etc.). Proven designs are those that use standard, durable construction materials, although designs that rely on plastic materials to a certain extent are also being investigated. Construction, maintenance and operation should use, to some degree, local labour requiring neither high skill levels in working nor complex machinery. The reference lifetime, given previously by scientists/researchers, was 20 years, however it is possible, with suitable research effort, to double this.
- New designs orientation: system studies should be oriented at synthesising new designs for the overall plant solution through novel combinations of sub-systems. Other systems for increasing the productivity of the still should be considered. Recommendations on this issue have been made by many authors such as cooling the glass using a PV fan, or by causing the brine to flow down the glass before reaching the solar still [26]. Other system studies concern multiple-function plants. Examples are the coupling of solar desalting systems with food production and space conditioning facilities [27]. These systems might also help to improve the attitude of users towards solar desalination technologies.

Programmes for these fields of research, devoted to water resource development and desalination, should be based on simulation models using technical parameters which are appropriate, in terms of desalination technology, to the needs of the arid lands living conditions. These models would mean further adaptation of existing technology to the specific needs of developing countries.

Adaptations could include design modifications that allow use of available local materials and locally manufactured components.

Primarily, research should focus on improving the economics of solar distillation, and thereby contribute to the solution of water supply problems, at least for small communities in arid areas.

Besides technical research field aspects, it would be valuable to consider the education aspect in remote arid areas. Education of the local community, on the basics of desalination, would create and spread awareness of the threat of water scarcity.

6.3 Devices for Converting Solar Radiation into Thermal Energy Suitable for Desalination

Solar energy collection devices are special kinds of heat exchangers that transform solar radiation energy to internal energy of the transport medium. The major component of any solar system is the solar collector. This is a device which absorbs the incoming solar radiation, converts it into heat, and transfers this heat to a fluid (usually air, water or oil) flowing through the collector. The solar energy thus collected is carried away from the circulating fluid, either directly to the hot water or to a thermal energy storage tank, from which can be drawn for use at night and/or on cloudy days.

In this section a review of the various types of collectors currently available is presented.

There are basically two types of solar energy collection devices: non-concentrating or stationary, and concentrating. A non-concentrating collection device has the same area for intercepting and absorbing solar radiation, whereas a sun-tracking concentrating solar collector usually has concave reflecting surfaces to intercept and focus the sun beam radiation to a smaller receiving area, thereby increasing the radiation flux. A large number of solar collector types are available on the market. A comprehensive list is shown in Table 6.1 [28], where the concentration ratio is defined as the aperture area divided by the receiver/absorber area of the collector.

Table 6.1 Solar heat collecting devices

Motion	Collector type	Absorber type	Concentration ratio	Temperature range (°C)
Stationary	Solar pond	Salt water layer	1	50–100
	Flat plate collector (FPC)	Flate Plate	1	50–90
	Evacuated tube collector (ETC)	Flat plate/circular tube	1	70–200
	Compound parabolic collector (CPC)	Tubular	1–5	60–240
Single-axis tracking	Linear fresnel reflector (LFR)	Tubular	15–45	60–250
	Parabolic trough collector (PTC)	Tubular	15–45	60–300
	Cylindrical trough collector (CTC)	Tubular	10–50	60–300
Two-axis tracking	Parabolic dish reflector (PDR)	Point	100–1,000	100–500
	Heliostat field collector (HFC)	Point	100–1,500	150–2,000

Note: Data partially extracted from Kalogirou [28].

6.3.1 Salt-Gradient Solar Ponds (SGSP)

The solar pond is one of the simplest devices for direct thermal conversion of solar energy [29]. Moreover, it is simultaneously a solar collector and a thermal storage device. Any pond converts insolation into heat, but most natural ponds quickly lose that heat, through vertical convection within the pond and through evaporation and convection at the surface. The solar pond artificially prevents both losses, because of its massive thermal storage capacity and measures taken to retard heat loss. Non-convective ponds employ a salt gradient to prevent the warm water at the bottom from rising to the surface. The salt concentration in such a pond is highest near the bottom and lowest near the surface. The salts most commonly used are NaCl and $MgCl_2$. Such ponds may be a reliable source of heat for a wide range of industrial and agricultural applications such as process heating, space heating, desalination and electricity generation. Solar ponds have several advantages over other solar technologies. They have a low cost per unit area of collector, inherent storage capacity and are easily built over large areas. It has been demonstrated that salinity-gradient solar ponds can be reliable heat sources at temperature levels of 50–90°C.

A typical salinity-gradient solar pond has three regions (see Fig. 6.7). The top region is called the surface zone or upper convective zone (UCZ). The middle region is called the non-convective zone (NCZ). The lower region is called the storage zone or lower convective zone (LCZ). The lower zone is a homogeneous, concentrated salt solution that can be either convective or temperature-stratified. Above it, the NCZ constitutes a thermal insulating layer that contains a salinity gradient. This means that the water closer to the surface is always less concentrated than the water below it. The surface zone is a homogeneous layer of low-salinity brine or fresh water.

If the salinity gradient is large enough, there is no convection in the NCZ even when heat is absorbed in the lower zone, because the hotter, saltier water at the

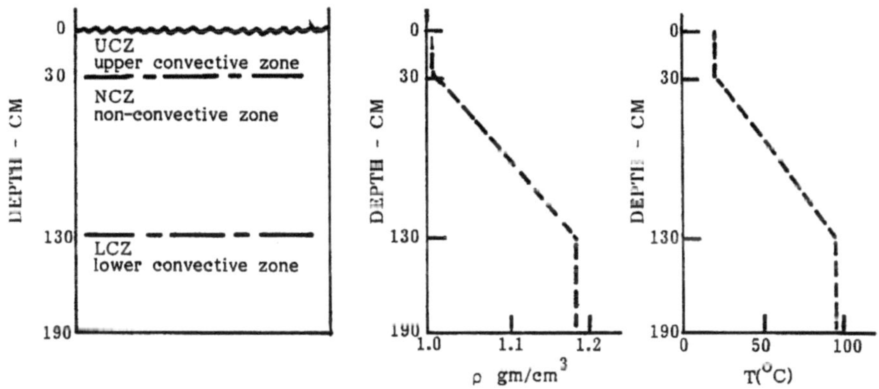

Fig. 6.7 Non-convective solar pond
Source: Lu et al. [29]

bottom of the gradient remains denser than the colder, less salty water above it. The thermal conductivity of water is moderately low, and if the gradient zone has substantial thickness, heat escapes upward from the lower zone very slowly. The insulating properties of the gradient zone, combined with the high heat capacity of water and the large volume of water, make the solar pond both a thermal collector and a long-term storage device.

The thermal efficiency of a solar pond, which is defined as the ratio of heat removal rate from the LCZ to solar radiation incident on the pond surface, is mainly affected by the clarity of pond water, pond configuration (especially the depth of the gradient zone) and temperature difference, ΔT, between the lower zone and surface zone. The greater the ΔT, the lower the thermal efficiency because of greater heat losses at higher pond temperatures. For this reason, solar ponds are more efficient for medium to low temperature thermal applications than for electric power generation, in which higher temperatures (usually above 85°C) are required for operating the generator efficiently. The thickness of the storage zone (LCZ) also has an effect on the thermal performance of the ponds. The daily temperature fluctuation in a solar pond with a thicker storage zone, is smaller than with a thinner storage zone. However, the pond with a thicker storage zone will have a longer start-up time. As an example, the El Paso solar pond [29] has a 1.2 m gradient zone and 1.35 m storage zone, and the bottom temperature increases at a rate of about 1°C per day during start-up, while the temperature fluctuation in the LCZ is 1–3°C between day and night.

The El Paso solar pond has a surface area of 3,000 m^2 and a depth of about 3.25 m. The UCZ, NCZ and LCZ are approximately 0.7, 1.2 and 1.35 m, respectively. The pond uses an aqueous solution of predominantly sodium chloride (NaCl). Figure 6.7 shows the typical density profile of a solar pond. The LCZ contains saturated or near-saturated brine with a concentration of about 26% by weight. The concentration in the UCZ (surface zone) is normally maintained at 1–4% salt by weight (10,000–41,000 mg/l). The operating temperature of the pond ranges from 70°C in winter to 90°C in early autumn. The highest temperature observed at the E1 Paso solar pond is 93°C, and the maximum temperature difference between the LCZ and UCZ is well above 70°C.

Figure 6.8 shows the percentage of total radiation, which hits the surface of a 2-m solar pond, reaching different depths of the pond. The Figure shows that about 30% of the incident radiation reaches the absorbing bottom surface of a pond. But since the bottom of a solar pond is an imperfect absorber, and heat losses occur from the top and bottom surfaces, the thermal efficiency rarely exceeds 25% [30].

6.3.2 Flat Plate Collectors (FPC)

Flat plate collectors are non-tracking (stationary) types of collectors. They are normally fixed in position and south-facing in the northern hemisphere (north-facing in the southern hemisphere) with an optimum tilt angle equal to the latitude of the location, with an angle variation of 10–15° depending on the application [28].

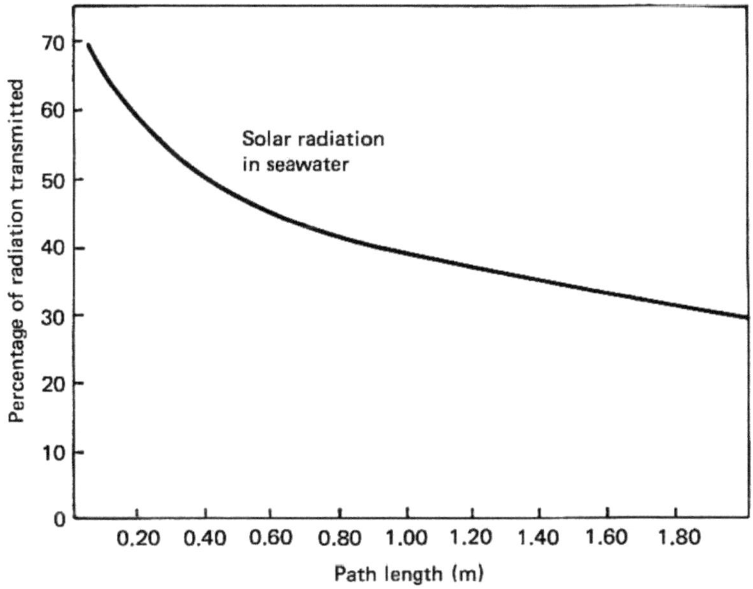

Fig. 6.8 Transmittance of water to solar radiation as a function of the thickness of the water layer Source: Kreider and Kreith, [30].

Figure 6.9 shows a cross section of a typical flat plate collector with a single glass cover. The absorber plate inside the collector converts sunlight into heat energy and transfers this thermal energy to a fluid, such as water. The fluid, in turn, removes the thermal energy from the collector and transfers it to a thermal storage facility. The front surface of the absorber plate has a black coating that absorbs solar radiation. The coating can be either a matt black paint or any one of a number of chemically deposited selective coating films. The back and sides of the absorber are insulated to prevent heat loss from the edges and back of the collector. The collector has a transparent glass pane covering the absorber plate. This glazing transmits solar radiation to the absorber plate and minimises heat losses from the hot absorber plate. Some collectors have more than one layer of glazing material. The type and number of glazings are dependent on the operating temperature of the collector and the type of coating on the absorber plate.

The materials used for the absorber plate include copper, aluminium, steel, glass and plastic, with copper and aluminium being by far the most common. Copper is the most popular material for collector tubes.

There are two ways in which absorbed solar energy leaves the absorber plate. One is by radiation and convection losses, and the other is via the heat transfer fluid in the collector tubes, whose specific purpose is to extract usable energy from the plate. In other words the usable energy output is given by the difference between the absorbed solar energy and the energy losses.

If the usable (collected) energy is Q_c, the area of the absorber plate is A_c, and incident solar radiation I, the collector efficiency can be defined as

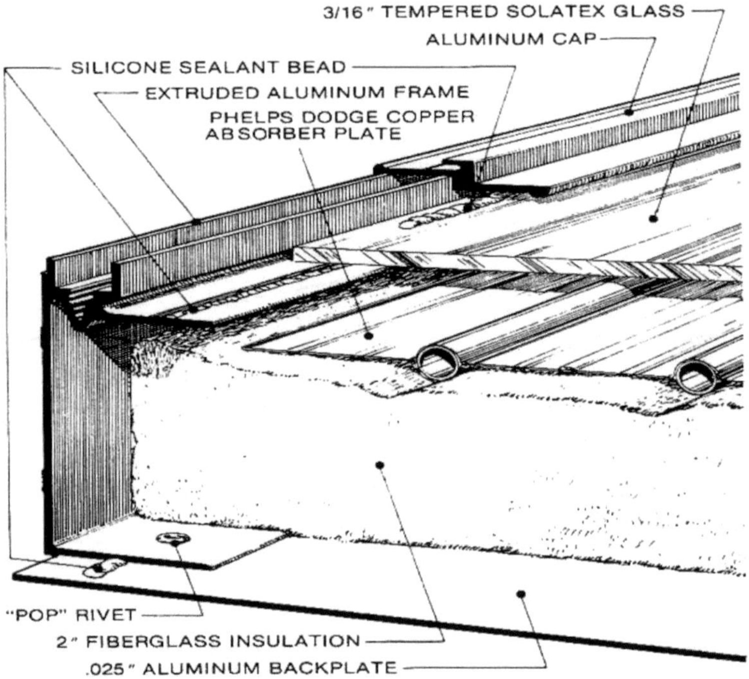

Fig. 6.9 Cross section of a flat plate solar collector
Source: Harris et al. [31].

$$\eta_c = \frac{Q_c}{A_c I} = \frac{\text{Absorbed solar energy}}{\text{Incident solar energy}} - \frac{\text{Energy losses}}{\text{Incident solar energy}} \tag{6.7}$$

If the energy losses can be assumed to be proportional to the difference between the absorber plate temperature, T_p and the ambient air temperature, T_a then

$$\text{Energylosses} = U_L(T_p - T_a) \tag{6.8}$$

where the proportionality constant U_L, is the thermal loss coefficient. The absorbed solar energy can be expressed as $I\,(\tau\alpha)$ where τ is the transmittance of the glazing and α is the absorptance of the absorber plate. The transmittance-absorptance product, $(\tau\alpha)$, is dependent on the incident angle of the direct beam radiation striking the glazing. An effective transmittance-absorptance product is used to compensate for the variation in $\tau\alpha$ throughout the day and is referred to as $(\tau\alpha)_e$.

The efficiency equation can then be expressed by the equation

$$\eta_c = (\tau\alpha)_e - U_L \frac{(T_p - T_a)}{I} \tag{6.9}$$

The use of the factor F_R (called the collector heat removal factor) enables the absorber plate surface temperature, T_p, in the above equation, to be replaced by the more convenient fluid inlet temperature T_i, thus

$$\eta_c = F_R(\tau\alpha)_e - F_R U_L \frac{(T_i - T_a)}{I} \tag{6.10}$$

F_R is a function of the fluid, the fluid flow rate, the collector-fluid interface and the interface geometry in order to relate T_i and T_p for a particular collector [30].

Note that Eq. (6.10) is in the form of a straight line: y = constant + mx. Collector efficiency curves are ordinarily plotted with η_c (instantaneous collector efficiency) on the vertical axis, as a function of $(T_i - T_a)/I$ plotted along the horizontal. This represents a straight line where the intercept constant $= F_R\ (\tau\alpha)_e$ and the slope, $m = F_R\ U_L$, while $x = (T_i - T_a)/I$. A typical flat plate collector having a single glass cover and non-selective absorber plate may have a y-intercept equal to 0.83 (dimensionless) and a slope of -8.47 W/(m^2.°C)

$$\eta_c = 0.83 - 8.47(\frac{T_i - T_a}{I}) \tag{6.11}$$

Flat Plate Collectors (FPCs) with selective absorber plates and double glazing usually achieve higher efficiency compared with non-selective, single glazing collectors.

Collector efficiency curves are plotted from data obtained with normal solar incidence i.e. with the collector facing directly into the sun. In actuality, the incident angle for FPC changes continuously as the sun moves across the sky. The effective transmittance-absorptance product, $(\tau\alpha)_e$, decreases as the angle of incidence increases from zero (i.e. from normal incidence). This causes the y-intercept of the efficiency curve to move down the y-axis, thus shifting the whole efficiency curve downwards.

6.3.3 Evacuated-Tube Collectors (ETC)

Evacuated-tube collectors have been proposed as efficient solar energy collectors since the early twentieth century [30]. Figure 6.10 shows cross sections of several glass evacuated-tube collector concepts. Temperatures of the order of 200°C can be achieved at 40% efficiency, in bright sunlight using evacuated tubular collectors without concentration. Although receiver geometry varies, all rely on vacuums of the order of $10^{-3} - 10^{-4}$ mmHg to eliminate conduction and convection losses from the absorber surfaces. Radiation heat loss is also minimised by using a low-emittance (selective) absorber surface. Methods used to extract heat from evacuated tubes include:

- Heat pipe: a metal absorber is mounted in a single envelope vacuum tube and the absorber is attached to a heat pipe that penetrates the vacuum space via a glass-to-metal seal.

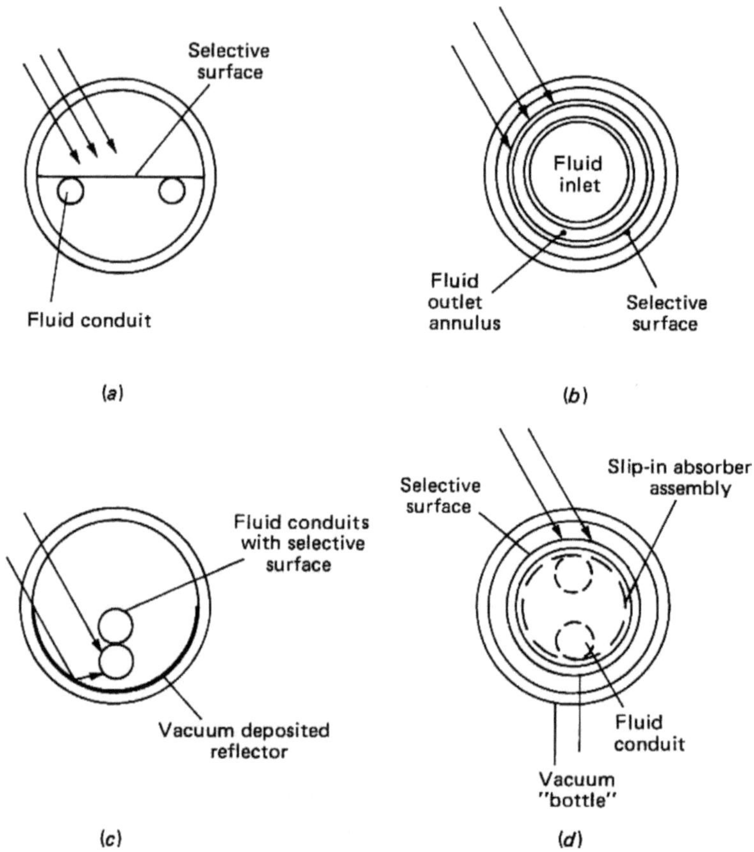

Fig. 6.10 Evacuated-tube solar energy collectors: (**a**) flat plate; (**b**) concentric tubular; (**c**) concentrating; (**d**) vacuum bottle with slip-in heat exchanger contacting rear surface of receiver
Source: Kreith and Kreider, [30].

- Flow through absorber tube: a single ended metal absorber pipe is mounted in a glass vacuum tube through a glass-to-metal seal. A central tube is used to deliver a heat removal fluid to the bottom of the metal absorber tube; this then flows up the annular space between the central tube and the larger metal absorber tube. Due to differential expansion the metal tube can only contact the glass at one end unless a vacuum bellows seal is used. An alternative configuration of this concept is to use two small diameter glass-to-metal seals at one end of a single envelope evacuated tube. A U-shaped heat removal fluid tube is introduced through the two seals and attached to the absorber.
- All-glass tubes: a Dewar type vacuum tube with the solar absorbing surface on the vacuum side of the inner glass tube. The absorbed heat is conducted through the inner glass tube wall and then removed by a fluid in direct contact with the inner glass tube, or by a heat removal fluid in a metal U-tube inserted in the inner tube, with a fin connecting the outlet arm of the U-tube to the inner glass tube.

- Storage absorber: Vacuum tubes greater than 100 mm diameter can function as both the absorber and the insulated hot water store. Tubes with 10–20 l of storage have been developed.

The water-in-glass concept, in which water is circulated directly through the inner glass tube, has good heat transfer from the glass absorber to the heat removal fluid, however, the operating pressure of the heat removal fluid is limited to a few metres of water head. The all-glass tube, with the U-tube heat removal system, has been successfully used for more than 20 years; however, it is expensive as a result of the plumbing and heat transfer fin in each evacuated tube. The efficiency of this system depends on the quality of the contact between the heat transfer fin and the glass absorber.

In a single glass tube solar collector, selective coating absorber plates enclosed in glass tubes, which are maintained under a vacuum, are used. High efficiency is maintained by a high vacuum and selective-coating absorber plates. The cross section of a single evacuated glass tube is shown in Fig. 6.11. An absorber plate, with a selective-coating surface, is inserted into the glass tube. Installed on the plate is an absorber tube through which passes the heat absorbing medium (water). The glass tube is maintained at a high vacuum of 10^{-4} mmHg. When the solar radiation passes through the glass tubes and is incident on the absorber plate, it is converted into heat, raising the temperature of the absorber plate. This heat is transferred to the heating medium (water) which flows through the absorber tube, generating hot water. This hot water is used for desalination. The vacuum construction and selective-coating surface plays an important role in transferring as much of the energy from the sun rays as possible to the heat medium.

The heat collection amount is defined as (incident energy from sun rays) – (heat loss). Consequently, technology which prevents the incident energy from escaping, i.e. raising the heat insulation performance, will serve to increase the performance of the solar collector. The heat loss from a solar collector is due to convection, conduction and radiation. Loss due to radiation can be reduced by using a selective coating. Vacuum insulation eliminates loss due to convection and conduction, hence can be said to be an ideal method for insulation of a solar collector.

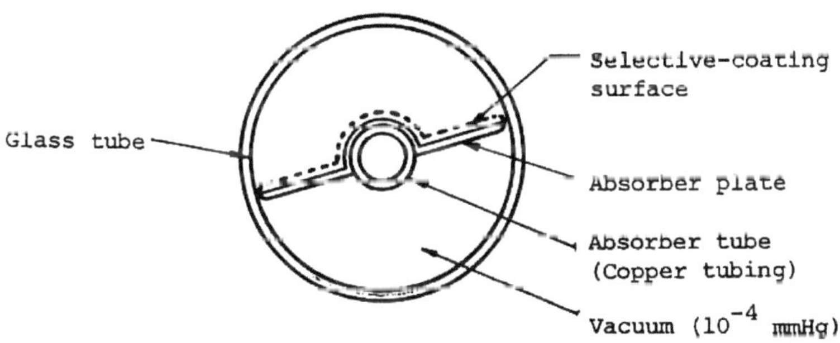

Fig. 6.11 Structure of a Sanyo evacuated tube collector
Source: Sanyo Electric Trading Co. [32].

Efficiencies can vary significantly with the irradiation and the type of collector. Standard values can range between 40 and 80%, with the lower values obtained at higher fluid temperature and lower solar irradiation.

6.3.4 Parabolic Trough Collectors (PTC)

Parabolic trough collectors are employed in a variety of applications, including industrial steam production, desalination and electrical power generation. Several PTC fields, with aperture areas from 500 to 5,000 m^2, have been built and operated throughout the world to provide process steam to industry. Most of these systems, however, supply process steam at 150–200°C [33]. High temperatures can be obtained from PTCs without any serious degradation of the collector efficiency, which can normally be maintained at values higher than 70%.

PTCs are the most mature solar technology to generate heat at temperatures up to 400°C for solar thermal electricity generation and process heat applications. The biggest application of this type of system are the Southern California power plants, known as solar electric generating systems (SEGS), which have a total installed capacity of 354 MWe. Another important application of this type of collector is installed at Plataforma Solar de Almeria (PSA) in Southern Spain, mainly for desalination purposes.

6.4 Possible Plant Configurations for Solar Thermal Seawater Desalination

Solar energy can be converted into thermal energy with different temperature levels using different energy collection devices. Figure 6.12 shows some possible configurations for solar thermal seawater desalination plants. Solar ponds can produce thermal energy at temperatures in the range 50–100°C. Flat plate collectors (non-evacuated) can produce thermal energy at temperatures in the range 50–90°C, while evacuated flat plate collectors can operate in a temperature range of 70–200°C. Concentrating collectors produce thermal energy in a wider range: 90–350°C.

Several pilot units have been tested so far. An example of a high-potential MED unit powered by solar collectors is the Abu-Dhabi ETC-MED plant. A description of this plant is presented in the following paragraphs, while another noteworthy example, the PTC-MED plant installed at the Plataforma Solar de Almeria, will be discussed in Chap. 10.

6.4.1 Performance Analysis of a High Potential Solar Thermal Desalination Plant: The Abu Dhabi Solar Desalination Plant

The solar desalination plant in Abu Dhabi was designed for an expected annual average fresh water production of 80 m^3/day. A simplified flow diagram of this plant

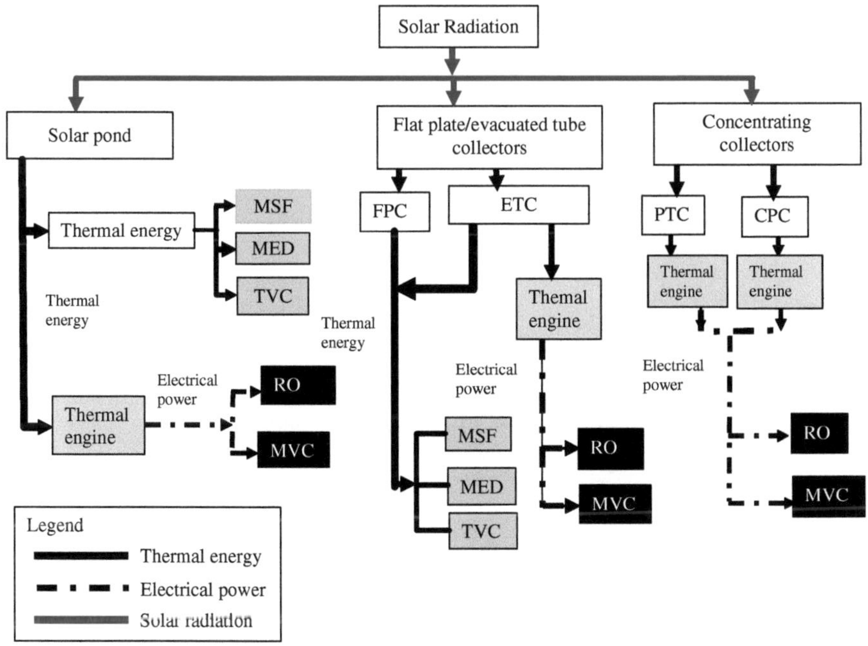

Fig. 6.12 Possible configurations for solar thermal seawater desalination plants

is shown in Fig. 6.13 and a photograph of the plant is shown in Fig. 6.14. Seawater intake is drawn from a nearby channel by a pump, which sucks the seawater from the channel bottom through a metallic screen cage (to rid the water of large objects), and pumps it into a seawater intake pit (settling tank). Sodium hypochlorite (NaClO)

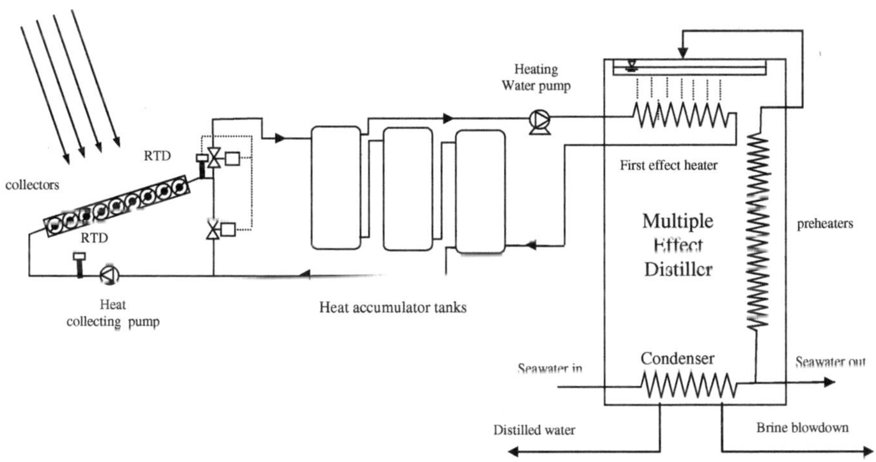

Fig. 6.13 A simplified schematic of the Abu Dhabi solar desalination plant

Fig. 6.14 The Abu Dhabi solar desalination plant

is injected into the seawater on its way to the intake pit, to disinfect it by killing the marine organisms it contains. A priming vacuum pump is used to ensure that the suction line is filled with seawater before starting the seawater intake pump. Clarified and disinfected seawater is then drawn from the intake pit, and pushed to the MED evaporator by a seawater pump. Most of this seawater flows through the condenser of the MED plant and when discharged from the condenser, part of it is taken as feedwater and the other part is rejected through the seawater outfall. The feedwater is drawn by the seawater feed pump and is supplied to a number of preheaters, connected in series. Prior to entering the first preheater, a scale inhibitor chemical (Belgard EV) is injected into the feedwater.

To create and maintain the vacuum inside the evaporator, two vacuum pumps are used: a two-stage water seal type vacuum pump with air ejector, and an oil seal type vacuum pump. Only one of the two vacuum pumps is needed to provide the required vacuum inside the evaporator; the power consumption of the oil seal vacuum pump is much lower than water seal version, but its operational reliability needed to be verified under actual operating conditions, at the time when the plant was first commissioned. Indeed, one of the objectives of the first year of the research program was to investigate the reliability of both vacuum pumps.

The heat-collecting water leaving the collector bank flows into the top of the heat accumulator. The heat accumulator is of the thermally stratified liquid type where, by virtue of density variation between the top and bottom layers, the higher temperature water is located in the upper region of the accumulator tank, while the lower temperature water occupies the lower region. The lower temperature water is drawn from the tank bottom and pumped through the collectors by the heat-collecting pump, which has a capacity of 80 m^3/h for a 26 m discharge head. The heat-collecting water is drawn from the top of the accumulator tank by the heating water circulating pump and is forced to flow into the heating tubes of the first effect of the MED evaporator. This evaporator is designed for a maximum distillate production of 120 m^3/day. By transferring heat to the cooler brine flowing and evaporating on the outside of the tubes, the heating water is cooled down and is then discharged into the accumulator.

6.4.1.1 The Solar Heat Collector Subsystem

The thermal energy required by the MED plant is entirely supplied by solar energy from a solar collector field and heat accumulator tanks. A bank of evacuated tube solar collectors, whose orientation with respect to the sun has been optimised to collect the maximum amount of solar radiation, is used to heat the collector fluid (water) to a maximum temperature of about 90°C. The total effective collector area of this bank is 1,862 m^2. Fresh water is used as a heat-collecting medium that circulates in closed circuit through the solar collector bank, the heat accumulator tanks and the MED evaporator.

The solar energy collecting system (SECS) has the function of collecting the solar energy, when it is available during the day using the collector bank, and storing this energy in the heat accumulator which supplies thermal energy to the evaporator with minimal fluctuations in supply temperature. This is desirable since steady state operation of the evaporator, close to optimum operating conditions, is highly desirable.

The basic unit in the collector bank is the Sanyo evacuated tube solar collector. This is a flat plate-type collector that employs selective coating absorber plates enclosed in glass tubes maintained under a high vacuum of 10^{-4} mmHg. Ten glass tubes with their absorber plates are incorporated within each collector. Along the centreline of each glass tube a single copper tube is located, which is attached to the middle of the absorber plate. The heat-collecting water flows through this centre pipe and absorbs the solar energy collected by the absorber plates.

The ends of each glass tube are sealed to a special stainless steel end cap, using a ceramic glass material with a coefficient of thermal expansion approximately the same as that of the glass tube itself. The difference in the thermal expansion between the copper tube and the glass tube is taken up by bellows installed between the end cap and the copper tube. Each collector consists of 10 individual tubes arranged in parallel. The heat-collecting water moves inside the centre tubes in a parallel-series arrangement, whereby in five of the tubes the flow is in one direction and in the other five it is in the opposite direction.

Attached to one end of the centre tubes is a header tube with an orifice located in the middle of the header tube. The opposite ends are connected to return bends which are used to connect pairs of centre tubes in series. Fourteen collectors are

Table 6.2 Specifications of a single collector

Item	Specification
Selective coating	Absorptivity $\alpha \geq 0.91$ Emissivity $\varepsilon \leq 0.12$
Absorber area	1.75 m^2
External dimensions	2,860 mm × 985 mm × 115 mm
Net weight	64 kg
Flow rate	700–1,800 l/h
Max. operating pressure	0.6 MPa

connected in series by coupling the different header tubes. Each collector has an absorber area of 1.75 m^2 and is coated with a black selective coating with an absorptivity, α 0.91 and an emittance, ε 0.12. The specifications of a single collector, as provided by the manufacturer (Sanyo), are shown in Table 6.2.

The collector bank consists of 1,064 collector units, making up a total collector area of $1{,}064 \times 1.75 = 1{,}862$ m^2. Twenty-eight collectors are combined to form a single-array pair of collectors, with its own support structure. Each array pair consists of 2 parallel arrays of collectors, with each array consisting of 14 collectors connected in series. Each array pair is 14.5 m long and 6.0 m wide and is oriented in a north/south direction at a slope of 1/50. Water is supplied from the main pipe on the south side and passes through the 14 collectors, connected in series, and exits into the main pipe on the north side.

Seventy-six array pairs are arranged in a U-shape to form the whole collector bank. All array pairs are connected in parallel and each is provided with 2 isolating valves (at inlet and exit), a drain valve and an air vent. The bank is divided into 6 blocks designated A, B, C, D, E and F. Blocks A and F consist of 12 array pairs, while the other blocks each consist of 13 array pairs. Figure 6.15 is a block diagram of the collector bank.

6.4.1.2 The Heat Accumulator Subsystem

The heat accumulator subsystem is designed to provide thermal energy to the evaporator during its 24 h/day operation. It consists of three carbon steel tanks with a total storage capacity of 300 m^3 and contains hot water at atmospheric pressure and at a temperature ranging from 64 to 90°C. The tanks are insulated with a 100 mm layer of fibreglass to minimise heat loss to the ambient air. All three cylindrical tanks have the same internal diameter (3.8 m) and wall thickness (9 mm). However, the tank heights are not identical; tank No. 1 has an effective height of 10 m whilst tank Nos. 2 and 3 have an effective height of 7.6 m. The heat-collecting water from the collector bank is introduced at the top of tank No. 1. The heat-collecting water to the collector bank is drawn from the bottom of tank No. 3. Heating water to the evaporator is drawn from the top of tank No. 1 and returns to the bottom of tank

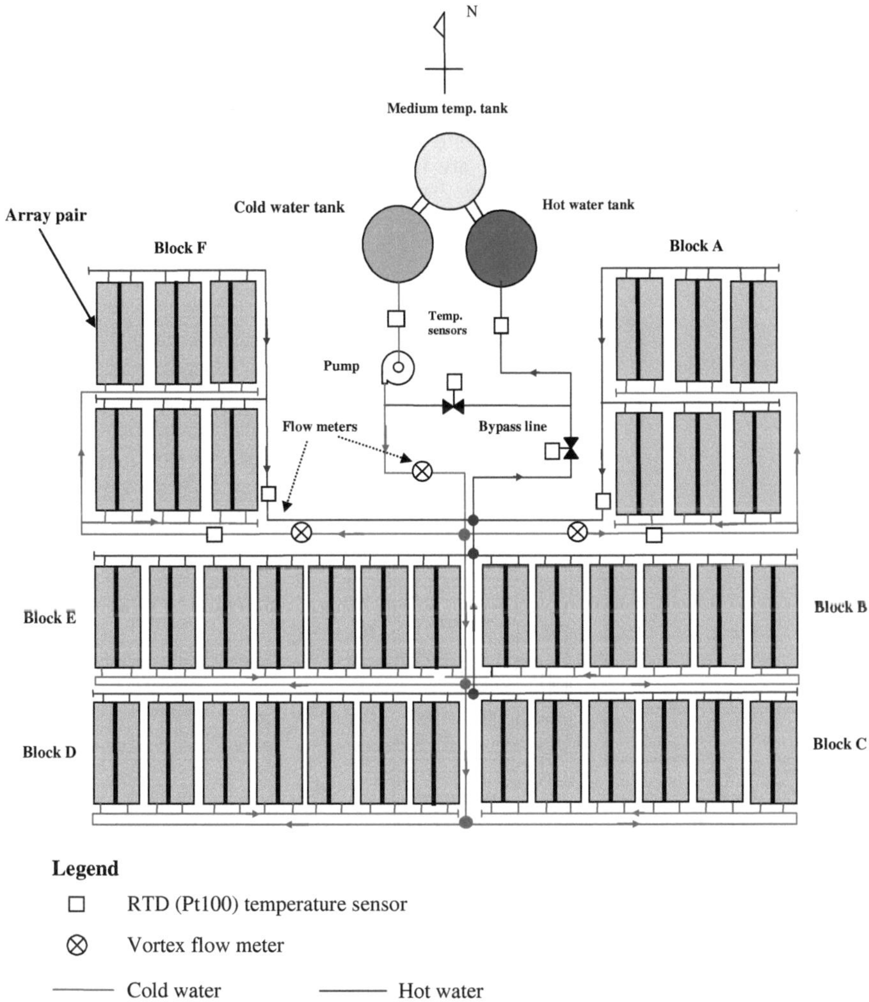

Fig. 6.15 Collector bank consisting of six blocks A, B, C, D, E and F

No. 3. The water is therefore stratified in such a way that the top water layers of tank No. 1 are always at the highest temperature, and the bottom layer of tank No. 3 at the lowest temperature.

The heat accumulator tanks have enough capacity to be able to provide the required thermal load (for the evaporator) for about 16 h following sunset, provided that the tanks were fully charged just before sunset. This feature makes it possible to operate the desalination unit during night time for most days. Only during extended overcast or hazy days, when sandstorms prevail, plant shut-down can be expected to occur at night, due to insufficient energy collection.

6.4.1.3 The MED Evaporator Subsystem

The MED evaporator (Fig. 6.16) has 18 effects, with the highest temperature effect (1) located at the top of the stack and the lowest temperature effect (18) located at the bottom. The 18 effects are actually arranged in a double-stack configuration where odd numbered effects 1, 3, 5, …, 17 are in one stack and even numbered effects

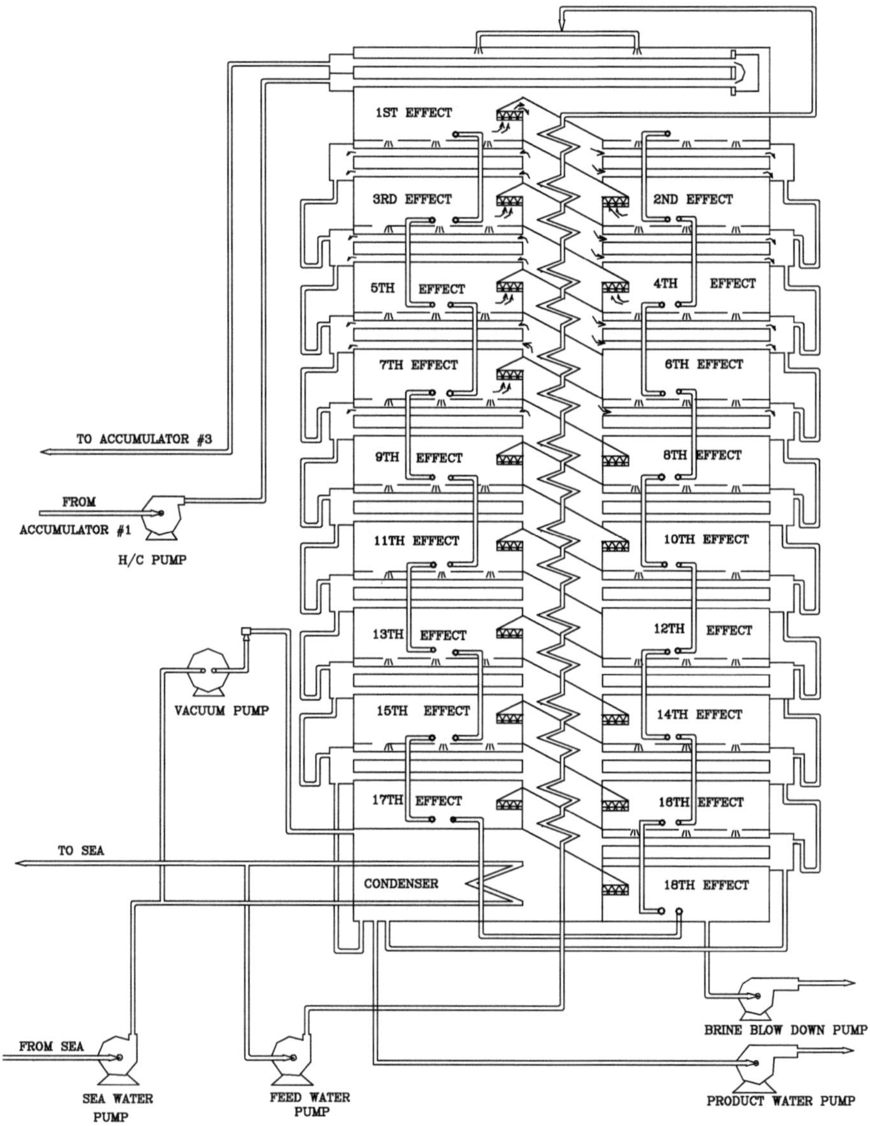

Fig. 6.16 The MED evaporator

2, 4, 6, ..., 18 are in the second. The double-stack arrangement is incorporated into a single evaporator vessel. In addition to the 18 effects, the evaporator has a final condenser designed to condense the vapour generated in the bottom and last effect (18). Heat input supplied to the first effect, by the heating water, is repeatedly used by evaporating a portion of the brine flowing into each effect. The evaporator operates under a vacuum that is produced by a vacuum pump connected to the final condenser. The absolute pressure to be maintained in the final condenser is designed to be 50 mmHg. The pressure to be maintained in each effect varies from just below atmospheric in effect (1), to about 50 mmHg in effect (18).

The feedwater flow rate amounts to 17.3 m^3/h; it flows through 17 preheaters before reaching effect (1), one preheater for each effect except for effect (18). These preheaters are designed to raise the feedwater temperature incrementally by flowing from the bottom effect (18) to the top effect (1). A horizontal tube, thin film multiple-effect distiller (MED) is used for desalination of seawater. The distiller is manufactured by The Sasakura Engineering Co., Ltd. This distiller type was chosen because of its high performance, its capability to accommodate large load fluctuations and its low consumption of electrical power. The maximum capacity of the distiller is 120 m^3/day.

Preheated feedwater is sprayed into the top of the first effect and descends down the evaporator stack, flowing as a thin film over the tube bundle in each effect. Part of the feedwater flashes into steam and thereby cooled by several degrees as it passes from one effect to the next. It is rejected at the bottom of the plant as cold, concentrated brine. In the top effect, heating water from the accumulator is used to partially evaporate the thin seawater film on the outside of the tube bundle. The generated vapour from one effect passes through demisters to the inside of the tubes in the next effect, where it condenses to form part of the product and simultaneously causes further evaporation from the seawater film on the tube bundle of this effect. The process is repeated from one effect to the next down the plant. The heat input from the accumulator is thus used over and over again, in successive evaporation/condensation heat exchanges in each effect, to produce more product and new vapour, thereby obtaining a maximum quantity of fresh water with minimum heat input. The vapour generated in the last effect (18) is condensed in a seawater-cooled condenser and part of the seawater is used as feedwater to the MED unit. The remaining seawater is rejected into the sea and carries most of the heat away from the process.

6.4.2 Design Features of the Plant

Table 6.3 lists the design specifications of the plant. At the planning stage, no detailed solar radiation data was available for Abu Dhabi City and the only data available was that for nearby Kuwait City. Therefore the data for Kuwait was used, with the annual mean daily solar radiation on horizontal surface taken as 5,000 kcal/m^2 per day. Based on the actual measurements, the average annual daily values were found to be slightly higher than this value (5,270 kcal/m^2 per day instead of 5,000 kcal/m^2per day).

Table 6.3 Plant design specifications

Design parameter	Assumed value/range
Solar radiation	5,000 kcal/m^2 per day (annual mean value on horizontal surface)
Ambient temperature	30°C (daytime mean temperature)
Rainfall	18.1–390.1 mm/year
Wind speed	5 m/s (for collector design) 30 m/s (for structures)
Relative humidity	Max. 100%, min. 10%, normal 25–90%
Seawater salinity	55,000 ppm TDS
System capacity	80 m^3/day as expected yearly average
Solar collector type	Evacuated glass tube collector 1,862 m^2 (effective absorbing area)
Heat accumulator	Thermally stratified vertical cylinder Capacity 300 m^3
Evaporator	Horizontal-tube, multiple-effect stack type (MES) Evaporator, capacity 120 m^3/day, specific heat consumption 43.8 kcal/kg product water

Nomenclature

A_c solar collector or solar pond area, m^2
F_R heat removal factor of solar collector
I solar irradiation on collector absorber surface or solar pond, kW/m^2
Q_c heat collected kJ/s
T_p collector absorber temperature, °C
T_a ambient temperature, °C
T_i inlet fluid temperature, °C
U_L collector heat loss, W/m^{2}K
G Radiation intensity, W m^{-2}
h Heat transfer coefficient ,W m^{-2} K^{-1}
he Equivalent mass transfer coefficient, W m^{-2} K^{-1}
m_d Specific fresh water production, kg s^{-1} m^{-2}
p Partial pressure of saturated water vapour, Pa
t Time
T Temperature, K

Greek Symbols

$(\tau\alpha)_e$ effective transmittance-absorptance product
α absorptance, numerical constant
η_C collector efficiency
τ transmittance
λ Latent heat of vaporisation for water, J kg^{-1}
η Efficiency of the solar still, dimensionless

Subscripts

c,wg refers to convective heat transfer between water and the cover
h refers to hour
g refers to glass cover
w refers to average water

Abbreviations

CPC compound parabolic collector
ETC evacuated tube collector
FPC flat plate collector
ITG integrated solar water desalination green house
LCZ lower convective zone
MED multiple-effect distillation
MES multiple-effect stack
MSF multi stage flash
MVC mechanical vapour compression
NCZ non-convective zone
PAR Photosynthetically Active Radiation
PR performance ratio
PSA Plataforma Solar de Almeria
PTC parabolic trough collector
RO reverse osmosis
SEGS solar electric generating systems
SGSP salt gradient solar pond
TVC thermal vapour compression
UCZ upper convective zone

References

1. Telkes M (1956) Research on methods for solar distillation. US Office of Saline Water, Washington, DC, Res. Dev. Prog., Rep. No. 13, PB 161388
2. Fath HES (1998) Solar distillation: a promising alternative for water provision with free energy, simple technology and a clean environment. Desalination 116: 45–56
3. Mink G, Aboabbous MM, Karmazsin E (1998) Air blown solar still with heat recycling. J. Sol. Energy 62: 309–317
4. Graeter F, Duerrbeck M, Rheinlaender J (2001) Muti- effect still for hybrid solar/fossil desalination of sea and brackish water. Desalination 138: 111–119.
5. Bourouni K, Chaibi MT, Tadrist L (2001) Water desalination by humidification and dehumidification of air: state of the art. Desalination 137: 167–176
6. Malik MAS, Tiwari GN, Kumar A, Sodha MS (1996) Solar distillation: a practical study of a wide range of stills and their optimum design, construction and performance, 1st edition, Pergamon Press Ltd., New York, USA

7. Kudish AI (1991) Water desalination. In: Parker BF (ed.) Solar energy in agriculture, Elsevier, New York
8. Tanaka K, Yamashita A, Watanabe K (1981) Experimental and analytical study of the tilted wick type solar still. In: (19th edition.) Proc. International Solar Energy Congress, Brighton
9. Sodha MS, Kumar A, Tiwari GN, Tyagi RC (1981) Simple multiple-wick solar still: analysis and performance. J. Sol. Energy 26: 127–131
10. Cooper PI, Appleyard JA (1967) The construction and performance of a three effects, wick type tilted solar still. Sun at Work 12: 1–4
11. Tanaka H, Nosoko T, Nagata T (2002) Experimental study of basin-type, multiple-effect, diffusion-coupled solar still. Desalination 150: 131–144
12. Tanaka H, Nakatake Y (2007) Numerical analysis of the vertical multiple-effect diffusion solar still coupled with a flat plate reflector: optimum reflector angle and optimum orientation of the still at various seasons and locations. Desalination 207: 167–178
13. Trombe F, Foex M (1961) Utilisation of solar still energy for simultaneous distillation of brackish water and air conditioning of hot houses in arid regions. U.N. Conf. On new sources of energy, Paper 35/S/64 Revised, Rome
14. Boutiere H (1971) Cultivation in arid areas and greenhouse-solar stills. COMPLES meeting, Athens, Greece
15. Bettaque R (1999) An integrated solar desalination system in controlled environment greenhouses. SunWord 23: 18–20
16. Selsuk MK, Tran VV (1976) An overview of solar still greenhouse performance and optimal design studies. Conf on Heliothermique and Development, ed. M.A Kettani and J.E. Soussou, Vol. II, 349
17. Strauch KH (1985) A closed system greenhouse with integrated solar desalination for arid regions. Acta. Hortic. 170: 29–36
18. Dumont M, de Cachard M (1984) A greenhouse with solar distillation. Plasticulture 61: 11–24
19. Chaibi MT, Jilar T (2004) System design, operation and performance of roof integrated desalination in greenhouses. J. Sol. Energy 76: 545–561
20. Zaragoza G, Buchholz M, Jochum P, Pérez-Parra J (2007) Watergy project: towards a rational use of water in greenhouse agriculture and sustainable architecture. Desalination 211: 296–303
21. Dunkle RV (1961) Solar water distillation: the roof type still and a multiple effect diffusion still. Int. Dev. Heat Transfer, ASME Proceedings (Part 5) 895–902
22. Mohamed AMO, Maraqa M, Al Handhaly J (2005) Impact of land disposal of reject brine from desalination plants on soil and groundwater. Desalination 182: 411–433
23. Someswara Rao N, Venkateswara Rao TNV, Babu Rao G, Venu Gopala Rao K (1990) Impact of reject water from the desalination plant on ground water quality. Desalination 78: 429–437
24. Gomkale SD (1988) Solar distillation as a means to provide Indian villages with drinking water. Desalination 69: 171–176
25. Cooper PI (1973) The maximum efficiency of single-effect solar stills. J. Sol. Energy 15: 205–217
26. Kaabi A, Smakdji N (2007) Impact of temperature difference (water-solar collector) on solar still global efficiency. Desalination 209: 298–305
27. Jensen MH, Eisa HM (1972) Controlled-environment vegetable production; results of trials in Puerto Penasco, Environmental Research La. Univ. of Arizona, USA
28. Kalogirou SA (2004) Progress in energy and combustion science. Prog. Ener. Combust. Sci. 30: 231–295
29. Huanmin Lu, Walton JC, Andrew HP (2001) Swift, Desalination coupled with salinity-gradient solar ponds. Desalination 136: 13–23
30. Kreith F, Kreider JF (1978) "Principles of solar engineering", McGraw-Hill Book Co., New York
31. Harris NC, Miller CE, Thomas IE (1985) "Solar Energy System Design", by John Wiley & Sons, Inc., New York

32. Sanyo Electric Trading Co. (1983) "Evacuated glass tube collector Model STC-BH250 R/L B", Technical Data, report No. WM-9177
33. Kalogirou SA (1998) Use of parabolic trough solar energy collectors for sea-water desalination, Appl. Energy 60: 65–88

Chapter 7
Membrane Distillation for Solar Desalination

Joachim Koschikowski, Marcel Wieghaus, and Matthias Rommel

Abstract Membrane distillation (MD) is a hybrid thermal/membrane desalination process in which pure water vapour from a salt solution passes through a hydrophobic membrane, driven by a difference in temperature, and condenses on the opposite side. This chapter starts with a detailed explanation of the principles behind membrane distillation. The four main types of MD technology are then discussed and the technical advantages and disadvantages of each technology are outlined, focusing on the crucial features for coupling MD with solar thermal energy. Heat and mass transfer phenomena are examined with regard to the influences of temperature polarisation and salt concentration on process performance. Finally, available semi-commercial MD systems are briefly presented with details on solar thermally driven MD systems for the autonomous desalination of brackish and sea water.

7.1 The Membrane Distillation Process

Membrane distillation (MD) is a separation technique which joins a thermally driven distillation process with a membrane separation process. The thermal energy is used for phase changing of liquid water into vapour. The membrane is permeable for vapour only and separates the pure distillate from the retained solution. MD offers significant advantages for the construction of desalination systems which are driven by solar or waste heat.

The principle setup of membrane distillation (MD) is based on a hydrophobic, microporous membrane (Fig. 7.1). Due to the high surface tension of the polymeric membrane materials, liquid water is prevented from entering the pores, while molecular water in the vapour phase can pass through.

J. Koschikowski (✉)
Fraunhofer Institute for Solar Energy Systems ISE, Freiburg, Germany; PSE AG – Projects in Solar Energy, Freiburg, Germany
e-mail: Joachim.koschikowski@ise.fraunhofer.de

A. Cipollina et al. (eds.), *Seawater Desalination*, Green Energy and Technology, 165
DOI 10.1007/978-3-642-01150-4_7, © Springer-Verlag Berlin Heidelberg 2009

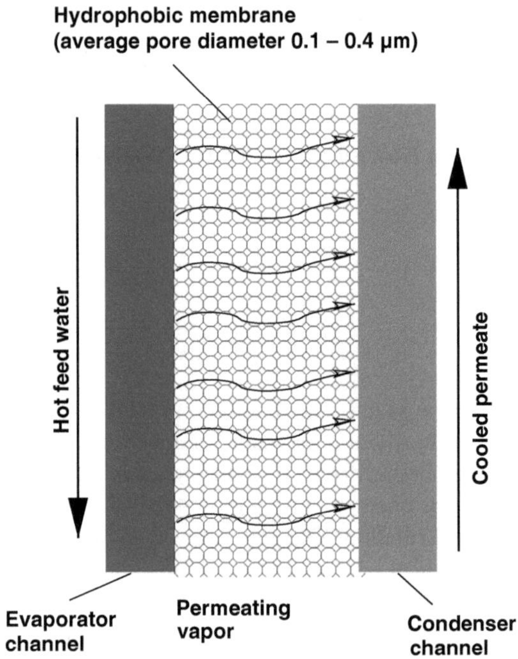

Fig. 7.1 Principle of direct contact membrane distillation (DCMD)

Indeed, a hydrophobic membrane is characterised by the fact that the surrounding liquid water cannot enter its pores (capillary depression). This effect depends on the relative intensity of cohesiveness between the liquid molecules, and the adhesive power between liquid molecules and the membrane material. These forces are responsible for the contact angle $\Theta = 180 - \Theta'$ between the liquid surface and the membrane wall. In the case of a non-wettable membrane, the contact angle is $\Theta > 90°$ and a convex meniscus, as shown in Fig. 7.2, is formed [1]. The hydrostatic pressure of the water columns on both sides of the membrane must remain less than the wetting pressure of the membrane, in order to restrict liquid water from passing through the pores.

Membranes for MD usually have a pore diameter of 0.1–0.4 μm and are made from PTFE, PVDF or PP polymers.

The driving force in an MD process is the vapour pressure difference across membrane interfaces. For direct contact membrane distillation (DCMD), this vapour pressure difference arises due to a temperature gradient between a hot feed stream in the evaporator channel, and a cooled permeate stream in the permeate channel.

Various technologically advanced configurations exist for the channel setup. For example, in air gap membrane distillation (AGMD), the vapour does not condense in a cold liquid stream on the membrane interface, but rather on a cooled plate separated from the membrane by an air gap. In vacuum membrane distillation (VMD)

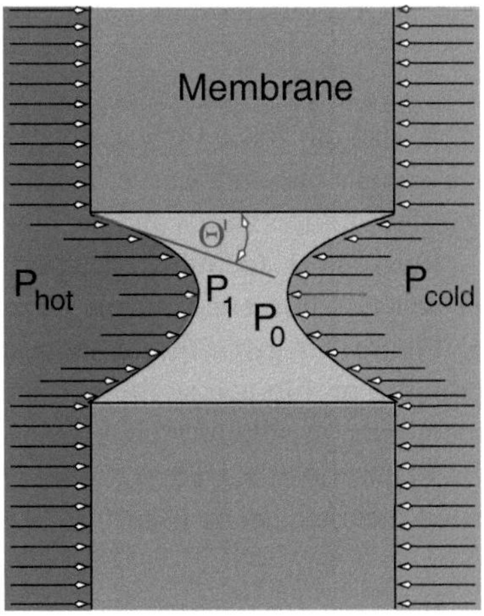

Fig. 7.2 Capillary depression of liquid water at a pore of a hydrophobic membrane

vapour is removed from the permeate channel by a vacuum pump, and condenses outside the membrane module.

In some process configurations, the heat necessary for the separation process is recovered by preheating, using the distillate stream, the feed inlet, which then undergoes a final stage of heating before entering the hot feed channel. In this way a good degree of thermal energy integration is achieved.

Today MD is not used in large scale desalination plants. There are, however, several advantages which make this a preferred technology for small plants, especially in remote applications where low temperature waste heat or a solar thermal heat supply is available, facilitating the energy self-sufficient operation of the unit.

The MD process presents several important advantages. The operating temperature of the process ranges between 60 and 90°C. At this temperature level solar thermal collectors show good performance, and waste heat from cogeneration is also easily available. Contrary to the continuous operation requirement for the reverse osmosis process, an intermittent operation of MD modules with dry periods is possible. The technical complexity of MD modules is low, compared to MSF or MED systems using vacuum stages. All of the construction materials for the modules can be polymeric, which prospectively leads to significant cost advantages compared to MED or MSF modules made from stainless metal alloys. The MD process can be designed with heat recovery in order to make it very energy efficient. High water purity can be achieved almost independently of the feedwater quality. The hydrophobic membranes used in MD are not very sensitive to scaling and fouling, as compared to the hydrophilic membranes used in reverse osmosis. Thus, an operation without chemical pretreatment is possible, if the feedwater composition is within certain limits.

7.2 Mass and Heat Transfer in Membrane Distillation

The vapour pressure difference between the evaporator and the condenser channel is the driving force for vapour permeation through the membrane. With respect to the vapour pressure curve, the mass transport is a function of the two membrane interface temperatures and the absolute temperature level.

Resistances for mass transport are a result of the membrane structure and the presence of non-condensable gases in the membrane pores.

Figure 7.3 gives the temperature profile T_1–T_0 and the corresponding vapour pressure profile across the hot and cold interface of a membrane in direct contact membrane distillation (DCMD).

To describe the transport phenomena in MD, classic heat transfer and gas permeation theories can be used.

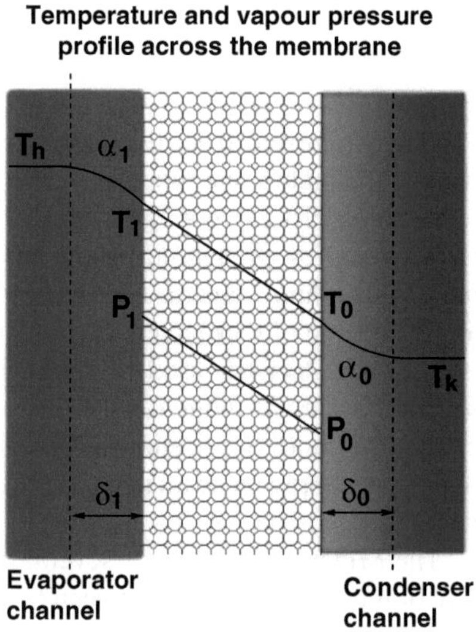

Fig. 7.3 Temperature profile and corresponding vapour pressure profile

7.2.1 Mass Transfer

In MD, it is assumed that mass transfer is based on convection and diffusion of water vapour through the microporous membrane. It can be described by the combination of Knudsen diffusion and Poiseuille flow, non-condensable gas locks in the pores excepted. Theoretically the superposition of both functions is necessary, since the average pore diameter of 0.2 μm is in the transition region of both models. Knudsen diffusion only can be used if the pore diameter is smaller than the mean free path of water molecules and the Poiseuille flow model is only valid if the diameter is 100 times larger than the mean free path. Gas permeation measurements reported by Fane et al., for different MD membranes demonstrate that Knudsen diffusion is dominant [2]. Based on this assumption, and on the knowledge of membrane geometry, the mass transfer can be expressed by the following equation:

$$N_w = \frac{-2r\varepsilon}{3\chi}\sqrt{\frac{8R\overline{T}}{\pi M}}\frac{1}{R\overline{T}}\frac{dp}{dx} \tag{7.1}$$

where N_w is the mass flux, the term $r\varepsilon/\chi$ describes the geometry of the pores, r is the average pore radius, ε is the porosity and χ is a factor describing the tortuosity of the membrane pores. The term $\sqrt{(8R\overline{T})/\pi M}$ is the average velocity of molecules, where $\overline{T}$ is the average temperature between the hot and cold membrane interface, R is the gas constant and M is the molecular weight. *dp/dx* describes the pressure gradient through the membrane. The geometry term, velocity term and temperature term are included within the coefficient regulating the Knudsen diffusion, *K.*

The simplified equation for mass transfer is expressed as follows:

$$N_w = -K\frac{dp}{dx} \tag{7.2}$$

The transport resistance caused by air bubbles inside the membrane can be described by molecular diffusion theory, considering the air as a stationary component [3]. The molar flux through the resistance of air can be calculated from the following equation:

$$N_w = \frac{-\varepsilon}{Y_{ln}}\frac{D}{\chi}\frac{dp}{R\overline{T}\,dx} \tag{7.3}$$

where Y_{ln} is the log mean mole fraction of air and *D* is the diffusion coefficient for vapour in air.

Combining the diffusion coefficient *D*, the pore geometry term ε/χ, R T and Y_{ln}, which is substituted by the fraction of partial pressure over average total pressure ($Y_{ln} = P_a/P_0$) in a molar diffusion characteristic, *J,* the simplified equation for the flux of vapour through the non dissolved gases in the pores can be described by:

$$N_w = -J\frac{1}{P_a}\frac{dp}{dx} \tag{7.4}$$

Here P_a is the average partial pressure of air in the membrane.

The combination of both equations for mass transfer and molecular diffusion results in:

$$N_w = \frac{-1}{dx}\frac{1}{\frac{1}{K}+\frac{P_a}{J}}dp \tag{7.5}$$

Finally, combining the constants for Knudsen diffusion, molecular diffusion, membrane thickness and average partial pressure of air into a transport coefficient, *C,* leads to the simplified equation:

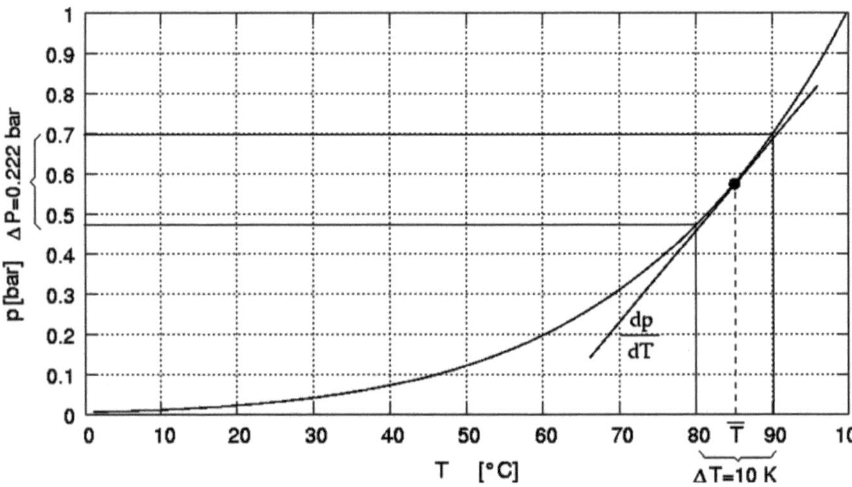

Fig. 7.4 Saturated steam curve, showing the relationship between pressure difference across the membrane, Δp, and temperature difference, ΔT, between the hot and cold membrane interface

$$N_w = C \cdot \Delta p \tag{7.6}$$

Usually C is determined by experimental investigation and is in the order of $3\times10^{-7}$to 4×10^{-6}kg/m^{2}s Pa, depending on membrane material and geometry [2].

As shown in Fig. 7.4, the pressure difference across the membrane, Δp, can be calculated from the temperature difference between the hot and cold membrane interface. Assuming a constant slope of the vapour pressure curve $(dp)/dT$ within a narrow interval of temperatures (i.e. $T_1 - T_0$) and considering the average temperature $\overline{T} = 0.5 \cdot (T_1 + T_0)$, Eq. (7.6) can be re-written as:

$$N_w = C \cdot \frac{dp}{dT} \cdot (T_1 - T_0) \tag{7.7}$$

The gradient $(dp)/dT$ can be calculated using the Clausius-Clapeyron equation with sufficient accuracy if $\Delta T_{1,0} \leq 10K$:

$$\frac{dp}{dT} = \frac{P \cdot \Delta h_v}{R \cdot \overline{T}^2} \tag{7.8}$$

where Δh_v is the latent heat needed for evaporation.

7.2.2 Heat Transfer

There are three stages of heat transfer in MD. The first is the heat transfer, Q_{S1}, from the hot bulk stream in the evaporator channel to the membrane interface. This is calculated as a function of the heat transfer coefficient, α_1, and the temperature difference between the bulk stream and membrane interface $T_h - T_1$:

$$Q_{S1} = \alpha_1(T_h - T_1) \tag{7.9}$$

This transfer is necessary since the evaporation occurs directly at the membrane interface, leading to significant temperature drops due to the need of latent heat for evaporation. This heat transfer should be as high as possible, since a big temperature drop at the membrane interface leads to significant temperature polarisation and reduction of the driving force. From this point of view, a turbulent flow profile should be achieved in the channel.

The second stage is the heat transfer through the membrane. It consists of three different mechanisms: first is the heat transfer, Q_{LM}, of the latent heat of vaporisation, Δh_v, which is transported with the vapour flux, N_w, through the membrane:

$$Q_{LM} = N_w \cdot \Delta h_v = C \cdot \frac{dp}{dT} \cdot (T_1 - T_0) \cdot \Delta h_v \tag{7.10}$$

The second mechanism is the heat flux through the membrane material and the third is the heat conduction through water vapour and air in the membrane pores. With respect to membrane porosity, the heat transferred by conduction through the membrane, Q_{SM}, can be assumed as:

$$Q_{SM} = \frac{\varepsilon_M k_V + (1 - \varepsilon_M) k_M}{\delta_M} \cdot (T_1 - T_0) \tag{7.11}$$

where δ_M is the thickness of the membrane, ε_M is the porosity of the membrane, k_V is the heat conduction coefficient of vapour and k_M is the heat conduction coefficient of the membrane material.

Heat conduction through the membrane should be minimised, as this fraction of energy cannot be utilised for evaporation and must be considered as heat loss. Air gap membrane distillation (which will be illustrated in the following paragraphs) is one approach aimed at the reduction of conductive losses, since the air gap between the cold membrane interface and the condenser surface provides efficient thermal insulation. Other approaches for direct contact membrane distillation aim at reducing heat conduction using multi-layer membranes with thin active layers and thermal insulation layers.

The third stage is the heat transfer, Q_{S0}, from the cold membrane interface to the cold bulk stream in the condenser channel:

$$Q_{S0} = \alpha_0 \cdot (T_0 - T_c)$$

where α_0 is the heat transfer coefficient between the cold membrane interface and the cold bulk stream and $T_0 - T_c$ is the temperature difference between the cold membrane interface and the cold bulk stream, as shown in Fig. 7.3.

To reduce temperature polarisation, very efficient heat transfer coefficients are important as the released latent heat leads to a large temperature increase at the cold membrane interface during condensation. Therefore, turbulent flow is also recommended in the condenser channel.

7.2.3 Temperature Polarisation

As mentioned in the previous section, temperature polarisation characterises the temperature distribution between bulk stream (T_h–T_c) and membrane interface temperatures, (T_1–T_0), which mainly depends on the ratio of heat transfer in the flow channels and through the membrane. The coefficient characterizing this relation is the temperature polarisation coefficient:

$$\tau = \frac{T_1 - T_0}{T_h - T_c} \tag{7.12}$$

It can be seen that, for a very efficient heat transfer between bulk streams and membrane interfaces ($T_1 \approx T_h$ and $T_0 \approx T_c$), the coefficient τ approaches 1. In the case of low heat transfer coefficients in the channels, compared to heat transfer coefficient through the membrane ($T_1 \approx T_0$), the polarisation coefficient τ approaches 0.

In MD systems, τ mainly depends on channel set-up, module design and operation. The temperature polarisation coefficient increases (i.e. temperature polarization decreases) with an increasing heat transfer coefficient, conversely, an increase of the mean temperature, which leads to an increase of vapour transport (see Eq. (7.7)), can also lead to significant reductions in temperature polarisation coefficients [3], i.e. an increase in temperature polarisation due to the higher temperature driving force required between the bulk and the membrane surface. Values below 0.1 for flat membranes in laminar flow operation, and above 0.9 for narrow hollow fibre membranes in turbulent flow configuration have been reported [4].

The temperature polarisation coefficient must be taken into account in order to calculate the mass transfer from the bulk temperatures, which are usually well known, compared to the membrane interface temperatures, which are almost immeasurable:

$$N_w = \tau \cdot C \cdot \frac{dp}{dT} \cdot (T_h - T_c) \tag{7.13}$$

It can be seen that temperature polarisation has a significant influence on flux and on the overall process efficiency, and therefore should be the focus of the MD process design.

7.2.4 Effect of Salt Concentration on MD Performance

Dissolved solids reduce the vapour pressure of water, leading to a reduction in process efficiency as the vapour pressure difference is the driving force in MD. Figure 7.5 shows the relative reduction in vapour pressure difference between salt water and distillate against salt concentration. The functions are provided for

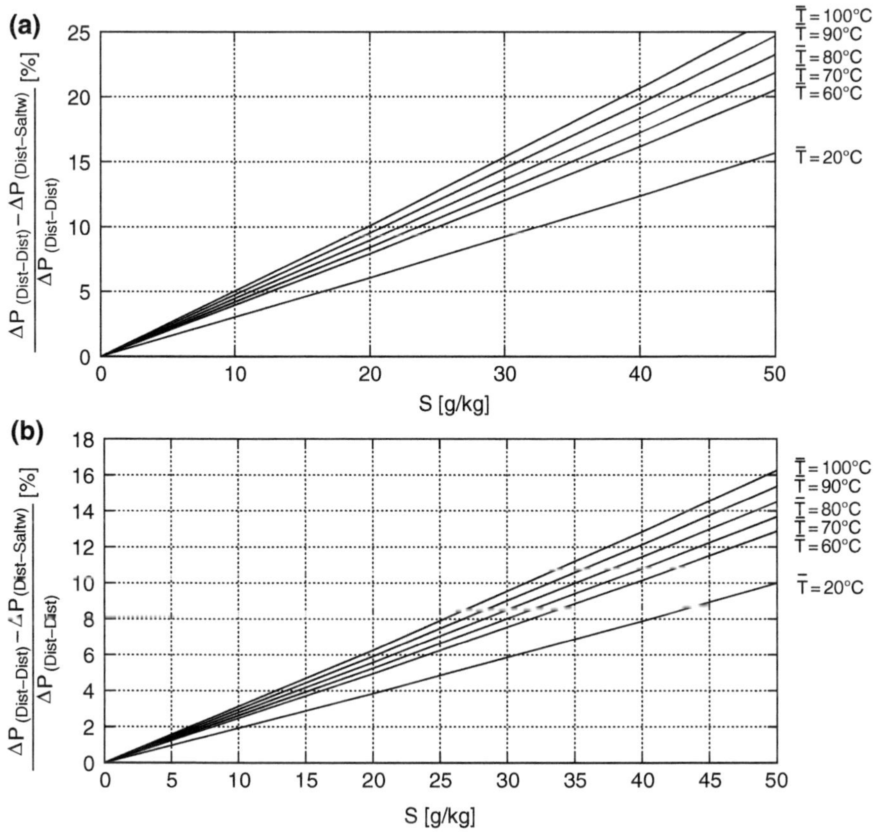

Fig. 7.5 Reduction of vapour pressure difference between distillate and salt water, against salt concentration, for different average feed temperatures at $\Delta T = 3\,K$ (**a**), and $\Delta T = 5\,K$ (**b**)

different average temperatures but it is assumed that the temperature difference across the membrane is always 3 K (Fig. 7.5a) and 5 K (Fig. 7.5b). It can be seen that the reduction in the pressure difference, or driving force, for a 35 g/kg salt solution at 80°C with $\Delta T = 3\,K$, is about 16%. For the same parameters, but a $\Delta T = 5\,K$, the reduction is only about 10%. These figures show that the influence of salt must be considered and gains even more importance for higher operation temperatures and lower trans-membrane temperature differences.

7.3 Different Types of Membrane Distillation Technology

The MD process can generally be subdivided into four different types:

- Air Gap Membrane Distillation (AGMD)
- Direct Contact Membrane Distillation (DCMD)

- Sweeping Gas Membrane Distillation (SGMD)
- Vacuum Membrane Distillation (VMD)

In all four types the hot liquid feed flows over one side of the membrane and the vapour formed on the evaporation surface passes through the membrane and reaches the other side, where vapour condenses according to different mechanisms related to the abovementioned configurations.

7.3.1 Direct Contact Membrane Distillation

In Direct Contact Membrane Distillation (DCMD) both sides of the membrane are in direct contact with a liquid stream. On the left of the membrane shown in Fig. 7.6a the hot liquid (i.e. hot seawater) flows in the evaporator channel, whilst on the right a cold liquid (i.e. cooled permeate or distillate) is circulated. Heat transfer (as well as mass transfer) occurs from the hotter to the colder side.

The liquid in the evaporator channel is constantly refilled and reheated, whilst the volume of the liquid in the permeate channel increases and heats up.

One of the main features of DCMD is that the gas gap between the membrane surface and the condensate stream is very narrow and only exists due to the hydrophobic nature of the membrane. This causes the temperature of the membrane surface in contact with the condensate to be very close to that of the condensate stream itself, thus allowing high temperature drops across the membrane, i.e. high driving forces for mass transfer.

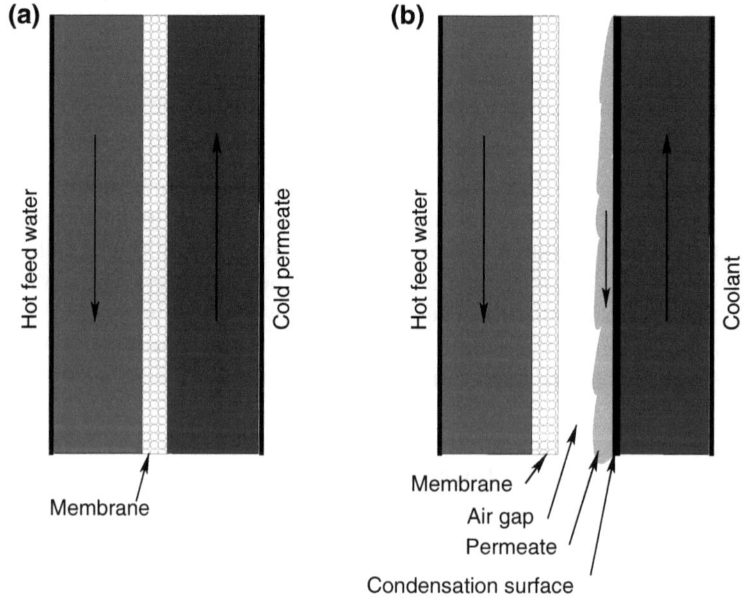

Fig. 7.6 Direct contact MD (**a**), Air Gap MD (**b**)

Conversely, the Direct Contact configuration causes a relatively high heat loss as the membrane is the only barrier for the transfer of sensible heat.

Therefore with DCMD, a relatively high specific flux, at the expense of a relatively high specific energy consumption, can be obtained.

Depending on the specification of the whole MD-Module design and the operating conditions (i.e. velocity of the bulk stream in the evaporator and condenser channel, thickness and material of the condenser foil and membrane) the effective difference in temperature across the membrane, and therefore the driving force, can be significantly higher than AGMD configuration, which is described in the next paragraph.

7.3.2 Air Gap MD

The modules in Air Gap Membrane Distillation (AGMD) consist of three separated channels: the evaporator channel (hot sea water), the condenser channel (cooling stream) and the air gap (or permeate channel). Condensation of the distillate takes place over a cold surface separated from the membrane by an additional gap of an inert gas, typically air (Fig. 7.6b). This air gap, between the membrane and the condensing surface, substantially reduces the heat loss through the membrane due to the thermal insulation of the air gap, the design and specification of which, i.e. air gap thickness, is of great importance. Generally speaking the effective difference in temperatures across the membrane decreases with an increasing air gap thickness. The temperature difference of the two bulk streams, however, increases with the thickness of the air gap. The advantage of AGMD, compared to DCMD, is the lower specific energy consumption but the disadvantage is the reduction of specific flux.

7.3.3 Sweeping Gas MD

In Sweeping Gas Membrane Distillation (SGMD) the microporous hydrophobic membrane separates an aqueous solution from a gas phase, which acts as a stripping agent. Vapour, formed by evaporation at the liquid/vapour interface, diffuses through the stagnant gas film within the membrane pores towards the sweeping gas. The left side (evaporation channel) is filled with a circulated hot aqueous solution, i.e. hot seawater (Fig. 7.7a). The right side (permeate channel) contains vaporised permeate as well as a stripping agent (i.e. dry air). In SGMD, the stripping agent in the permeate channel flows continuously and transports the vapour into a condenser which is located outside the module. The dimensions of the external condenser must be relatively large due to the high gaseous fraction, and the relatively high volume flow. A lower volume flow would result in an increased gas temperature and vapour fraction in the sweeping gas, and therefore in a decrease in the effective temperature drop across the membrane (the driving force for the mass transfer). The stripping

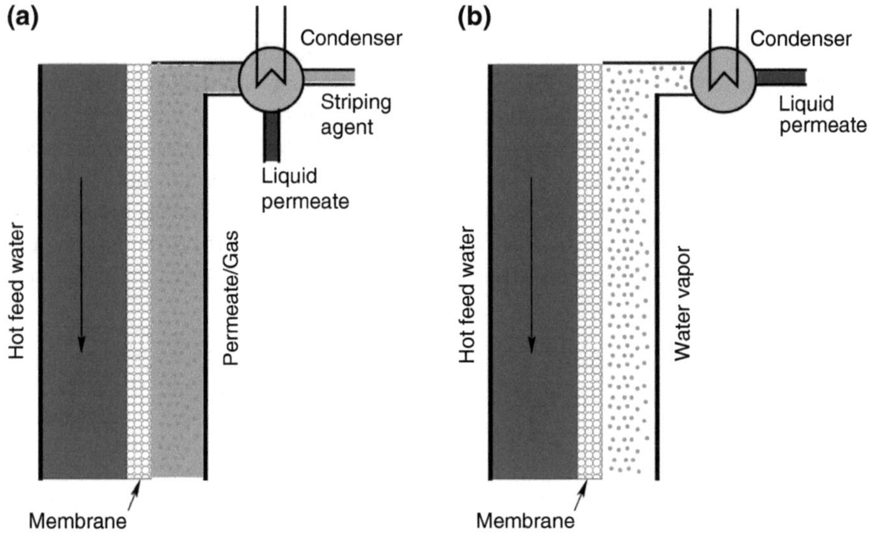

Fig. 7.7 Sweeping Gas MD (**a**), Vacuum MD (**b**)

agent can also be cooled via an additional channel located next to the permeate channel (analogous to AGMD).

The advantage of SGDM, compared to AGMD, is a lower resistance of mass transport due to turbulence achieved by the circulating gas phase and therefore a higher specific vapour flux [5].

7.3.4 Vacuum MD

Vacuum Membrane Distillation (VMD) also makes use of an air gap. As opposed to the types described above, this process is driven by a vacuum. The applied vacuum keeps the pressure on the permeate side below the equilibrium vapour pressure, thereby contributing to generating the driving force. As with SGMD, the water vapour condenses in an external condenser. VMD can be used for the removal of volatile components in an aqueous solution (i.e. volatile organic compounds (VOC's) like ethanol). Additionally, it can be applied when the feed contains non-volatile salts as, for example, in sea water.

The main advantage of VMD is the simultaneous removal of non-soluble inert gases which block the membrane pores and consequently decrease the flux. Even at lower temperature differences, the flux could be increased by the applied vacuum.

The operation of MD at lower temperatures is beneficial to the reduction of general scaling effects. On the other hand, the applied vacuum also decreases the CO_2 concentration of the feed, which causes a higher scaling index.

Vacuum MD processes have lower specific (thermal) energy consumptions. However, the process complexity and electrical energy consumption is significantly higher [6].

7.4 Membrane Distillation Systems

7.4.1 Overview of Membrane Distillation Systems

As already mentioned, membrane distillation is a technology which is not commonly used for desalination today. There are no MD systems, produced on a large economical scale, currently available on the market; they are all in between the research and pilot stage. The following brief overview of the MD market does not claim completeness.

In the 1970s and 1980s, when development of membrane technology became very popular, MD was also the subject of comprehensive investigation and development. The American company GORE developed a spiral wound MD-module for desalination purposes [7]. The German company ENKA developed a hollow fibre MD-module, mainly for industrial water processing. GORE technology was used later by German companies Sick and SEP for building solar driven desalination units, based on spiral wound MD-module technology. In pilot production, the units produced 500–800 l of distillate per day [8, 9].

The Japanese company Takenaka developed, together with the Promotion Centre for Water Re-use in Tokyo, a solar driven MD-system based on one MD-flat plate module and a 12 m^2 field of vacuum tube collectors. The unit was commissioned in 1994 and had a maximum productivity of 40 l/h [10].

The Swedish company Scarab Development today provides flat plate MD-modules, based on the work by the Swedish National Development Co originally done in 1983. Different pilot units were constructed for industrial purposes using this module, for the production of ultra pure water for the semiconductor industry. Production ratios of 12–20 kg/h are reported for high temperature gradients across the membrane [11]. Currently the SCARAB module is being tested with solar energy supply, within the frame of the MEDESOL project [12].

Today's largest MD-system is the MEMSTIL system, developed by TNO in the Netherlands. The MD-modules are of flat plate type. Pilot plants have been installed by the Kepel Seghers Company in Singapore and by Eon in Rotterdam. The design capacities of the waste heat-driven units are 80 and 50 m^3/day, respectively [13, 14].

7.4.2 Preliminary Design Considerations for Solar Membrane Distillation Systems

In general, all available constructions of membrane modules are also applicable to the design of MD-modules. Hollow fibre, flat plate and spiral wound modules exist in MD. Hollow fibre modules are more often used for vacuum membrane distillation. Flat plate modules are favourable for large capacities because very large membrane areas can be realised. Spiral wound modules with circular flow can be built with very long channels, thus enabling efficient energy recovery in very compact module geometries. The disadvantage of the spiral wound construction is the limit in size as a result of maximum channel width. Several modules can be connected in parallel, but nevertheless it makes sense to use this technology only for low capacities, of below 20 m^3/day.

In designing a solar-powered desalination system, the question of energy efficiency is very important, since the investment costs mainly depend on the area over which the solar collectors are to be installed. The power consumption of the auxiliary equipment (e.g. the pump), which will be supplied by photovoltaic units (PV), also has a significant influence on total system costs. Therefore, system design has to focus not only on distillate output, but also on a highly optimised heat recovery arrangement, to minimise the consumption of thermal energy. Heat recovery can be achieved by an external heat exchanger or by an *ad-hoc* internal heat recovery configuration, where the feedwater is used as coolant for the condenser channel.

An example of the internal construction principle of a MD module with integrated heat recovery is shown in Fig. 7.8. The setup has three different channels: the condenser channel, the evaporator channel and the distillate channel. The condenser and the distillate channel are separated by an impermeable condensation foil, while the evaporator and the distillate channel are separated by a hydrophobic, vapour-permeable membrane. The hot water (e.g. 80°C inlet temperature) is directed along this membrane, passing through the evaporator channel from its inlet to its outlet, and cooling down as it flows (e.g. 28°C evaporator outlet temperature). The feedwater (e.g. 20°C inlet temperature) passes through the condenser channel in counter-flow, from its inlet to its outlet, warming up as it flows (e.g. 72°C outlet temperature). The partial vapour pressure difference, caused by the temperature difference across both sides of the membrane, is the driving force for the vapour passing through the membrane. The heat of evaporation is transferred to the feedwater by condensation along the condenser channel. As a consequence, the heat of evaporation is (partly) recovered by the process. Because the energy required for evaporation is obtained from the brine, the brine temperature decreases. The liquid

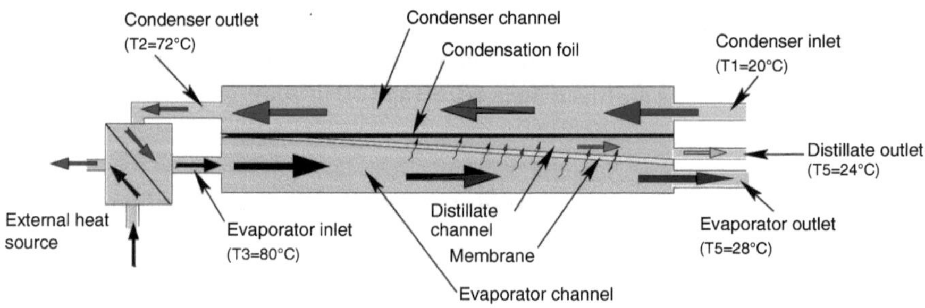

Fig. 7.8 Principle of an MD module setup with integrated heat recovery

distillate is collected at the distillate outlet at a temperature between the feed inlet and the brine outlet temperature levels. The heat input, which is necessary for the required temperature gradient between the two channels (e.g. 8 K), is supplied to the system between the condenser outlet and the evaporator inlet. Thus, the thermal energy consumption of the system is given by the flow rate and the temperature increase in the feedwater between these two points. Heat recovery has a significant influence on the energy consumption of the MD system. In thermal desalination processes, the "Gained Output Ratio" (GOR) is a central parameter for the evaluation and assessment of the heat recovery function. The GOR can be calculated as the ratio between the latent heat needed for evaporation of distillate and the energy input supplied to the system from external sources.

7.5 Examples of Prototypal Solar MD Units

Due to the low thermal capacity of MD modules, the feed flow rate and process temperature can be changed very quickly, without causing instabilities in the desalination operation. For this reason MD modules can also be connected directly to a corrosion-free solar thermal collector without heat storage.

Set out below is information relating to units developed by the Fraunhofer Institute for Solar Energy Systems (Freiburg, Germany).

The technical specifications of the MD modules used for the construction of such compact systems are:

- hydrophobic membrane, mean pore size 0.1–0.4 μm
- height 450–800 mm
- diameter 300–400 mm
- membrane area 7 –12 m^2
- feed temperature at evaporator inlet 60–85°C
- maximum feed volume flow 500 l/h
- specific thermal energy consumption 100–200 $kWh/m^3{}_{distillate}$ (GOR 3-7)
- distillate output 10–40 l/h
- all components are made of polymer materials

Two different design approaches were adopted for the design of solar powered MD systems. The first refers to a "Compact System", used to produce small capacities of fresh water, of between 0.1 and 0.5 m^3/day. The second refers to a "Two-Loop System" for capacities larger than 2 m^3/day.

7.5.1 Solar MD Compact Systems

In the Compact System (Fig. 7.9) the cold feedwater is pumped by a PV powered pump into the condenser channel. The pre-heated feedwater, after leaving the condenser channel, enters the solar thermal collector, where its temperature is increased

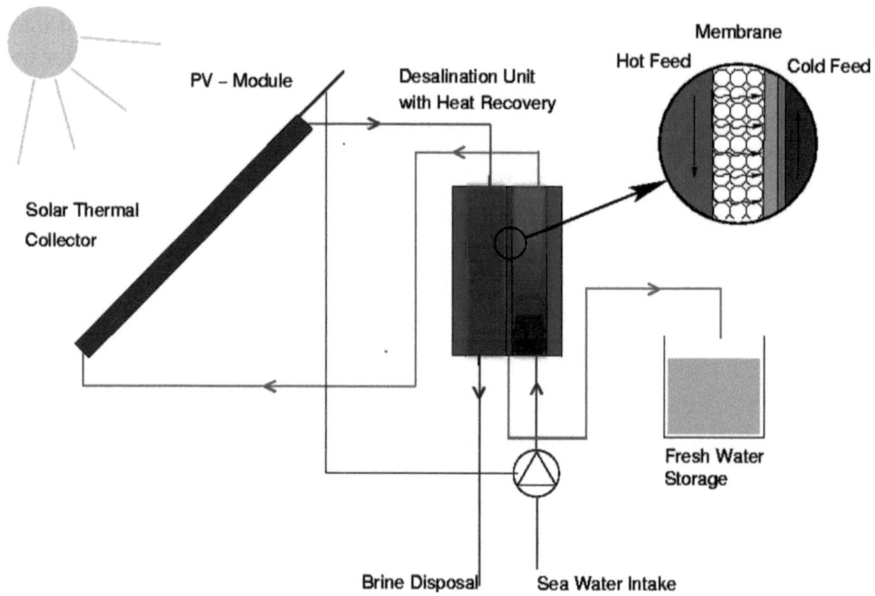

Fig. 7.9 Schematic of the compact system

by 5–10 K. The feed water then leaves the collector to enter the evaporator channel. The advantage of this configuration is its simplicity and efficiency compared to the placement of an intermediate heat exchanger to separate the solar collector's circuit from the MD system. The disadvantage is that common flat plate collectors cannot be used; instead the collectors must be equipped with seawater-resistant riser and header tubes. For compact systems, the collectors consist of tubes made from a seawater-resistant alloy.

The main components of the above mentioned compact system are a 0.5 m^3 feed storage, 1 MD module, a 7 m^2 solar thermal flat plate collector, a pump and a PV module for the electrical power supply of the pump and control system. While the feed storage is mounted above the collectors, most of the hydraulic components are installed in a closed housing underneath the mounted collectors. The distillate produced is collected in a freshwater tank. The brine rejected from the evaporator outlet of the MD module is re-circulated to the feed storage. In this way, the salt concentration, as well as the feedwater temperature in the feed storage, increases during the day while the content of water decreases due to distillate production. The feed storage is refilled automatically when a certain level or temperature is reached. In fact the negative effects of high temperature and salinity of the feed on the system performance require the feed tank content to be refreshed every 5–12 h of operation

For system control, the DC-driven pump is connected to a maximum power point converter (MPP) and supplied by an 80 Wpeak PV module. Thus, the flow rate in the system is directly controlled by the solar irradiation on the PV module. This simple control mechanism has a very low technical complexity and can be very effective if the characteristics of the pump, the PV module, the hydraulic system, and the thermal performance of the collectors fit well together. For low solar irradiations the thermal energy gained from the collectors is small, so a low flow rate is needed to achieve a reasonable operating temperature for the MD process. During hours of operation with high irradiation (i.e. noon), a high flow rate is needed to keep the operating temperature at a specific maximum of e.g. 85°C. For optimal performance, however, a more complex control system should be implemented.

Since 2004, five compact systems were constructed, installed and operated in five different countries within two EU-projects (MEMDIS: NNE5/2001/819 and SMADES : ICA3-CT-2002-10025). The systems were installed in Pozo Izquierdo (Grand Canary), Alexandria (Egypt), Irbid (Jordan), Morocco and Freiburg (Germany). In December 2007, an improved Compact System with a maximum daily capacity of 120 l was set up in Tenerife, Spain (Fig. 7.10). The new Compact System was designed also taking into consideration transportability, ease of installation and low maintenance [15, 16].

In order to give an idea of the reliability of such systems and on their production efficiency, Fig. 7.11 presents some long term measurements recorded in the Grand Canary unit. The daily distillate gain is plotted against the cumulative daily gain of solar energy. The plotted measurement period started in mid June 2005 and ended in mid June 2006. As can be seen from the measurements for the year, there is no decrease of specific energy demand during the observed period. For example, a daily solar gain of 7 kWh/m^2 enabled an average distillate production of 60 l/day in June

Fig. 7.10 Compact system installed in December 2007 in Tenerife, Spain

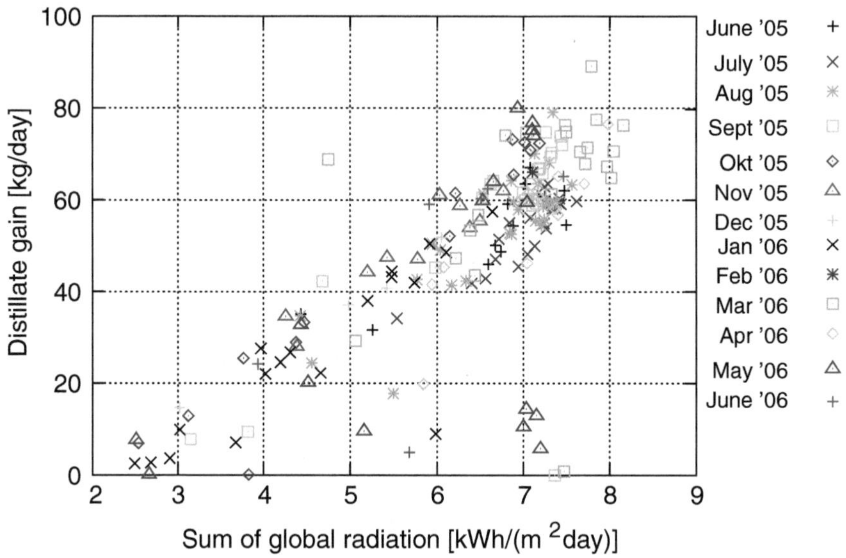

Fig. 7.11 Long term measurement: daily permeate production against the sum of daily solar irradiation

2005, as well as in June 2006. Small variations from this value occur depending on the actual water level, the temperature in the feed storage tank, and as a result, if and when the system started to refill.

Finally it is worth mentioning that a maximum distillate production of 180 l/day was achieved (in a laboratory installation) using feedwater with a salinity of 400 ppm. The productivity with salt water (~35,000 ppm) as feed, however, is 10–20% less and a maximum of only 150 l/day was reached [17].

7.5.2 Solar MD Two-Loop Systems

The concept of the "Two-Loop System" is different from the Compact System. From an economical point of view, it is favourable for daily capacities above 1,000 l of distilled water. A schematic of the setup is given in Fig. 7.12.

The four main design differences with reference to the Compact System and their effects on the operation of the unit are:

- A thermal storage tank and an electrical battery is used to enable an extended operation time of the MD modules, even after sunset. → Consequences: Larger distillate production per MD module → Lower specific module costs

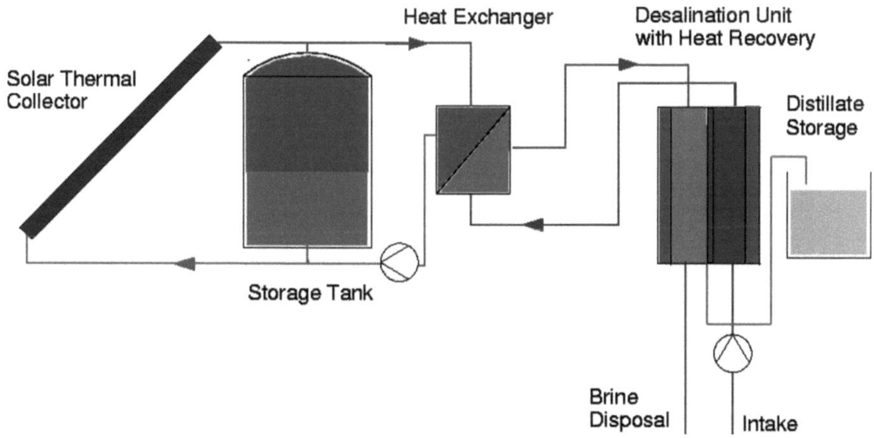

Fig. 7.12 Principle set up – Two-loop system

- The system consists of two loops. The desalination loop is operated with seawater and is separated from the collector loop (operated with tap water) by a corrosion-resistant heat exchanger → Consequence: Usage of economic standard components in the solar collectors, without the need for cost-intensive saltwater-resistant materials
- In the desalination loop, five MD-modules operate in parallel. The MD modules are exactly the same as in the compact systems. → Consequence: Enables a standard multi-module production
- A control unit is used to control the system operation. At times of low irradiation, the desalination unit is operated with heat directly from the collector field. If enough irradiation is available, the surplus energy can be stored in the storage tank and, after sunset, the desalination process can be continued using heat from the tank. → Consequence: The operation conditions can be adjusted to optimal performance conditions for the MD module.

Two-Loop Systems can be powered completely by solar thermal energy and PV. No additional energy supply is necessary.

Two Two-Loop Systems were installed in Aqaba in December 2005 and in Grand Canary in February 2006. Figure 7.13 shows the collector field of the Two-Loop System in Grand Canary (Spain) and Fig. 7.14 shows the desalination unit, of the same Two-Loop System, with five MD modules. Numerical simulations were carried out for system design and development of an adapted control strategy for two different pilot plants. The design capacity for the Aqaba system was 0.7–0.9 m^3/day, and for the Grand Canary system was 1–1.5 m^3/day. Table 7.1 gives key data for both systems, determined by simulation using weather data sets for Aqaba and Grand Canary.

Fig. 7.13 Collector field of the Two-loop system in Grand Canary, Spain

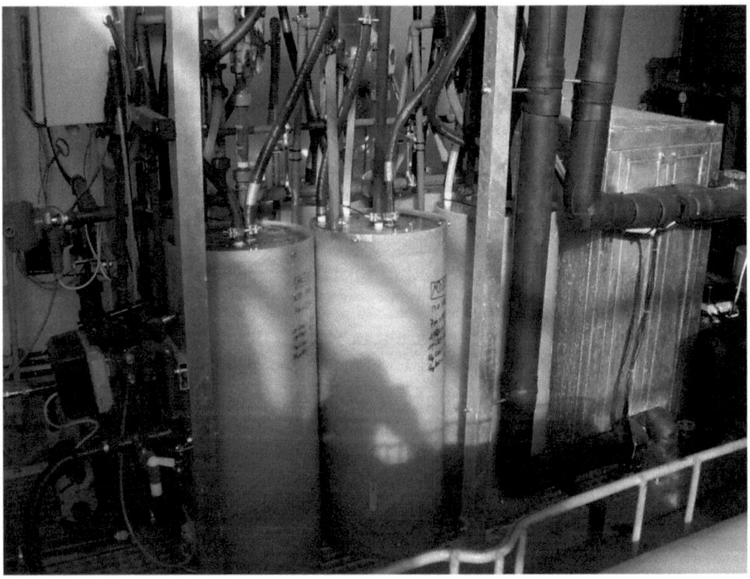

Fig. 7.14 Desalination unit with five MD modules and heat exchanger (on the right)

Table 7.1 Key data of the two "two loop systems" designed by simulation computations

	Aqaba	Grand canary
Design capacity [m^3/day]	0.7–0.9	1–1.5
Collector area [m^2]	72	90
Collector type	Standard flat plate	Flat plate double glassed AR-coated
Heat storage capacity [m^3]	3	4
Number of MD modules	4	5
PV area [kWp]	1.44	1.92

Nomenclature

C	Membrane constant	kg/(Pa s m^2)
D	Diffusion coefficient	m^2/s
Δh_v	Latent heat	kJ/kg
J	Molar diffusion characteristic	mol/(s m)
K	Coefficient for the Knudsen equation	mol/(Pa s m)
k_V	Vapour Thermal Conductivity	kW/(m °C)
k_M	Membrane Thermal Conductivity	kW/(m °C)
M	Molecular weight	kg/mol
N_w	Mass flux	kg/(m^2 s)
P_1	Partial pressure of vapour at cold membrane interface	Pa
P_0	Partial pressure of vapour at warm membrane interface	Pa
P_a	Average partial pressure of air in the membrane	Pa
Q_{LM}	Latent heat transfer through the membrane	W/m^2
Q_{S0}	Heat transfer from cold membrane interface to bulk stream	W/m^2
Q_{S1}	Heat transfer from warm bulk stream to membrane interface	W/m^2
Q_{SM}	Sensible heat transfer through the membrane	W/m^2
R	Gas constant	J/(mol K)
r	Pore radius	m
$\overline{T}$	Average temperature	°C
T_0	Temperature at cold membrane interface	°C
T_1	Temperature at warm membrane interface	°C
T_c	Temperature of cold bulk stream	°C
T_h	Temperature of warm bulk stream	°C
α_0	Heat transfer coefficient cold membrane interface to bulk stream	W/(m^2K)
α_1	Heat transfer coefficient warm bulk stream to membrane interface	W/(m^2K)
δ_0	Size of boundary layer on the cold membrane interface	m
δ_1	Size of boundary layer on the warm membrane interface	m
ε	Porosity	–
Y_{ln}	Log mean mole fraction of air	–
χ	Tortuosity	–

Abbreviations

AGMD	air gap membrane distillation
DC	direct current
DCMD	direct contact membrane distillation
MD	membrane distillation
MED	multi effect distillation
MPP	maximum power point
MSF	multi stage flash
PP	polypropylene
PTFE	polytetrafluorethylene
PV	photovoltaic
PVDF	polyvinylidenfluoride
SGMD	sweeping gas membrane distillation
VMD	vacuum membrane distillation

References

1. U. Ditscher, Untersuchung des Wassertransports über die Gasphase durch eine poröse Membranmit hydrophober Gerüststruktur unter dem Einfluss einer Temperaturdifferenz, Inaugural-Dissertation an der Mathematisch- Naturwissenschaftlichen Fakultät der Universität zu Köln, Köln 1990
2. A.G. Fane, R.W. Schofield, C.J.D. Fell, The efficient use of energy in membrane distillation, Desalination 64, 231–243, Elsevier, Amsterdam 1987
3. T.K. Sherwood, R.L. Pigford, C.R. Wilke, Mass Transfer, McGraw-Hill, New York 1975
4. R.W. Schofield, A.G. Fane, C.J.D. Fell, Heat and mass transfer in membrane distillation, Journal of Membrane Science 33, 299–313, Elsevier, Amsterdam 1987
5. C.H. Lee, W.H. Hong, Effect of operating variables on flux and selectivity in sweeping gas membrane distillation for dilute aqueous isopropanol, Journal of Membrane Science 188, 79–86, Elsevier, Amsterdam 2001
6. N. Couffin, C. Cabassud, V. Lahoussine-Turcaud, A new process to remove halogenated VOCs for drinking water production: vacuum membrane distillation, Desalination 117, 233–245, Elsevier, Amsterdam 1998
7. W.L. Gore et al., United States Patent Number 4,545,862 United States Patent and Trademark Office, 1985
8. G. Wiedner, W. Heinzl, Entwicklung, Bau und Erprobung einer solarbetriebenen Meerwasserentsalzungsanlage nach dem MD-Verfahren, Bericht Projektphase I zum BMBF-Projekt FKZ: 0328762A, M
9. C. Bier, Bau und Erprobung einer solarbetriebenen Meerwasserentsalzungsanlage nach dem Verfahren der Membrandestillation, Phase II Schlussbericht zum BMBF-Projekt FKZ: 0329085 A, M
10. A Solar Desalination System Using the Membrane Distillation Process, Technical Brochure No. 46, CADDET Center for renewable energy, United Kingdom 1996
11. L. Chuanfeng, Polygeneration of electricity, heat and ultra pure water for the semiconductor industry, Master thesis heat and power technology, Department of Energy Technology, Royal Institute of Technology, Stockholm 2004
12. MEDESOL consortium, Critical assessment of the state-of-the-art and bibliographic review on membrane distillation technology, solar collector technology and low-fouling heat transfer modified surfaces, Deliverable 1 of MEDESOL project, Almeria 2007

13. J.H. Hanemaaijer, Salt to fresh water using MEMSTIL membrane distillation, Presented at Euromembrane, Gardini, Naxos 2006
14. G.W. Meindersma, C.M. Guijt, A.B. de Haan, Desalination and water recycling by air gap membrane distillation, Desalination,187, 291–301, Elsevier, Amsterdam 2006
15. S. Basel, Entwicklung eines kompakten solaren Meerwasserentsalzungssystems, Diplomarbeit FH Augsburg, angefertigt am Fraunhofer ISE, Freiburg, 2005
16. M. Ebermeyer, Aufbau und Vermessung einer solarthermisch angetriebenen Wasserentsalzungsanlage, Diplomarbeit FH Biberach, Fraunhofer ISE, Freiburg, 2005
17. M. Rommel, M. Hermann, J. Koschikowski, Final project report, JOR3-CT98-0229 SODESA, Fraunhofer ISE, Freiburg, 2002

Chapter 8
Photovoltaic Reverse Osmosis and Electrodialysis

Application of Solar Photovoltaic Energy Production to RO and ED Desalination Processes

Jürgen Rheinländer and Dieter Geyer

Abstract Arid zones with less than a thousand inhabitants, remotely situated from large scale sources of power and potable water, are candidates for implementation of small scale Reverse Osmosis (RO) and Electrodialysis (ED) desalination processes using power supplied by Photovoltaic generation (PV). An ED/PV combination may only compete with an RO/PV combination for the desalination of unpolluted brackish water with low salinity. Energy and cost efficiencies depend on raw water quality, climate, accessible energy sources, number of consumers and demand per head, potentials for cogeneration (water and power) and hybridisation, industrial and educational environments. A selection of 10 plant configurations, including different processes and a wide range of capacities, is presented with typical design information and exemplary analysis of life cycle performance for specific sites. Results show that plant capacity should not be too small, power recovery should be provided wherever feasible, and any opportunity for cogeneration should be exploited. Hybridisation reduces levelised costs of electricity and water and increases the reliability of a minimum water supply under emergency conditions.

8.1 Introduction

Administrations of countries with arid regions show growing concern about means of raising the standard of living in small districts remote from the country's major centres of population. Among the most important elements of standards of living, anywhere in the world, are reliable water and power supplies.

There are several reasons to consider using power supplied from Photovoltaic (PV) electricity generation for the operation of Reverse Osmosis or Electrodialysis desalination plants. For example:

J. Rheinländer (✉)
Center for Solar Energy and Hydrogen Research Baden-Württemberg, Industriestr. 6, D-70565 Stuttgart, Germany
e-mail: rheinlaender1j@aol.com

A. Cipollina et al. (eds.), *Seawater Desalination*, Green Energy and Technology,
DOI 10.1007/978-3-642-01150-4_8, © Springer-Verlag Berlin Heidelberg 2009

- The location of the user community is remote from the nearest connection point to the national (or regional) power grid, and the district is so small that transport of fuel for a local diesel power station would be expensive and unreliable. Both the grid extension and the local diesel power station option would be less economic than PV.
- As above, the location of the user community is remote from the nearest grid connection point. A local diesel power station is evaluated as more economic than the PV option, but there are strong reasons to give priority to PV, e.g. natural resources to be protected or high environmental standards set for a tourist site, limiting noise and pollution.
- The district is equipped with a local grid, powered from a diesel power station. Reliability of this power station is poor and accidental shut downs, for several days, are frequent. Powering the desalination plant from both the local grid and a PV system could help to secure a minimum water supply.
- The district is small and dependent on long distance water trucking from a central water treatment plant. Reliability of vehicles, drivers and fuel supply may be limited, as well as hygiene standards for equipment. Evaluation of true total cost shows that local, small-scale desalination using power from PV would be more reliable and economic.
- The district is small and dependent on brackish but bacteriologically clean underground water. Here the use of ED for desalination might require less power than RO, and therefore the possibility of direct use of direct current (DC) for the ED process may be the major advantage over connection to a local or regional alternating current (AC) supply.

Cases where these and other reasons are relevant favour the integration of PV power generation with RO or ED desalination. They may be found at coastal sites dependent on sea water, as well as at rural inland sites dependent on brackish water. All three technologies, RO, ED and PV, are commercially available in very small units, and compact plants, pre-assembled to a large extent, and easy to transport, can be built for small villages or even single domestic or commercial buildings.

The coupling of RO or ED processes with PV energy systems, is a straightforward technique, as it involves the use of electrically driven conventional desalination technologies with well established RE technologies. Chapters 3 and 4 give a description of the relevant desalination processes. In the following paragraphs a description of past and present experiences with RO-PV and ED-PV units is shown. A summary of several different scenarios is also presented to highlight advantages and disadvantages of the proposed couplings. Chapter 10 gives a description of relevant examples of pilot operating units.

PV power plants may be designed in units with power capacities ranging from a few Watts (W) to several Megawatts (MW). The output from PV conversion of solar energy is direct current (DC). A PV system may be connected to a power grid or operated as a standalone ("island") system. When connected to a national or local power grid, the DC output from a PV system is inverted into AC, as most grids are AC grids with centrally controlled frequency. When operated as an island system, the DC output may be supplied unaltered, if the end-user process is designed for DC consumption.

With both the grid-connected PV unit and the island system, an electrochemical energy storage unit, referred to as an ACCU (accumulator) in the schematics presented in this chapter, may be integrated in order to uncouple the generation of PV electricity from the user/process demand. In grid connected systems an ACCU is not common, as grid power capacities are usually much larger than those of PV units, and the grid compensates for the unsteadiness of the solar energy source.

In island systems, an ACCU may have an important role in stabilising the power supply to a user process requiring steady operation, as indeed, is strongly recommended for RO. The other main purpose of the ACCU could be the extension of daily operation times of a desalination system into night time hours, aiming at increasing the plant's capacity and economic performance. This implies the design of a PV generator capacity with a solar multiple (SM) of the nominal power demanded by the desalination system.

In communities not connected to grid power, inhabitants will usually demand both potable water and electricity to run their domestic, industrial and public appliances. Therefore the project of water desalination for such a remote place, in most cases, will include simultaneous power supply to the desalination plant as well as to the community.

There are places where, in the past, sufficient underground water with low salinity was available and power supply was established via grid connection or from a diesel power station. With growing populations and increasing stress on water use and on hygiene standards, desalination may become indispensable. Such a situation demands for comprehensive analysis of the energy situation on site, usually concluding with a cogeneration solution for power and water. With a choice between extension of the grid, upsizing of the diesel power engine or integration with renewable energy (RE), the RE option may be the optimum solution.

8.2 Experience from Experimental and Demonstration Plants

In the 1980s, not long after the start of commercial markets for both RO desalination and PV power generation, the first projects combining them to use RE for desalination emerged, generally with public financial support. Several reports were published on design and implementation of these plants and on the experiences of their operation:

The "Desalination Guide Using Renewable Energies" (1998) edited by the Center for Renewable Energy Sources (CRES) in Greece, on behalf of the European Commission, presented a first comprehensive review and comparison with other combinations of RE with desalination. An update of this information was published by García-Rodríguez [1], including a list of 20 RO plants driven by PV, the largest designed for 50 m^3/day, and 8 ED/PV plants for a maximum of 10 m^3/day. This paper presented a valuable and extensive collection of references on all possible combinations of RE with desalination processes published up to 2003.

At the same time, Tzen and Morris [2] reviewed the status of technologies for desalination and decentralised power supply in regard to the most promising couplings such as PV with RO, wind energy with mechanical vapour compression distillation, and geothermal energy with multiple effect distillation. Five years later, Tzen et al. [3], made a worldwide count of small scale desalination systems (up to the capacity of 50 m^3/day) and found 32 systems combining RO with PV and 6 systems combining ED with PV. Most of these systems were installed for research and demonstration purposes and operated under non-commercial conditions.

The concentrated brine which leaves the RO-vessels at pressure levels a few bars lower than feed pressure, carries potential energy that should be recovered whenever possible. As PV electricity is very expensive, power recovery from RO is perhaps the most significant way of reducing water production cost. Recognising the importance of energy efficiency when using PV, Kunczyinski [4] tested the three commercially available small-scale energy recovery mechanisms in a SWRO/PV plant designed for 19 m^3/day: the ERI PX-15 pressure exchanger; the Spectra Watermakers' Clark Pump; and the Danfoss axial piston motor. The plant achieved an energy consumption level as low as 2.6 kWh/m^3 and with the three systems together, accumulated over 70,000 h of operation running entirely on solar energy.

8.2.1 Estimated and Actual Costs

Karagiannis and Soldatos [5] collected information on water production costs from a large number of studies and performance reports for desalination processes and their power supplies.

Manolakos et al. [6] conclude that even a water cost of about 8 €/m^3 could be competitive for small Greek islands, where this value is not much above that of water hauling by boat. Helal et al. [7] performed a study on the economic feasibility of PV/RO desalination for a production capacity of 20 m^3/day, for remote areas in the United Arab Emirates. They compared a fully solar-driven RO plant (22 kW_{peak}) against one fully diesel-driven plant (10 kW) and one diesel-assisted PV system (11 kW_{peak}), where all plants produced the same total annual quantity of water. No ACCU was used and in the hybrid system the annual contribution of the PV source covered about 1/3 of the electricity demand. All RO systems were equipped with energy recovery from the concentrate.

The conclusions from this study are quite remarkable as it was made based on conditions in a country known to have access to very cheap fossil fuel resources, leaving no chance for competition by RE in the minds of energy experts:

- Optimal design selection depends primarily on the cost of primary energy and on the cost of solar panels. The solar-driven plant configuration becomes most favourable at panels costs of below 6 €/$W_{peak}$and a primary cost of (thermal) energy greater than 40 €/MWh (including cost of transport and storage).

- For small capacity RO plants in remote areas, the labour cost becomes a significant cost fraction of the water cost: around 0.75 €/m^3.
- For the input data used in this study, results showed that the fully solar-driven alternative is very competitive, having a specific water cost of 5.5 €/m^3. This cost can be reduced via incentives to encourage the use of solar panels, such as reduction of interest on capital expenditure, exemption from taxes and reduction of land cost.

8.2.2 Lessons Learned

Helal et al. [7] report that an important consideration in using PV panels in hot environments is the performance deterioration of crystalline silicon solar cells with increasing temperature, where efficiency drops by about 0.4% per °C. This effect should be taken into account when sizing a solar array. When selecting inverters high priority should be given to those with low sensitivity to high ambient temperatures.

Though capacities of the motors involved may be small, the use of 1-phase motors may not be advisable when powered by PV through a DC/AC inverter; the inrush currents at start-up may be too high. Experience with 3-phase motors was more positive [3].

As the presence of skilled technicians on sites with small autonomous desalination plants can not be expected, because of cost and of the difficulty of finding trained people willing to live remote from industrial centres, a high degree of automation may be preferred. However, automation is prone to defects and plant failures. Tele-diagnostics may help, in the future, for sites accessible by wireless telecommunication (FZI [8]).

Referring on the benefit of batteries (ACCU) in stabilizing the operation of a PV driven RO plant (refer Sect. 8.1), the choice will need to be oriented towards local conditions. Therefore the decision between batteries or accelerated wear of membranes should be based on the expected local replacement cost. In some places suppliers of membranes are located close by, due to the existence of a large conventional RO plant, in other places it may be easier to find large battery importers.

The majority of the research and development performed so far on RO/PV and ED/PV systems is characterised by the ambition to achieve autonomous RE driven systems, rather than hybrid systems integrated with other power and water generation plants. Certainly autonomy is the biggest challenge in this context, but implementation and performance have shown that water production costs remain high compared to what the target customers can afford. One of the biggest reasons for the small number of attempts to build hybrid and integrated plants may be the barrier between research environments and field application conditions, which require mutual cooperation and understanding between power and water authorities. Also the reluctance of industry to deal with equipment for the production of a "few litres/hour" of water, at locations which can only be accessed by expensive travel, may explain the lukewarm interest of manufacturers so far.

Somehow these barriers need to be removed to find solutions closer to industrial size and with economic competitiveness. The cost of water supply is the main criteria affecting the feasibility of water desalination projects involving renewable energy. The main factors impacting water cost at a site, for a given raw water quality and local availability of energy sources, are:

- Selection of technology
- Size of system
- Specific energy consumption
- Opportunity for hybridisation and cogeneration

The main purpose of the following paragraphs is to present a few examples of the impact of these factors and of typical configurations approaching optimum solutions in many practical cases. The cost related input to the computation performed for the following examples is based on values published in the above mentioned reports and other experience with RO, ED and PV to date.

The upper limit of plant size here considered is around 50 m^3/day throughout the year, as a process of this capacity would only be justified for a population of more than 500 people. Places with larger populations or big tourist resorts, with higher water consumption even though having a smaller number of inhabitants, would either be connected to a regional power grid or have a high capacity power station.

Due to the low desalination capacity range of the solutions considered within this chapter, there are no major related environmental issues. Optimisation of water production cost will always include reduction of energy consumption (even in the case of purely fossil fuel sources) and thus reduction of CO_2-emissions. The overall balance of emissions (including production of components) of PV power supply is not zero, but is smaller than for most relevant power alternatives.

8.3 Sample Case Studies

Benefits to be expected from different plant sizes and configurations and from feasible technical improvements of the desalination process and of its power supply are best explained and demonstrated with simplified case studies for the more common configurations. This chapter includes the results from studying 10 sample cases, 4 for sea water RO (SWRO), 3 for brackish water RO (BWRO) and 3 for brackish water ED (BWED). The ACCU capacities are designed for a 24 h full load operation from the ACCU alone, in all cases.

Particularly for sea water RO, the recovery of the potential energy carried by the concentrate flow leaving the membrane vessels, is the principal method of reducing demand from the rather expensive PV energy source. Four technologies are commercially available, and the lower limits of their usual ranges of application are characterised by a minimum concentrate flow.

- Pelton Turbine (PT), above 10 m^3/h
- Pressure exchanger (PX), above 2 m^3/h
- Axial piston motor (APM), above 0.2 m^3/h
- Pressure intensifier (PI), "Clark pump", above 0.03 m^3/h

For PV driven RO plants, the Pelton Turbine would not be a reasonable choice, as a process with a 10 m^3/h capacity is beyond the actual capacities of PV-RO units presented in this chapter.

The following assumptions relevant to plant economics are used for all cases:

- life cycle of the project: 20 years
- discount rate: 6% per annum
- local cost of fuel (gasoil, including transport and storage): 0.5 €/kg
- no cost inflation

8.3.1 Photovoltaic Sea-Water Reverse Osmosis (SWRO/PV)

RO plants are usually operated with high pressure pumps requiring AC, though, in principle, pumps, instrumentation and control units designed for DC could be used. The following principal combinations of PV electricity generation with RO power supply exist, where the first listed are the preferred options for autonomous units with small desalination capacities , and the latter those for larger plants with hybrid power supply and integrated with power and water networks:

1. PV – DC pump – RO – (DC to village)
2. PV – ACCU – DC pump – RO – (DC to village)
3. PV – DC/AC inverter – AC pump – RO – (AC to village)
4. PV – DC/AC inverter – ACCU – AC pump – RO – (AC to village)
5. PV – DC/AC inverter – ACCU – grid connection – AC pump – RO
6. PV – DC/AC inverter – grid connection – AC pump – RO

The option "AC to village" is not included in those cases with grid connection as this is assumed to have been the original purpose of the power grid. For the sake of brevity not all of these combinations are considered in the following paragraphs.

8.3.1.1 Base Case: Small Standalone SWRO/PV System with 10 m^3/day Capacity

A frequently proposed small-scale configuration combining desalination with RE is an RO system, with power supply from a PV generator, including an ACCU for the uncoupling of power use from generation (combination no.2). The main equipment components required for the desalination of sea water are shown in Fig. 8.1.

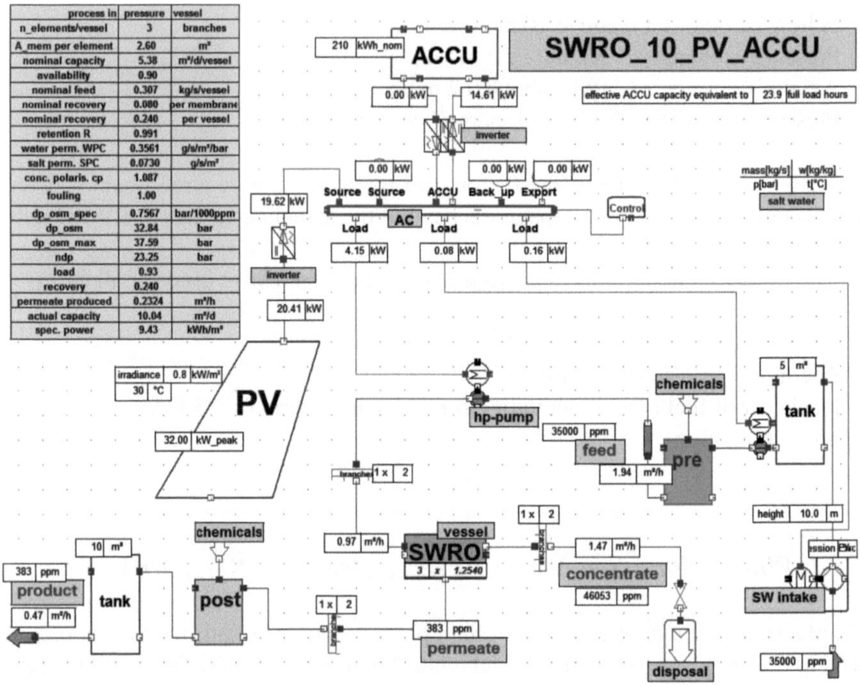

Fig. 8.1 Schematic of a 10 m^3/day sea water reverse osmosis powered by PV

This is a design flow diagram for performance prediction from the Process Simulation Environment *IPSEpro* (SimTech [9]) using the process model library *RESYSpro* [10].

The example is designed for the desalination of 10 m^3/day of sea water with 35,000 ppm salinity, and continuous power supply. This would be sufficient, at most remote sites in arid countries, to supply about 200 people with 50 l of potable water per day. Typical quantities and sizes of the components for such a capacity are:

- 2 RO vessels 2.5 inches (6.25 cm diameter) 3 m (120 inches) long
- 3 membrane elements per vessel each with a 2.6 m^2 area
- 0.083 m^3/h nominal permeate flow capacity of each membrane element
- high pressure pump, rising feed water pressure from 3 to 58 bar
- pre-treatment unit with feed pump, e.g. for pH adjustment, chlorination, adsorption onto active carbon (depending on raw water quality)
- post-treatment unit for adjustment of residual salinity (optional)
- 5 m^3 storage tank for raw water, to decouple sea water pumping from desalination process
- 10 m^3 storage tank for product water (i.e. 1 day production reserve)
- sea water intake (for small systems usually a beach well), particle filter
- Concentrate disposal (return to sea)

- PV generator with 32 kW_{peak} capacity (about 300 m^2 module surface)
- 2 DC/AC (19 and 14.5 kW) inverters and 1 AC/DC (14 kW) inverter
- ACCU with capacity for a 24 h full load operation
- AC busbar with power control for balancing power flows between PV, ACCU and pump motors

A power recovery unit is not included in this example, though an APM (axial piston motor) or a PI (pressure intensifier) would be suitable for the range of flow rates expected. Though both PV and ACCU supply DC, in most practical cases AC motors are selected, as in most markets they are more easily available and cheaper than DC motors. Also subsequent connection of the desalination system to a (local) power grid would be easier.

Concerning the use of inverters, it is worth noting that modern PV inverters allow the substitution of all AC/DC inversions and the bus-bar shown in Fig. 8.1, by one unit which includes DC/DC converters for battery charging and discharging.

The key performance parameters for operation of all components under design conditions are:

- pressure in feed to RO vessels: 58 bar;
- net driving pressure (average along line of 3 membranes): 23 bar;
- permeate production: 0.47 m^3/h;
- concentrate disposal: 1.47 m^3/h;
- water recovery: 24%;
- process availability: 90%;
- water production capacity (taking into account process availability): 10 m^3/day;
- total power required by process: 4.4 kW;
- specific energy consumption (including all auxiliary pumping): 9.43 kWh/m^3;
- power generated by PV system (AC output from inverter): 19.6 kW.

The nominal capacity of the ACCU is determined based on the assumption that the process requires 105 kWh electric per day for continuous operation. Allowing for a 50% depth of discharge, the ACCU's nominal capacity should be 210 kWh for 1 day of operation from the ACCU alone.

The nominal capacity of the PV generator is determined with reference to the expected irradiance for winter days at the site: e.g. 0.503 kW/m^2 irradiance on the tilted PV panel surface during 8 h of a typical winter day. Running process simulation for this irradiance yields 14 kW output from PV after DC/AC inversion, or 112 kWh of electricity from the 8 sunshine hours. This quantity is just over the daily 105 kWh required for the RO process. 35 kWh is used directly during sunshine hours and the remainder is stored in the ACCU for the 16 night time hours (the small surplus generated is consumed by inverter losses).

Of course in practice there will be many days in winter with much lower daily irradiance, and therefore not enough electricity will be available for the process on these days. On the other hand, demand for water basically follows irradiance and

Table 8.1 Typical meteorological conditions relevant to temperature and solar irradiance in Aqaba (Jordan)

Operating mode	Hours/year	Irradiance kW/m^2	Ambient temperature °C
Summer day	1,476	0.574	34
Summer night	1,476	–	25
Winter day	960	0.503	21
Winter night	1,920	–	14
Spring/autumn day	1,220	0.562	29
Spring/autumn night	1,708	–	20

ambient temperatures during the four seasons of the year. At sites not suffering from extended periods of sunshine absence, the suggested sizes for the PV generator and the ACCU should adequately meet the needs of the consumers.

On a typical summer day (referring, for example, to the data in Table 8.1) with 0.574 kW/m^2 irradiance and 34°C ambient air temperature during the 12 sunshine hours (higher temperature slightly reduces PV panels efficiency), the PV generator produces 12 h × 13.5 kW = 162 kWh AC (inverter output). 52 kWh are used directly for the process during sunshine hours, and the remaining 110 kWh are stored in the ACCU for the other 12 h of operation, until the next sunrise. Few kWh are lost to the inverters.

For this and the following examples in this chapter, a rough prediction of technical and economic life cycle performance can be done with the help of*RESYSpro*. When the purpose of analysis is only a pre-feasibility study of a RE-desalination project, this tool runs a simplified annual time series simulation. Six operating modes are taken into consideration, and irradiance and ambient temperature values are allocated to each of these modes, for a given site. For this example, a coastal site near Aqaba (Jordan) at a latitude of 29° is chosen, where total annual irradiance on flat surfaces, south facing and tilted at an angle of 29° to the horizontal, amounts to 2,159 kWh/year. The data, obtained from an analysis of meteorological data collection for the site, for a typical meteorological year (TMY), is shown in Table 8.1.

The irradiance values for daytime are averages over the assumed hours of sunshine: 12 h per day in summer, 8 in winter and 10 in spring and autumn. Therefore the differences, from season to season, between the average irradiance values are not large. The energy performance of the integrated system is calculated for these 6 operating modes separately and the key results, such as water production, are summed for all hours of the year. Consideration of ambient temperature is made due to its strong impact on PV efficiency.

The total annual energy required for desalination of 3,666 m^3/year is 38,427 kWh and the total energy generated by the PV system is 50,087 kWh/year. Thus 30% excess energy is generated, mostly during summer months, and cannot be utilised because seasonal storage capacity is not available. However this loss is necessary to secure sufficient PV capacity for the winter months, as explained above in the discussion of the design settings for PV and ACCU.

The levelised costs of electricity (LEC) and water (LWC) produced by the system are derived by: (1) allocating costs to equipment and the regular replacement thereof, materials and fuel consumed (if appropriate), civil works and labour for operation and maintenance; (2) making assumptions on plant life duration and other economic parameters, such as the capital discount rate; (3) performing a life cycle cost analysis. The economically relevant inputs used for the pre-feasibility analysis of the example system presented include:

- investment cost of desalination equipment: 16,491 € (=> 1,641 €/(m^3/day));
- investment cost of power supply equipment: 207,631 € (=> 6,581 €/kW_{peak});
- total investment cost of plant: 224,122 € (=> 21,952 €/(m^3/day));
- annual operating and maintenance (O&M) cost, including staff: 14,327 €/year;
- (average) annual replacement cost (RO membranes and ACCU): 4,576 €/year.

Relating cost to water delivery yields:

levelised electricity cost (LEC): 0.78 €/kWh;
levelised water cost (LWC): 10.47 €/m^3.

The LWC is the price at which the water should be sold to the public, if the entire project was planned to payback at the end of its 20 year life. If a higher price is obtained, there will be a gain for the investor, and the payback period will be shorter than the life cycle.

8.3.1.2 Variations to the Base Case

With reference to the base case above, three different variations are considered.

Adding a Pressure Intensifier

In the preceding example, the RO process reaches a specific power consumption of almost 10 kWh/m^3. As power generation by PV is very expensive, there is strong motivation to use energy recovery. The potential for saving is about 50% of the power supplied to the high pressure feed pump. From the four commercially available power recovery technologies (Pelton turbine, pressure exchange, axial piston motor (APM) and pressure intensifier (PI)) only the APM is available for the range of mass flows between 5 and 40 m^3/day, and for very small flows, of less than 1.2 m^3/day, the PI is the only option.

A group of parallel Clark pumps is fed with 76% of the main feed water stream at a pressure of about 11 bar. A concentrate stream, with practically the same mass flow, helps to increase pressure in the feed flow up to 58 bar, which is the pressure required at the inlet to the RO vessels. In a pipe parallel to the Clark Pump line, the high pressure feed pump pressurises the remaining 24% of feed up to the same pressure level. Under design operating conditions, the power consumed by the smaller high pressure pump is less than ¼ of the power calculated for the process without the above power recovery. The total power load is reduced from 4.4 to 1.7 kW and the specific energy consumption is reduced from 9.4 to 3.6 kWh/m^3. The biggest benefit from power recovery is the reduction of the required sizes for PV generators and ACCUs. With power recovery, PV panels of 12 kW$_{peak}$, and an ACCU with a nominal storage capacity of 80 kWh, will suffice.

The results from simulating performance of this system, for the same site conditions as earlier, demonstrate the significant reduction in the cost of water production by using power recovery:

LEC = 0.88 €/kWh;
LWC = 5.68 €/m^3.

Economies of Scale for a SWRO/PV Plant of 50 m^3/day

Assuming that the district is inhabited by approximately 1,000 people and that the desired standard of potable water supply per person is 50 l/day, a plant designed for 50 m^3/day would be required.

For energy recovery at this capacity, there are commercially available pressure exchange units, with a proven track record in many larger RO plants. The power recovery efficiencies of pressure exchangers and pressure intensifiers are quite similar.

The following typical quantities and sizes of components for such a capacity differ from those of the base case:

- 7 RO vessels 2.5 inches (6.25 cm diameter) and 4 m (160 inches) long;
- 4 membrane elements per vessel each with a 2.6 m^2 area;
- high pressure pump rising feed water pressure from 3 to 62 bar;
- 20 m^3 storage tank for raw water, to decouple sea water pumping from desalination process;
- 50 m^3 storage tank for product water (i.e. 1 day production reserve);
- PV generator with 54 kW$_{peak}$capacity (about 500 m^2 module surface);
- 2 DC/AC (32 and 23 kW) and 1 AC/DC (8 kW) inverters;

As a result of the increased size of the plant, with the same site conditions as before, a strong reduction of water production cost is obtained, leading to the following values :

LEC = 0.70 €/kWh
LWC = 3.87 €/m^3

Adding a Diesel Engine for Cogeneration

Communities of several hundreds of people, who require desalination of water, in general need power supply for other applications such as light, telecommunications, refrigeration for food and medicines,. Therefore it is generally worthwhile analysing the opportunities for simultaneous generation of power for the village.

The benefits expected from such cogeneration are:

- Better exploitation of the installed power generation capacity
- No ACCU needed
- Better plant component operating efficiencies due to flexibility of load management
- Higher productivity of the water desalination plant due to the possibility of operation, even in cases of partial failure of power generation equipment
- Lower LEC and LWC

This case is a study of hybridising power supply to the 50 m^3/day system previously discussed, and the increase of the power generation capacity up to a minimum direct electricity supply for the village.

The capacity at the grid connection node for power supply to the village is set to a modest value of 50 kW (equivalent to about 50 W/head).

The following values of power demanded by the village are allocated to the 6 operating modes:

50/30 kW during summer day/night
30/10 kW during winter day/night
40/20 kW during spring/autumn day/night

The assumptions on village power are that the predominant power needs originate from the refrigeration of valuable food or medicines, rather than from supply of light and telecommunications. Therefore the loads are highest during hot summer days and lowest for cold winter nights. No ACCU is required, as excess PV output during sunshine hours is fed into the village power supply, and the power load of the desalination process during the night is covered by the diesel engine output.

The results from simulating performance of this system, for the same site conditions as before, show the significant reduction in the cost of water production by using power recovery, due to economies of scale and the synergetic effect of cogeneration of water and power:

LEC = 0.27 €/kWh
LWC = 2.26 €/m^3

8.3.1.3 Comparison of Settings and Results in Case Studies Using SWRO with PV

The key results from the preceding four case studies are summarised in Table 8.2 and demonstrate the benefits to be expected from power recovery, economies of scale and hybridisation applied to the fictive desalination project. The general conclusions from these case studies are:

- If possible, by contracting water supply to a larger number of customers, desalination plants should be designed as large as possible.
- Wherever cogeneration of power and water is justified by simultaneous demand for both, this opportunity should be exploited.
- Hybrid power supply from fossil and renewable energy sources can reduce LEC and LWC, if the fuel is locally available at a competitive price.
- The hybrid system will allow maintaining a minimum supply of water and power, even if the diesel engine is out of service for several days.

Table 8.2 Comparison of settings and results of 4 SWRO/PV case studies

	Units	Base case	Power recovery	Larger capacity	Hybrid PV/DG
SWRO capacity	m^3/day	10	10	50	50
Power recovery	type	–	PI	PX	PX
Design power for RO	kW	4.4	1.7	7.9	7.9
PV peak capacity	kW	32	12	54	54
ACCU capacity	kWh	210	80	380	–
Diesel gen. cap.	kW	–	–	–	50
Spec. energy	kWh/m^3	9.4	3.6	3.4	3.4
Solar fraction	%	100	100	100	27
CO_2 emission	tons/year	–	–	–	281
Water to village	m^3/year	3,666	3,666	18,323	18,323
Power to village	MWh/year	–	–	–	249
Total investment cost	€'000	224	103	396	357
Desalination	€'000	16	18	52	52
Power from PV	€'000	208	85	344	292
Power from DG	€'000	–	–	–	13
O&M cost	€'000/year	14.3	9.3	27.5	31.1
Fuel cost	€'000/year	–	–	–	42.7
Replacements (avg.)	€'000/year	4.6	2.1	7.8	2.5
Life cycle project cost	€'000	441	239	814	1,247
LEC	€/kWh	0.78	0.88	0.70	0.27
LWC	€/m^3	10.47	5.68	3.87	2.26

8.3.2 *Photovoltaic Brackish-Water Reverse Osmosis (BWRO/PV)*

In many places, in arid countries, far from the coastline, underground water of potable quality has been pumped from wells for a significant period of time. Due to population growth and increased demand of water per head, many of these wells have become brackish and treatment, in addition to filtering, is now required. If such places are remote from the country's water supply centres, and if the number of people living there is small, a desalination plant combining a brackish water RO system with power supply from a PV generator, including an ACCU for uncoupling the use of power from generation, could be an option. As in the previous case of PV-SWRO, the main equipment components required are shown in Fig. 8.2.

8.3.2.1 Base Case: Standalone BWRO/PV System with 10 m^3/day Capacity

As in the previous SWRO/PV base case, this example is designed for the desalination of 10 m^3 of water per day, assuming a continuous power supply. Typical quantities and sizes of components for such a capacity are similar to the previous, apart from:

- unit for concentrate disposal, which could be, for example, an evaporation basin
- PV generator with 5 kW$_{peak}$ capacity (equivalent to about 50 m^2 module area)
- 2 DC/AC (3 kW) inverters and 1 AC/DC (1 kW)

A schematic of the standard case process is depicted in Fig. 8.2.

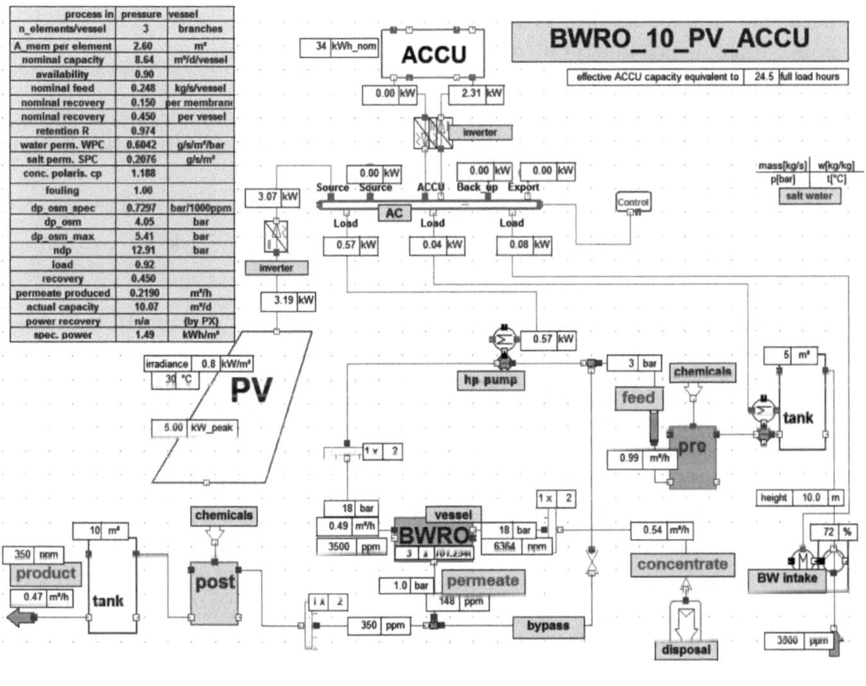

Fig. 8.2 10 m^3/day brackish water reverse osmosis powered by PV

As the expected total power demand for the pumps is less than 1 kW, there is an option to install DC motors thereby avoiding the need for inversion of the PV output into AC. Brushless DC motors could be selected to reduce the O&M requirements. The higher specific cost of DC motors compared to that of AC motors would be balanced by the saving on the cost of inverters. The only disadvantage of the decrease in voltage in the ACCU, due to the falling state of charge, would be a slight reduction in water production.

When brackish water passes through the RO process, the permeate flow leaves the RO vessels with a rather low residual salinity: 150 ppm TDS in this example. As 300–500 ppm is the typical range of standard salinities for potable water, it is possible to implement a bypass line for a small quantity of feed water to be mixed with the permeate flow. The salinity of the product water is therefore adjusted to a value of 350 ppm. This helps to reduce power consumption of the high pressure pump.

The significantly lower power requirements of BWRO (compared to SWRO) mean that integration of a power recovery unit into this configuration is not required, though an APM or a PI could match the range of flow rates expected. The key performance parameters for the operation of components at design conditions are:

- salinity of raw water (TDS): 3,500 ppm
- irradiance on tilted PV panel surface: 0.8 kW/m^2
- pressure in feed to RO vessels: 1.8 MPa
- net driving pressure (average along line of 3 membranes): 1.3 MPa
- permeate production: 0.47 m^3/h
- concentrate disposal: 0.54 m^3/h
- water recovery: 45%
- availability of process: 90%
- water production capacity (taking into account process availability): 10 m^3/day
- total power required by process: 0.69 kW
- specific energy consumption (including all auxiliary pumping): 1.49 kWh/m^3
- power generated by PV system (AC output from inverter): 3.1 kW

The nominal capacity of the ACCU is designed to cover 1 day of full load operation using the ACCU alone. The nominal capacity of the PV generator is designed with a value that allows 24 h of full load operation using the electricity generated during sunshine hours, of a typical winter day, at the selected site. In this case the site is taken to be Hassi Khebbi, Algeria at a latitude of 29°, where annual total irradiance on south facing flat surfaces tilted at 29° to the horizontal amounts to 2,109 kWh/year. The winter day irradiance is 0.538 kW/m^2 on tilted PV panel surfaces. This irradiance yields 2.2 kW output from PV after DC/AC inversion or 17.6 kWh electricity, from 8 sunshine hours. This quantity is just over the daily 16.6 kWh required for the desalination unit specified above. 5.5 kWh will be used directly during sunshine hours and the rest is stored in the ACCU for the 16 night time hours (the small surplus generated is consumed by inverter losses).

On a typical summer day, with 0.532 kW/m^2 irradiance and 32°C ambient air temperature during the 12 sunshine hours, the PV generator produces 12 h × 2.0 kW = 24 kWh AC (inverter output). 8.3 kWh of these are directly used for the process during sunshine hours, and the remaining 15.7 kWh are stored in the ACCU. The ACCU can absorb a maximum of 17 kWh and a few kWh are lost to the inverters.

The energy performance of the integrated system is calculated in a similar way to the case of the SWRO/PV above. The total annual energy demanded for desalination of 3,674 m^3/year is 6,067 kWh and the total energy generated by the PV system is 7,880 kWh/year. Thus 30% excess energy is generated, mostly during the summer months, and cannot be utilised as long term storage capacity is not available. However this loss is necessary to secure sufficient PV capacity for the winter months, as explained above in the discussion on the design settings for PV and ACCU.

The economically relevant inputs used for a pre-feasibility analysis of the example system presented include:

investment cost for desalination equipment: 1,122 €/(m^3/day)
investment cost for power supply equipment: 8,165 €/kW_{peak}
annual O&M cost including staff: 8,569 €/year
annual replacement cost (RO membranes and ACCU): 1,120 €/year (average)

Relating cost to water delivery:

LEC = 1.21 €/kWh
LWC = 3.87 €/m^3

8.3.2.2 Benefits of Increased Scale and Power Recovery

In general, the specific power consumption for BW desalination is much lower than for SWRO. Therefore the power equipment in these plants is smaller and the water production cost lower. Naturally hybridisation and cogeneration can be beneficial to BWRO plants in the same way as they are to SWRO systems.

As the method of analysis and the interpretation of performance results in BWRO cases are very similar to those of SWRO cases, the details are not explained here but the key results are shown in Table 8.3.

Moreover, if a configuration for 10 m^3/day with 5 kW PV and a 34 kWh ACCU capacity were to be equipped with energy recovery too, the peak power capacity required would drop to about 3.5 kW. The 35 m^2 of PV panel area would then be small enough to allow installation of the panels on the roof of two standard 6 m transport containers.

One of the greatest advantages of such a solution would be the opportunity to assemble the entire desalination system (except the storage tanks) into containers, in a workshop in a large city, and place the PV generator on the top. The expensive travelling and subsistence costs of expert technicians would be reduced to the

Table 8.3 Comparison of settings and results of 3 BWRO case studies

	Units	Base case	Increased capacity	Power recovery
BWRO capacity	m^3/day	10	50	50
Power recovery	Type	–	–	PI
Design power for RO	kW	0.7	3.1	2.2
PV peak capacity	kW	5	22	16
ACCU capacity	kWh	34	152	106
Spec. energy	kWh/m^3	1.5	1.35	0.95
Solar fraction	%	100	100	100
Water to village	m^3/year	3,674	18,288	18,288
Total investment cost	€'000	52	183	146
O&M cost	€'000/year	8.6	22.9	22.7
Total life cycle cost	€'000	163	489	438
LEC	€/kWh	1.21	0.82	0.92
LWC	€/m^3	3.87	2.33	2.09

installation of the raw water intake, storage tanks and connections to the village water network. The cost saving of central prefabrication and functional testing could be considerable. The only bottleneck for implementation might be the difficulty in accessing the site for the container-carrying trucks.

8.3.3 Photovoltaic Electrodialysis (ED/PV)

Electrodialysis is a physical phenomenon capable of transporting ionic compounds such as salts from one solution, the diluate, to another solution, the concentrate, by applying a directed electric current. It is therefore suitable to be used for the desalination of water by DC. Principles of the ED process are described in Chap. 3.

Here, the simplest case of a batch desalination process is carried out by circulating the solution through the stack until the conductivity of the diluate reaches the target value. In parallel to this change, the power consumption rises because of the increase in voltage drop over the cell. It is more practical to run an ED process continuously in the "feed and bleed mode", adopting a plant configuration as shown in Fig. 8.3 for the standard case of a 50 m^3/day unit analysed below.

The pretreated raw water is fed into the diluate and concentrate storage tanks at the same rates of product and waste removal from the system. By mixing the diluate with feed and concentrate, the salinities in the storage tanks are kept at 1,000 and 10,000 ppm, respectively. Thus the inlet flows to the ED stacks carry constant salinities. The stack design is made to achieve 200 ppm in the diluate, and to release a concentrate with almost 18,000 ppm. The mass flow ratio, between diluate and concentrate feeds into the stacks, is set to 10. A selection of design values is shown in the data table of Fig. 8.3. The DC power consumption for desalination of 50 m^3/day is 2.1 kW (excluding pumping power). Thus the specific energy consumption in this case is 0.89 kWh/m^3. As the ED process itself is operating with DC power, it can run directly from the output of a PV unit, if the pumps too are designed for DC

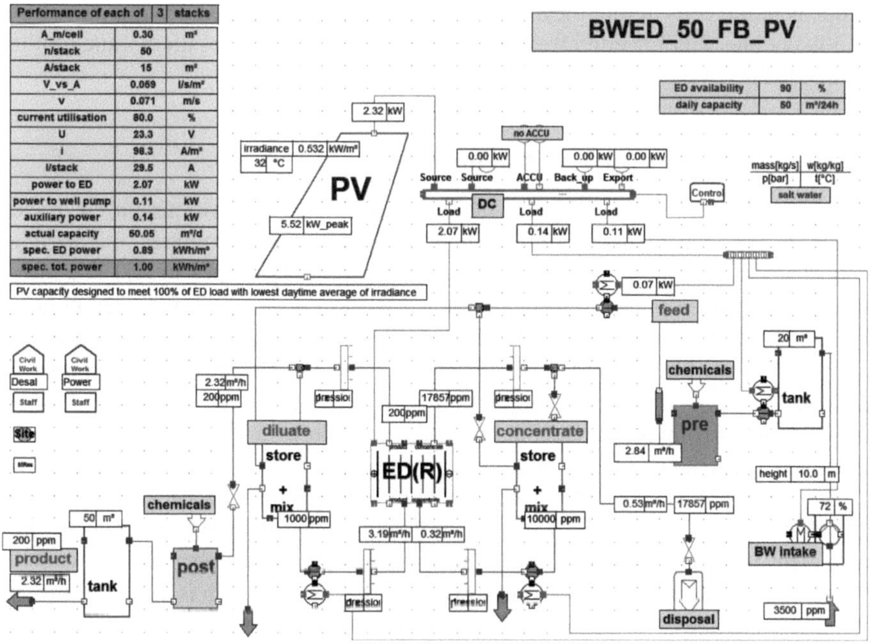

Fig. 8.3 50 m^3/day brackish water electrodialysis powered by PV

operation. When operated with a grid power supply, an AC/DC inverter is required to run the ED stacks. As the output of low salinity water is proportional to the current supplied, the process allows for variations in PV power input which mirrors variations in solar irradiance.

Obviously the main advantage of ED over RO is observed in the desalination of brackish water with very low salinity. On the other hand, it has to be borne in mind that the ED process does not prevent bacteria from passing through the system and into the product water. Bacteriological treatment has to be performed before if the raw water source is not clean. Furthermore, the raw water must be pretreated to prevent materials from entering the membrane stack that could either harm the membranes or clog the narrow channels in the cells. To prevent thick layers of ions accumulating on the membrane surfaces, the operating principle of reversing electrode polarity is applied. To allow for this, the diluate and concentrate channels throughout the stack are identical in construction. Several times an hour, the polarity is reversed and the flows are simultaneously reversed so that the diluate channel becomes the concentrate channel and the concentrate channel becomes the diluate channel. The result is that the ions are attracted in opposite directions across the membrane stack. Immediately after the reversal of polarity and flow, the diluate is disposed of, until the stack and lines are flushed out and the desired water quality is restored. This flush takes only 1 or 2 min, and then the unit can resume production of potable water. The reversal process is useful in breaking up and flushing out scale, slime and other deposits in the cells before they build up and create a problem. Flushing allows the unit to operate with fewer pretreatment chemicals and minimises membrane fouling (Buros [11]).

8.3.3.1 Standalone ED/PV System for 50 m^3/day Capacity

One principal advantage of ED over RO is its lower energy consumption for desalination of water with low salinity. Therefore the best opportunities for application of small standalone ED/PV systems are remote inland sites in arid countries, where these systems can compete with small BWRO plants. The main components required for a small standalone BWED plant with power supply from PV are shown in Fig. 8.3. DC power is supplied directly to the ED stack(s) and DC-powered pumps. Therefore water production is directly linked to power input from PV. No energy storage is involved. Decoupling of production from demand is achieved by separate storage tanks for raw and product water.

This example is designed for the desalination of 2.3 m^3 of water per hour using the DC power output of a 5.5 kW_{peak}PV generator, absorbing 0.532 kW/m^2 irradiance. This is the average for 12 sunshine hours per day in summer for Hass Khebbi, the same site selected for the BWRO examples above. 2.3 m^3/h is equivalent to 50 m^3/day, if the power supply is continuous and with 90% availability. As the system does not include an ACCU, and there are no other sources of electricity, the production for a typical summer day is less than 28 m^3/day, for a winter day is 19 m^3/day, and for the whole year the yield is predicted to be 7,890 m^3. This would be sufficient, at most remote sites in arid countries, to supply about 430 people with 50 l of potable water per day.

Typical quantities and sizes of components for such a capacity are:

- 3 ED stacks each with a 15 m^2 membrane area
- 2.32 m^3/h total nominal diluate flow
- 2.1 kW DC power consumption for nominal water production
- pretreatment unit with feed pump for pH adjustment, chlorination, adsorption onto active carbon (depending on raw water quality)
- post-treatment unit for adjustment of residual salinity (optional)
- 20 m^3 storage tank for raw water, to decouple sea water pumping from desalination process
- 50 m^3 storage tank for product water (1 day production reserve)
- water intake with particle filter
- concentrate disposal (e.g. evaporation basin)
- PV generator with 5.5 kW_{peak}capacity (about 55 m^2 module surface)
- DC busbar with power control for balancing power flows between PV, ED stacks and pump motors

The key performance parameters for the operation of components under design conditions are (also reported in Fig. 8.3):

- Salinity of raw water (TDS): 3,500 ppm
- operating pressure in ED stacks, permeate and concentrate buffers: 0.2 MPa
- nominal diluate production: 2.32 m^3/h
- concentrate disposal: 0.53 m^3/h
- water recovery: 81%
- process availability: 90%
- average water production capacity: 21 m^3/day
- total power requested by ED and pumping: 2.32 kW
- specific energy consumption (including all auxiliary pumping): 1.0 kWh/m^3
- power generated by PV system: 2.32 kW

For this example, the inland desert location of Hass Khebbi, Algeria at a latitude of 29°, is again selected. The total annual energy generated by the PV system is 8,981 kWh/year, and completely used for the desalination of 7,890 m^3/year. The economically relevant inputs used for a pre-feasibility analysis of the example system presented include:

investment cost for desalination equipment: 1,387 €/(m^3/day)
investment cost for power supply equipment: 6,011 €/kW_{peak}
annual O&M cost including staff: 20,548 €/year
annual replacement cost (ED membranes): 938 €/year (average)

Relating cost to water delivery:

LEC = 0.64 €/kWh
LWC = 3.86 €/m^3

8.3.3.2 Benefits of ACCU and Hybridisation

As only DC-powered units are involved in a BWED system, if DC motors are used for the pumps, the integration of an ACCU can easily help to extend daily production of a BWED/PV system and reduce LWC by better exploitation of the investment in the ED equipment. When present, the nominal capacity of the ACCU is designed to cover one day of full load operation.

Naturally hybridisation and cogeneration can be beneficial to BWED plants in the same way as they are to SWRO and BWRO systems. Therefore there are no additional descriptions of these configurations included in this chapter.

As the method for analysis and the interpretation of performance results in the case of BWED are very similar to those applied to the SWRO cases, the details are not explained here but the key results are shown in Table 8.4.

The main conclusion of this study on ED applied to brackish water, is that BWED driven by PV electricity may well become competitive compared with BWRO, and both can be expected to compete with traditional potable water trucking over long distances, from large water supply centres to remote places.

Table 8.4 Comparison of settings and results of 3 BWED case studies

	Units	Base case	ACCU	Hybrid PV/DG
BWED capacity	m^3/day	50	50	50
Design power for ED	kW	2.1	2.1	2.1
PV peak capacity	kW	5.5	16	16
ACCU capacity	kWh	–	112	–
Diesel gen. cap.	kW	–	–	50
Spec. energy	kWh/m^3	1.0	1.0	1.0
Solar fraction	%	100	100	10
CO_2 emission	tons/year	0	0	285
Water to village	m^3/year	7,890	18,280	18,288
Power to village	MWh/year	–	–	249
Total investment cost	€'000	103	165	178
O&M cost	€'000/year	20.5	23.4	27.6
Total Life cycle cost	€'000	349	468	667
LEC	€/kWh	0.64	0.77	0.23
LWC	€/m^3	3.86	2.23	1.63

8.4 Final Remarks

- Very small desalination plants are often not economically viable.
- Increasing process efficiency and reducing power demand is essential for powering desalination using PV.
- With RO desalination, power recovery should be included whenever possible.
- PV as the main source of power for desalination can be competitive at very remote sites in arid countries, where grid connection is not possible and local fossil fuel costs are comparable to those on the world market.
- RO with PV and ACCU is generally the preferred configuration, but ED with PV and ACCU can compete in brackish water desalination.
- Cogeneration of water and power can secure better exploitation of plant capacity and improve operational flexibility and process efficiency.
- Hybridisation of power supply, by combining PV and diesel powers, can reduce LEC and LWC, and increase the reliability of a minimum water supply under emergency conditions.

Abbreviations

AC	alternating current
ACCU	energy storage unit
APM	axial piston motor
BWED	brackish water electrodialysis

BWRO	brackish water reverse osmosis
DC	direct current
DG	diesel generator
ED	electrodialysis
EDR	electrodialysis reverse
LEC	levelised electricity cost
LWC	levelised water cost
PI	pressure intensifier
PT	Pelton turbine
PV	photovoltaic
PX	pressure exchanger
RO	reverse osmosis
SWRO	sea water reverse osmosis
TMY	typical meteorological year

References

1. García-Rodríguez L (2003) Renewable energy applications in desalination: state of the art. Solar Energy 75: 381–393
2. Tzen E, Morris R (2003) Renewable energy sources for desalination. Solar Energy 75: 375–379
3. Tzen E, Theofilloyianakos D, Kologios Z (2008) Autonomous reverse osmosis units driven by RE sources – experiences and lessons learned. Desalination 221: 29–36
4. Kunczinsky Y (2003) Development and optimisation of 1000–5000 GPD solar power SWRO. IDA World Congress on Desalination and Water Reuse, Bahamas, 2003
5. Karagiannis I, Soldatos P G (2008) Water desalination cost literature: review and assessment. Desalination 223: 448–456
6. Manolakos D, Mohamed E S, Karagiannis I, Papadakis G (2008) Technical and economic comparison between PV-RO system and RO-Solar Rankine system. Case study: Thirasia island. Desalination 221: 37–46
7. Helal A M, Al-Malek S A, Al-Katheeri E S (2008) economic feasibility of alternative designs of a PV-RO desalination unit for remote areas in the United Arab Emirates. Desalination 221: 1–16
8. FZI (2008) Telediagnosis for industrial plants. Forschungszentrum Informatik Karlsruhe, Germany, www.fzi.de/ids/projekte.php?id=28
9. SimTech: IPSEpro V4.0 User Documentation (2003)
10. Rheinländer J (2007) De-central water and power supply integrating renewable energy – technical and economic performance prediction. Springer and NATO, Solar Desalination for the 21st Century, 111–126
11. Buros O K (2000) The ABCs of Desalting. International Desalination Association (IDA), Topsfield, USA, 2nd edition 2000.

Chapter 9
Wind and Wave Energy for Reverse Osmosis

Eftihia Tzen

Abstract The idea of using Renewable Energy Sources (RES) to drive desalination processes is fundamentally attractive, as a considerable number of in-depth studies and real-life applications demonstrate. Renewable energy systems convert naturally occurring energy (sunlight, wind, etc.) into usable electrical, mechanical or thermal energy. Most of these systems are well established and reliable, with a significant number of applications all over the world. The selection of the most suitable technological combination for RES/desalination is an important factor in the success of a project. Wind energy turbines to drive Reverse Osmosis (RO) units is the second most used combination, following that of photovoltaic (PV)/RO systems. Only a few studies and applications have been done on the use of wave energy to drive RO units. Wave energy is a relatively new technology with only a small number of applications being used for electricity production. In this chapter, an overview of wind and wave energy technologies and their coupling with RO units for seawater desalination is presented. Additionally, a description of existing applications, economic data, as well as market potential, is provided.

9.1 Advanced Wind Energy Technology

9.1.1 Technology Description

Wind electricity technology has come a long way since the prototype of just 25 years ago. Two decades of technological progress have resulted in today's wind turbines being state-of-the-art, modern, modular technology- based and rapid to install. Modern wind turbines have improved dramatically in their power rating, efficiency and reliability. Today, the lessons learnt from the operation of wind power plants, along with continuing R&D, have made wind electricity very close in cost to power

E. Tzen (✉)
Centre for Renewable Energy Sources (CRES), Wind Energy Department,
19th km Marathonos Ave, 19009 Pikermi, Greece
e-mail: etzen@cres.gr

A. Cipollina et al. (eds.), *Seawater Desalination*, Green Energy and Technology,
DOI 10.1007/978-3-642-01150-4_9, © Springer-Verlag Berlin Heidelberg 2009

from conventional utility generation in some locations. Generation costs have fallen by 50% over the last 15 years, moving closer to the cost of generation from conventional energy sources.

In Europe, wind potential, which is the most vital parameter for a wind park installation, varies enormously across the continent, as totally different climates can be found from the temperate ocean of the British Isles and Ireland to the Mediterranean climate of Greece or the Spanish Levant. Furthermore, the effect caused by the presence of considerable differences in temperature over short distances, coinciding with specific mountainous features, can substantially modify the possibilities of exploiting wind power within a particular area. The European wind atlas, presented in Fig. 9.1, shows the extent of the variation of the resource across the continent [1].

The most important parameter of the information provided by wind maps is the average wind speed, which is often used as a general indication of the wind power potential at a particular site. The wind conditions in Europe are determined by three general elements: water and land distribution, the occurrence of great mountain barriers and the considerable difference in temperature between the polar air of the North and the subtropical air from the South [2].

Although the typical mean annual wind speed required for a viable development is around 7 m/s, turbines can be found on sites characterised by mean speeds of about 5 m/s.

Wind turbines are basically classified by the position of the spin axis, which can be vertical (VAWT Vertical Axis Wind Turbine) or horizontal (HAWT Horizontal Axis Wind Turbine). Modern wind turbines are typically of horizontal type (Figs. 9.5 and 9.6). The principal elements of a horizontal axis turbine are: the rotor, the multiplier-gear, the electrical generator, the tower and the control system. The rotor consists of the blades, the hub and the spin axis (see Fig. 9.2) [3]. The surface swept by the blades forms the collection area of the turbine. Wind turbines are also classified as fixed or variable pitch, depending on whether the blades embedded in the hub are permanently fixed or not. The inside of the hub houses the hydraulic devices which change the pitch of the blades, if applicable.

In a typical wind turbine, the kinetic energy of the wind is converted to rotational motion by the rotor. The rotor turns a shaft, which transfers the motion into the nacelle (the large housing at the top of a wind turbine tower). Inside the nacelle, the slowly rotating shaft enters a gearbox that greatly increases the rotational shaft speed (see Figs. 9.3 and 9.4). The output (high-speed) shaft is connected to a generator that converts the rotational movement into medium-voltage electricity. The electricity flows down heavy electric cables inside the tower to a transformer, which increases the voltage of the electric power to the distribution voltage. The distribution-voltage power flows through underground lines to a collection point where the power may be combined with that of other turbines [4].

The amount of energy which the wind transfers to the rotor through the blades, depends on the density of the air, the rotor area and the wind speed. The wind speed is extremely important for the amount of energy a wind turbine can convert to electricity.

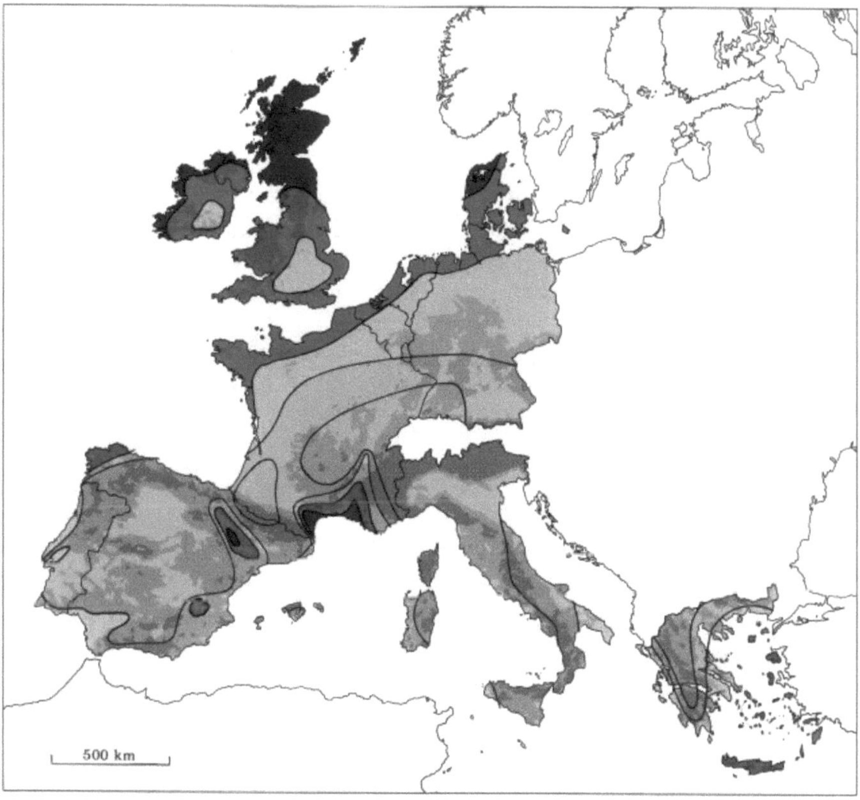

Wind resources[1] at 50 metres above ground level for five different topographic conditions										
	Sheltered terrain[2]		Open plain[3]		At a sea coast[4]		Open sea[5]		Hills and ridges[6]	
	$m s^{-1}$	Wm^{-2}	$m s^{-1}$	Wm^{-2}	$m s^{-1}$	Wm^{-2}	$m s^{-1}$	Wm^{-2}	$m s^{-1}$	Wm^{-2}
	> 6.0	> 250	> 7.5	> 500	> 8.5	> 700	> 9.0	> 800	> 11.5	> 1800
	5.0-6.0	150-250	6.5-7.5	300-500	7.0-8.5	400-700	8.0-9.0	600-800	10.0-11.5	1200-1800
	4.5-5.0	100-150	5.5-6.5	200-300	6.0-7.0	250-400	7.0-8.0	400-600	8.5-10.0	700-1200
	3.5-4.5	50-100	4.5-5.5	100-200	5.0-6.0	150-250	5.5-7.0	200-400	7.0- 8.5	400- 700
	< 3.5	< 50	< 4.5	< 100	< 5.0	< 150	< 5.5	< 200	< 7.0	< 400

Fig. 9.1 European wind atlas
Source: RISO [1]

The maximum attainable power from a wind turbine is estimated by the following formula:

$$P_{WT} = \frac{1}{2}\ \rho_{air}\ C_p \cdot \pi r^2 \cdot V^3 \tag{9.1}$$

where, P_{WT}: wind turbine output, kW; r: rotor radius, m; V: wind speed, (m/s); C_p: power coefficient; ρ_{air}: air density at the site, kg/m^3

The actual wind turbine power output is determined by its power curve. The power curve of a wind turbine is a graph that indicates the electrical power output for the turbine at different wind speeds. Power curves from commercially

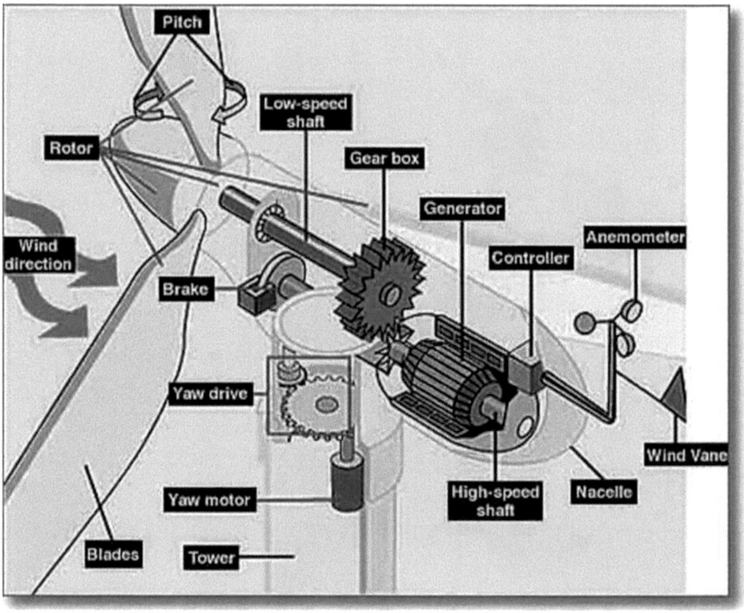

Fig. 9.2 Internal view of a wind turbine, with gear box
Source: EERE [3]

available wind turbines are typically constructed on the basis of averaging intervals of 1–10 min. A wind turbine is characterised by the *cut-in* wind speed (i.e. the speed at which it starts to generate power), the *rated* wind speed (i.e. the speed at which it starts to generate power at the rated level,) and the *high-wind cut-out* wind

Fig. 9.3 Erection of the rotor of a W/T
Source: CRES

Fig. 9.4 View of the gear box of a W/T
Source: CRES

Fig. 9.5 A gearless horizontal type 500 kW W/T
Source: CRES

speed (i.e. the speed at which the unit is shut down for safety reasons). The cut-in speed is typically around 4 m/s. The cut-out speed typically lies between 25 and 30 m/s. The energy efficiency of a wind turbine is defined by the following:

- Capacity Factor (%): the ratio of the actual annual energy output divided by the theoretical maximum output, corresponding to turbine operation at its rated (maximum) power during the whole year. In practice, capacity factors usually range from 20 to 40%.
- Availability (%): the percentage of time during the year that the wind turbine is able to operate.

Fig. 9.6 Erection of a W/T
Source: CRES

Wind energy has the tremendous advantage that it can be harnessed at a very wide range of scales; installations can vary from a few kW wind generators up to MW-scale (megawatt scale) wind parks. Wind turbines can be used as standalone applications, or they can be connected to a utility power grid. They can also be combined with a photovoltaic (PV) system or a diesel generator and batteries, thus forming "hybrid" systems, which are typically used in remote locations where connection to a utility grid is not available.

9.1.2 Commercial Applications and Economics

Although focus has been given to large utility-connected wind parks, there are many other applications, such as autonomous wind energy systems for water pumping, water desalination, as well as local power supply systems in conjunction with diesel generators.

The commitment towards wind energy by national and international governments stems from environmental concerns and the need to reduce greenhouse gas emissions. However wind energy has developed to such an extent that for many sites and many countries it is becoming the most economic new electricity supply option. The economics of a wind scheme depend upon technical, resource and cost parameters. The latter two vary from country to country. In general the cost of the equipment represents the highest portion of the total investment cost. Figure 9.7 presents the cost breakdown of a typical wind power project.

The costs for electrical and civil engineering infrastructure, transport, management and administration, land, etc. can vary between 25 and 35% of the cost of the installed turbines. The annual operating and maintenance cost accounts for around

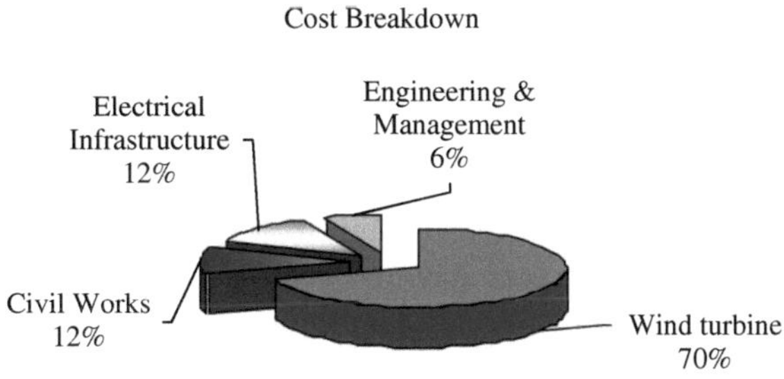

Fig. 9.7 Cost breakdown of W/T projects

1.5% of the initial investment. Also, to the above mentioned cost, a percentage of around 1% per annum of the initial investment should be added, for the cost of insurance [2].

In general, turbine costs vary from a low of 637 €/kW to a high of 1,000 €/kW [5]. However, higher costs have also been reported (Table 9.1).

The cost of offshore installations is much higher and can reach a cost of 2,375 €/kW (UK). The levelised cost of wind electricity depends mainly on the wind potential of the installation site and the size of the wind turbine used.

In Canada, for instance, the cost of energy from wind varies from 0.049 to 0.078 €/kWh, in Greece from 0.026 to 0.047 €/kWh, and in Japan from 0.057 to 0.070 €/kWh for wind turbines of 500–1,000 kW and from 0.040 to 0.057 €/kWh for wind turbines larger than 1,000 kW [5].

Table 9.1 Wind turbine costs

Country	Turbine cost (€/kW)	Total installed cost (€/kW)
Germany	982	1,289
Denmark	845	980
Italy	950	1,200
USA	871	1,121
Japan	637	1,210

9.1.3 Current Market

In 2006, cumulative installed wind power capacity worldwide increased by 26%. According to the IEA (International Energy Agency), thirteen member countries[1] added more than 100 MW of new capacity, and three countries added more than a GW each of new capacity: the United States (2.4 GW), Germany (2.2 GW) and Spain (1.58 GW). Canada, Portugal, the United Kingdom and Japan added 490 MW

[1] IEA member countries are: Australia, Canada, Denmark, Finland, Germany, Greece, Ireland, Italy, Japan, Korea, Mexico, Netherlands, Norway, Portugal, Spain, Sweden, Switzerland, UK and USA.

or more. Total generating capacities of each country varied greatly, from Germany with 20,622 MW to Switzerland with about 12 MW. In the UK, during 2006, more than 630 MW of wind capacity was added, generating nearly 4.6 TWh of electricity.

The prospects for the global wind industry are promising. Even in a conventional scenario, the total worldwide installed wind power capacity could quadruple from 30 GW in 2003 to 160 GW by 2012. According to the IEA, the overall wind power capacity for the year 2006 reached 75 GW [5].

Regarding the wind turbine industry, the average rated capacity of new wind turbines installed in 2006 continued the trend towards larger installations. The average rating in 2005 increased to 1.6 MW, while in 2006 the average rating rose to nearly 1.7 MW. In addition to MW-scale wind turbines, intermediate sizes of 660–850 kW are being manufactured in several countries, for single-turbine installations or small wind power plants (Italy, United States). Small-sized turbines, less than 100 kW, are also being manufactured in Italy, Mexico, Spain, France, Sweden, UK, Denmark and the United States.

9.2 Advanced Wave Energy Technology

9.2.1 Technology Description

The oceans cover 75% of the world's surface and as such, ocean energy is a global resource. There are different forms of renewable energy potentially available in the oceans: inter alia waves, currents, thermal gradients, salinity gradients and tides. Ocean waves represent a form of renewable energy created by wind currents passing over open water. Ocean wave energy is captured directly from surface waves or from pressure fluctuations below the surface. Ways to exploit these high energy-dense resources are being investigated worldwide.

Considerable progress has been made over the past decade in this sector in Europe, resulting in some technologies being at, or near, commercialisation; others still require further research. Compared with other forms of offshore renewable energy, such as PV, wind or ocean current, wave energy is continuous but highly variable, although wave levels at a given location can be confidently predicted several days in advance. Ocean energy technologies are not yet economically competitive when compared to "conventional" RES technologies, such as wind energy technology.

Wave power varies considerably in different parts of the world, and wave energy cannot be harnessed effectively everywhere (see Fig. 9.8) [6]. Areas rich in wave power include the western European coast, the coasts of Canada and USA and the southern coasts of Africa, Australia and South America. According to resource studies, the area of the north-eastern Atlantic (including the North Sea) has a wave power resource of about 290 GW and the Mediterranean of 30 GW [7].

For assessment of wave resources, several methods have been established which mainly provide the height of the waves, and in some cases, their period and direction.

In Table 9.2, these methods and the respective types of information provided are shown [9].

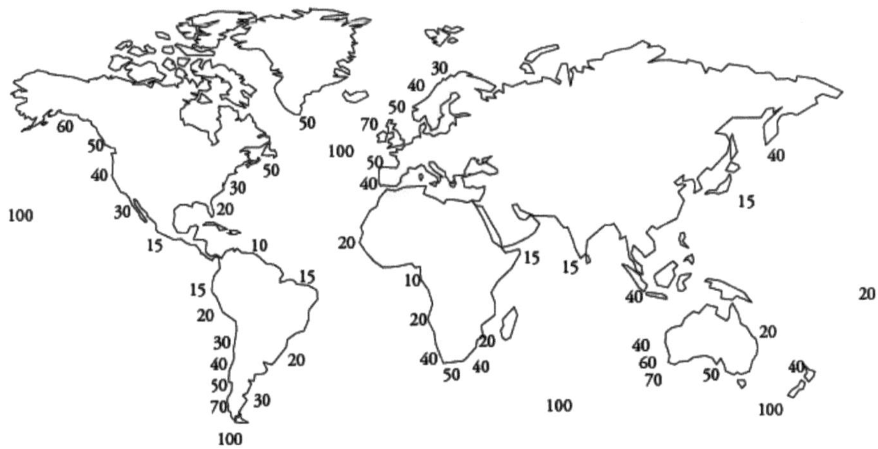

Fig. 9.8 Approximate global distribution of wave power levels. Power expressed in kW/m Source: ETSU [8]

Table 9.2 Methods of measuring ocean waves

Method	Type of data provided		
	Height	Period	Direction
Buoys	Yes	Yes	Some
Satellite	Yes	No	No
Visual observation	Yes	Yes	Yes
Hindcasts	Yes	Yes	Yes

The common measure of wave power, P, is

$$P = \frac{\rho \cdot g^2 \cdot H^2 \cdot T}{32\pi} \tag{9.2}$$

where, P: wave power, watt per meter (W/m) of crest length; ρ: density of seawater (around 1,025 kg/m^3); g: acceleration due to gravity (9.8 m/s^2); T: period of wave (s); H: wave height (m).

Several governmental and private research programmes have been established, mainly in the UK, Portugal, Ireland, Norway, Sweden and Denmark, aimed at the development of industrial/commercial wave power conversion devices [10].

A variety of technologies have been proposed to capture the energy from waves. Some of the more promising designs are undergoing demonstration testing on a commercial scale.

Wave energy devices are usually classified by the distance of the installation from the shore. Thus, there exist *Shoreline* devices, *Nearshore* devices and *Offshore* devices. Offshore systems are situated in deep water, typically of more than 40 m.

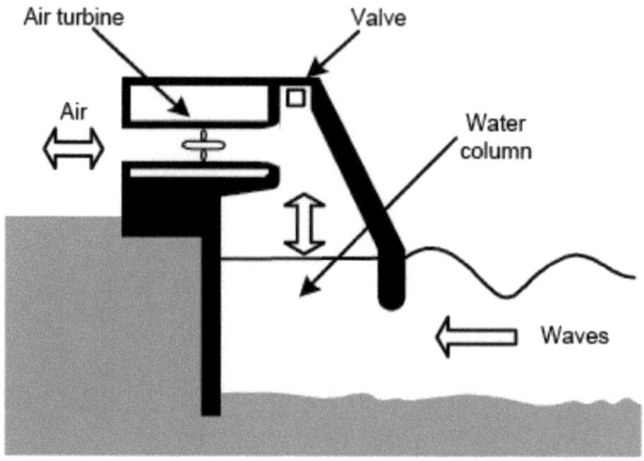

Fig. 9.9 OWC device

9.2.1.1 Shoreline Devices

Shoreline devices are fixed to or embedded in the shoreline, having the advantage of easier installation and maintenance. Furthermore, shoreline devices do not require deep-water moorings or long lengths of underwater electrical cable. However, they usually experience a much less powerful wave regime. Some of the most well-known shoreline devices are presented below.

Oscillating Water Column (OWC)

The OWC device comprises a partly submerged concrete or steel structure, open below the water surface, inside which air is trapped above the free surface of the water (see Fig. 9.9). The oscillating motion of the internal free surface, produced by the incident waves, makes the air flow through a turbine that drives the electrical generator [11]. The axial-flow Wells turbine, invented in the late 1970s, has the advantage of not requiring rectifying valves. It has been used in almost all prototypes. Several OWC prototypes have been built on the shorelines of Norway, China, Scotland (LIMPET, 500 kW nominal power) and Portugal (Pico Island, 400 kW nominal power) [12] or incorporated into a breakwater (in the harbour of Sakata, NW Japan) [11–13]. Another wave energy plant based on an OWC device is the one which was installed in Vizhinzam, India in 1990. The plant is shown in Fig. 9.10 [14].

Pendulor

The Pendulor device consists of a rectangular box which is open to the sea at one end. A pendulum flap is hinged over this opening, so that the action of the waves causes it to swing back and forth (see Fig. 9.11). This motion is then used to power a hydraulic pump and a generator. A 15 kW prototype has been tested in Muroran, Japan [11].

Fig. 9.10 The OWC in India

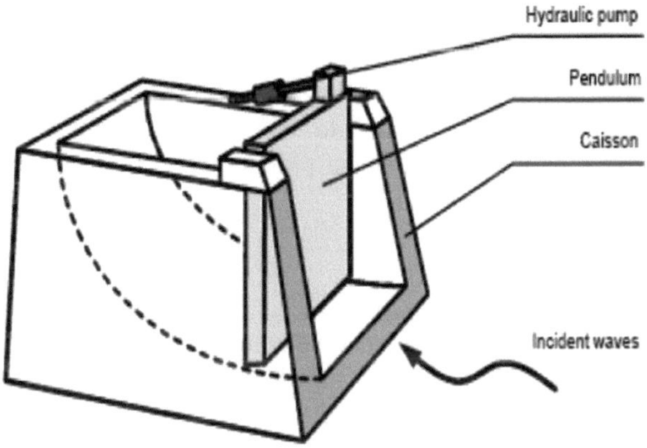

Fig. 9.11 Pendulor device

Tapchan Device

Another shoreline device mentioned in published literature is the *Tapchan device (tapered channel system)*.This device consists of a reservoir built into a cliff a few meters above sea level (see Fig. 9.12) [11]. A gradually narrowing channel, with wall height above mean water level, leads into the structure. Incoming waves increase in height as they move up the channel, eventually overflowing the lip of the channel and pour into the reservoir [10]. Water stored in the reservoir is then used to move a turbine and produce electricity.

9.2.1.2 Nearshore Devices

These devices are deployed at moderate water depths, in a range of 20–30 m, at distances of approximately 500 m from the shore.

Fig. 9.12 A Tapchan device in Norway, 1986

Wave Energy Point Absorber

Point absorbers are floating structures with components that move relative to each other due to wave action. The wave motion is converted to high pressure hydraulic energy by a floater driving a piston system, anchored to the seabed (Fig. 9.13) [11]. The piston system pressurises seawater to around 200 bar, which is transferred to the shore to drive a hydraulic motor and produce electricity, or used for pressure driven desalination units (e.g. RO) [7].

WaveRoller

The WaveRoller device is a plate, submerged at moderate water depths and anchored in vertical position on the seabed. The movement of the bottom waves moves the plate and the kinetic energy produced is collected by a piston pump. This energy can

Fig. 9.13 Energy point absorber

be converted to electricity, by the use of a generator linked to the WaveRoller plate or by a closed hydraulic system in combination with a generator/turbine system [7]. The device is developed by a company in Finland.

9.2.1.3 Offshore Devices

Offshore devices exploit the more powerful wave regimes available in deep water, of more than 40 m depth. More recent designs concentrate on small, modular devices, yielding high power output when deployed in arrays. Some of the more promising offshore wave converters are presented below.

Wave Dragon

The Wave Dragon is a floating offshore wave energy converter of the overtopping/run-up type (Fig. 9.14) [15]. Two wave reflectors focus the waves onto a patented double curved ramp. The waves run up the ramp and over the top into the reservoir. The resulting hydraulic head is utilised for power production through a number of propeller turbines. The Wave Dragon was developed by a group of companies in Denmark. Moreover, the company announced the development of a multi-MW demonstration project. It involves a two-stage development, financing, construction and operation of up to 77 MW of wave-generated electricity in Wales [7].

Pelamis

Pelamis is a semi-submerged, articulated structure composed of cylindrical sections, linked by hinged joints [16]. The wave-induced motion of these joints is resisted by hydraulic rams that pump high-pressure oil through hydraulic motors via smoothing

Fig. 9.14 The wave dragon
Source: NREL

Fig. 9.15 Pelamis device
Source: NREL

accumulators (see Fig. 9.15) [17]. The hydraulic motors drive electrical generators to produce electricity. Power from the joints is fed down a single umbilical cable to a junction on the seabed. Several devices can be connected together and linked to shore via a single subsea cable. Figure 9.15 shows a Pelamis device [17].

Pelamis was developed by a company in the UK. In 2004, a 750 kW commercial scale prototype was installed at the European Marine Energy Centre in Orkney. The prototype was 12 m long, 3.5 m in diameter and contained three 250 kW power modules. Since then three 750 kW Pelamis devices have been installed at the Aguçadora Wave Park in Portugal [7]. Other plans for wave farms include: a 3 MW array of four 750 kW Pelamis devices in the Orkneys, off the northern coast of Scotland, and the 20 MW Wave hub development off the north coast of Cornwall, England [18].

Archimedes Wave Swing (AWS)

The AWS consists of an upper part (the floater) of an underwater buoy that moves up and down in the wave, and a lower part (the basement or pontoon) which stays in position. The periodic changing of pressure in a wave initiates the movement of the upper part. The floater is pushed down under a wave crest and moves up under a wave trough. To be able to do this, the interior of the system is pressurized with air and serves as an air spring. The air spring, together with the mass of the moving part, is resonant with the frequency of the wave (see Fig. 9.16) [19]. The mechanical power required to damp the free oscillation is converted to electrical power by means of a Power Take Off system (PTO) (Fig. 9.17) [20]. The AWS was originally developed by a company in the Netherlands. In 2005 a 2 MW prototype was installed in Portugal. The system was tested for 7 months. During this period the system supplied the 15 KV local grid and demonstrated its control and reliability.

Power Buoy

The system uses an oceangoing buoy to capture and convert wave energy into electricity, via a patented power take-off. The produced power is transmitted to shore

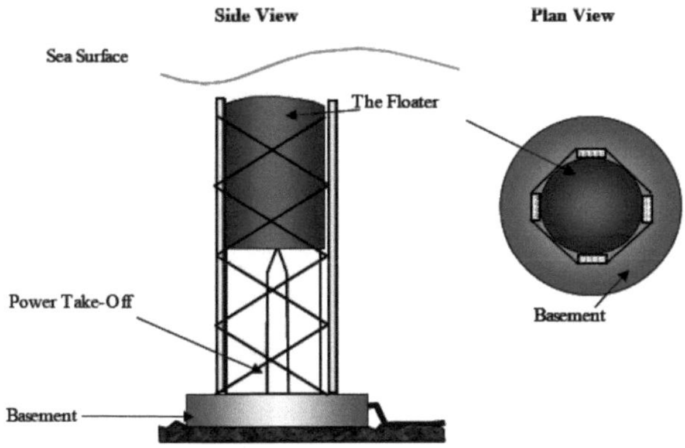

Fig. 9.16 Outline of AWS

via an underwater power cable. In case of extreme waves, the system automatically locks up and stops power production. The PowerBuoy is a point absorber and was developed by a company in the USA [21]. In June 2004, a 40 kW unit was installed off the coast of Oahu, Hawaii, while in October 2005 another 40 kW demonstration unit was put in operation in Atlantic City, New Jersey [22].

Other wave energy absorber devices are the Wavebob and the AquaBuOY (Fig. 9.18) [23]. The Wavebob comprises a wave energy absorber and a hydraulic power take-off system driving synchronous alternators. The absorber is an axisymmetric, compound, self-reacting oscillator operating primarily in the heave mode.

In AquaBuOY, the energy transfer takes place by converting the vertical component of wave kinetic energy into pressurised seawater, by means of two-stroke hose pumps. Pressurised seawater is directed into a conversion system consisting of a turbine driving an electrical generator. The power is transmitted to shore by means of a secure, undersea transmission line.

Fig. 9.17 Archimedes wave swing
Source: NREL

Fig. 9.18 AquaBuOY
Source: NREL

The offshore wave energy industry has also concentrated on small-scale devices. One of the best known is the McCabe wave pump. The device consists of three narrow, rectangular steel pontoons, which are hinged together across their beam. These pontoons are aligned so that their longitudinal direction heads into the incoming waves and they can move relative to each other amongst the waves. The basic aspect of the device is the damper plate attached to the central pontoon, which increases the inertia of the central pontoon and ensuring that it stays relatively still. Therefore, the fore and aft pontoons move relative to the central one, by pitching about the hinges. Energy is extracted from the rotation about the hinge points by linear hydraulic rams, mounted between the central and two outer pontoons, near to the hinges. Control of the characteristics of the hydraulic system allows the device to be tuned to the prevailing sea state and so optimise energy capture.

9.2.2 Commercial Applications and Economics

As presented in the previous paragraph, several wave energy converters have been developed, most of them as prototype demonstration units. However, there are plans for the installation of large wave farms, of capacity higher than 3 MW.

The predicted electricity generating cost from wave energy converters has shown a significant improvement in the last 20 years, and has reached an average price of 0.08–0.1 €/kWh. Compared with the average electricity price in the European Union, which is approximately 0.04 €/kWh, the price of electricity from wave energy is still high, but is forecast to decrease with further technological development [7].

9.2.3 Current Market

Several EU countries, Japan, India, Canada, Russia, USA and others have been involved in the development of wave energy conversion. During the last 5 years interest in wave energy in Europe has been growing significantly. Several wave energy companies have been involved in the development of new wave energy schemes, such as the previously described Pelamis, Archimedes Wave Swing and Wave Dragon.

Among the shoreline devices, the Oscillating Water Column (OWC) is presently the most established technology. An example of this is the Land Installed Marine Powered Energy Transformer (LIMPET), installed by a Scottish wave energy developer on the island of Islay, Scotland. According to the company, the current Limpet device was installed in 2000 and produces power for the national grid.

A company based in Greece has developed a wave energy point absorber device (nearshore device). The device has completed full-scale tests in depths of 10–20 m, and the company provides its systems to the market for electricity production, or for fresh water production using desalination [24].

In the offshore devices category, two examples are the Pelamis wave power and the PowerBuoy. In Septmeber 2008, Pelamis has installed the first commercial wave farm in Agucadoura, located off the coast of northern Portugal. The farm employs three wave energy converters, snakelike, semi-submerged devices that generate electricity with hydraulic rams driven by waves. This first phase of the new renewable energy farm is rated at 2.25 MW with 3 machines, and the second phase will add an additional 25 machines to bring the capacity to 21 MW, enough to power 15,000 homes [25]. Furthermore, the installation of a 1.25 MW wave farm-utilising PowerBuoy technology is in development on the northern coast of Spain. The same company has announced the installation of a 100 MW wave farm in the UK [7].

R&D on wave energy is underway in several countries worldwide, even in countries with poor wave energy resources (e.g. Sri Lanka and Mexico). Regarding European countries, the UK, Ireland and Portugal seem more active in the field of wave energy. The first step in improving collaboration between wave energy developers, academia and the electricity industry was taken by the European Commission in 1999 with the formation of the European Thematic Network on Wave Energy, under the 5th Framework Programme [19].

9.3 Wind/Reverse Osmosis Technologies Combination

Renewable energy systems, mainly wind and solar energy, can be coupled with desalination systems in order to provide the necessary energy input, which in itself may become a significant contribution in remote (off-grid) and arid areas. In addition, the desalinated water can be used as a temporary energy storage, thus providing a means for the "regulation" of one of the most important inherent characteristics of RES, i.e. their intermittence. When the system is grid-connected, the desalination plant can operate continuously as a conventional plant and the renewable energy

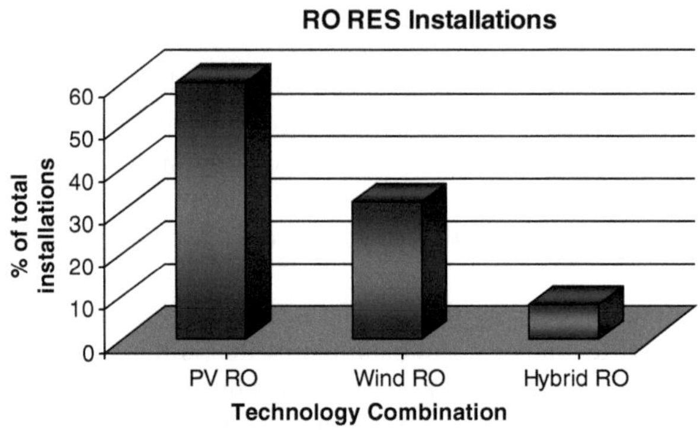

Fig. 9.19 RO/RES installations

source merely acts as a fuel substitute. In such cases, the desalination load could also be used to moderate the amount of energy injected into the grid. This can be a very useful concept for hybrid electricity systems, with high use of intermittent renewable electricity sources.

Wind energy to drive Reverse Osmosis (RO) units is the second most used combination of RES/desalination technologies. As shown in Fig. 9.19, the majority of the installed RES/RO plants use the coupling of photovoltaics with RO, followed by wind/RO.

9.3.1 Wind/RO Technologies Description

The utilisation of wind turbines to drive Reverse Osmosis units is technically feasible. The major drawback to the combination of desalination processes and wind energy is the fluctuation of power supply generated by the wind turbines. In general, desalination systems were traditionally designed to operate with a constant power input; unpredictable and non-steady power input forces the desalination plant to operate in sub-optimal conditions. Up until now, only a few standalone, battery-less wind/RO units have been developed and studied.

However, such RES desalination systems operating under variable power conditions have not yet proven their performance and economic advantage. A battery system is typically used for medium-term storage, whereas flywheels can be used for short-term storage [26].

An important factor, especially in standalone systems, is the control and full automation of the system. In both grid and standalone systems, a special energy management system should be designed.

Standalone wind energy systems, to drive RO for seawater desalination, can be used in conjunction with other conventional or renewable power sources (e.g. diesel, photovoltaic); these are known as hybrid systems. A standalone Wind/RO system consists of the following equipment:

- Wind generator
- Charge controller
- Battery bank
- Inverter
- RO unit

The battery bank is used for power stability and as an energy supply during periods when wind energy is not sufficient to drive the desalination unit. Charge controllers are used for the protection of batteries from overcharging. The inverters are used to convert the DC current from the battery output, to AC for the load. A diesel generator (DG) backup can also be used to charge the battery bank or to drive the RO unit directly. A typical diagram of a hybrid standalone PV/Wind/DG/RO unit is presented in Fig. 9.20.

For the Reverse Osmosis process, energy recovery devices are used to recycle the energy in the pressurised reject brine, thereby improving the overall efficiency of the system. Large RO units utilise energy recovery devices, which can reduce energy consumption down to 2–2.5 kWh/m^3. Their use increases the initial cost of the system but effectively reduces the energy requirement.

Autonomous systems are mainly small scale. [2] The majority of small scale installations have so far been developed within research projects. In general, the aim of

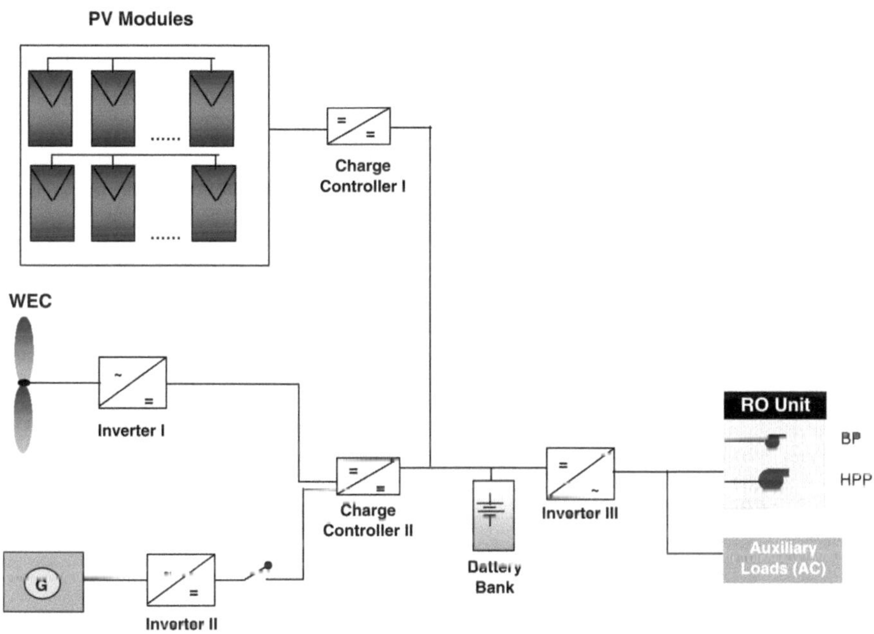

Fig. 9.20 A typical hybrid RO system

[2]For RES desalination, "small" systems are characterised by a capacity of up to 50 m^3/day, "medium" in the range of 50–150 m^3/day and "large" above 150 m^3/day.

each project is the evaluation of the combination of the technologies and the development of compact and reliable systems. The designs usually also allow the parallel production of water and electricity.

9.3.2 Wind/RO Applications

A number of units coupling wind turbines with RO desalination have been designed, installed and tested. Most of these installations were built within research or demonstration projects.

As early as 1982, a small system was set up in France (Ile du Planier), with a 4 kW turbine driving a 0.5 m^3/h RO desalination unit. The system was designed to operate either via direct coupling or with batteries. Other Wind/RO plants installed during the 1980s include a 6–9 m^3/day RO with a 6 kW Wind Turbine (W/T) on the island of Suderoog, Germany and in Mersa Matruh, Egypt, for the production of 25 m^3/day fresh water [27]. Table 9.3 presents some of the most well-known Wind/RO plants around the world.

Table 9.3 Wind energy driven reverse osmosis desalination plants

Location	RO capacity, (m^3/h)	Electricity supply	Year of installation
Ile du Planier, France	0.5	4 kW W/T	1982
Island of Suderoog, Germany	0.25–0.37	6 kW W/T	1983
Island of Helgoland, Germany	40	1.2 MW W/T and diesel	1988
Fuerteventura, Spain	2.3	225 kW W/T and 160 KVA diesel, flywheel	1995
Pozo Izquierdo, Gran Canaria, Spain SDAWES	8 units ×1.0	2 × 230 kW W/T	1995
Therasia Island, Greece APAS RENA	0.2	15 kW W/T, 440Ah batteries	1995/1996
Tenerife, Spain; JOULE	2.5–4.5	30 kW W/T	1997/1998
Island of Syros, Greece; JOULE	2.5–37.5	500 kW W/T, standalone and grid connected	1998
Keratea, Greece PAVET Project	0.13	900 W W/T, 4 kWp PV, batteries	2001/2002
Pozo Izquierdo, Gran Canaria Spain, AEROGEDESA project	0.80	15 kW W/T, 190Ah batteries	2003/2004
Loughborough University, UK	0.5	2.5 kW W/T, no batteries	2001/2002
Milos island, Greece[a] OPC programme	2 × 41	850 kW W/T, grid connected	2007
Island of Irakleia, Greece[a] OPC programme (Figs. 9.23 and 9.24)	3.3	30 kW W/T offshore, batteries	2007
University of Delft, Netherlands	0.2–0.4	Windmill, no batteries	2007/2008

[a]Both plants were developed within the Greek Ministry of Development's "Competitiveness" Programme.

Fig. 9.21 The RO unit at Pozo Izquierdo
Source: ITC

During the last decade, several interesting wind or hybrid RO units for seawater desalination have been installed and tested. A 15 kW wind turbine to drive a 0.80 m^3/h RO unit was installed in Pozo Izquierdo, Grand Canary, by the Instituto Tecnologico de Canarias, within the AEROGEDESA project (Figs. 9.21 and 9.22). The system started its operation in 2004 [28]. The project concerns the electrical coupling of a commercial wind turbine to an RO unit for seawater desalination, operating under a constant regime and managing storage and available wind energy use through a battery bank of minimal capacity [29]. A freshwater storage tank of 250 m^3 is also available. The whole system is fully automatic. The salinity of the

Fig. 9.22 The W/T at Pozo Izquierdo
Source: ITC

Fig. 9.23 View of the offshore Wind/RO unit in Irakleia island, Greece

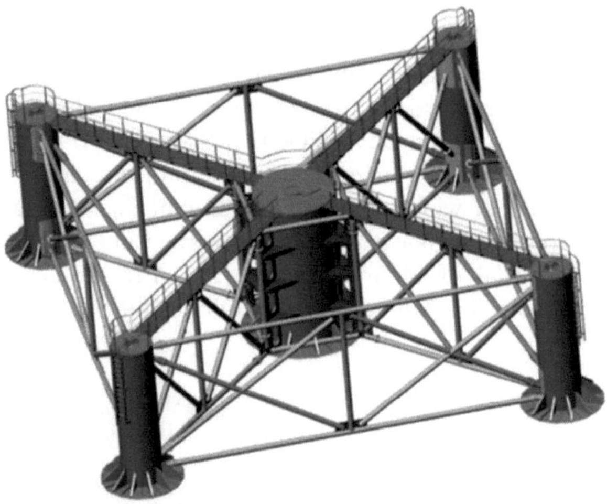

Fig. 9.24 Structure of the offshore unit

feed water is 35,500 ppm TDS while the salinity of the product water is less than 500 ppm TDS. The nominal operating pressure of the RO unit is 55 bar and the recovery ratio is 37%. The energy consumption of the RO unit is 7 kWh/m^3. The hours of operation of the RO unit depend on the availability of the wind; the wind speed in the area ranges from 4.5 to 13 m/s and on average the unit operates 15 h a day during winter and 20 h a day during summer. The unit water cost is estimated at 3–5 €/m^3.

In Greece, the Centre for Renewable Energy Sources (CRES) installed an autonomous hybrid (Wind/PV) RO unit for seawater desalination (Figs. 9.25-9.28).

Fig. 9.25 CRES 900 W W/T

The system is at pilot scale, operating in a closed water loop as seawater is not available at the test site [30].

The RO unit has a 130 l/h capacity and is driven by a hybrid system consisting of 4 kWp PVs and a 900 W wind turbine (W/T) (Fig. 9.27). The system also includes charge controllers, a battery bank and two inverters to drive the booster and the high-pressure pump for the RO unit (Fig. 9.28). The system has been in operation since 2001.

The most well-known, larger wind/RO plants are the plants on the islands of Syros, Greece, and Tenerife and Fuerteventura, Spain. The first two plants were installed as part of the European Commission (DG XII) JOULE research programme. The objective of the project was to develop the concept of a family of

Fig. 9.26 CRES 4 kWp PV unit

Fig. 9.27 CRES RO autonomous unit

Fig. 9.28 Battery bank of CRES hybrid unit

modular seawater desalination plants, adaptable to a broad range of regions and making use of the locally available wind energy resources. The family concept was based on a limited number of standardised modules. These modules (advanced technology wind energy converters, hybrid power units, variously sized seawater desalination units, designed to be capable for off-grid or grid-connected operation, etc.) were equipped with a highly flexible power conditioning and management system [31]. Additionally the concept allows for the parallel production of water and electricity, according to the operator's needs.

The plant on the island of Tenerife was the first to be built and consisted of a 30 kW W/T coupled to two RO modules. The plant on Syros is based on a 500 kW

Fig. 9.29 500 KW W/T on Syros
Source: [31]

W/T and eight RO modules with an output between 60 and 900 m^3/day of freshwater. The smaller unit on Tenerife was used to test some of the concepts, prior to the construction of the larger plant on Syros. At the Syros project, the 500 kW W/T is coupled to a grid management system which allows for electricity frequency and voltage control, and self-adaptation of the wind energy converter to the weak electricity grid of the island (Figs. 9.29 and 9.30). The electricity from the wind turbine is buffered in the energy storage system, which includes a diesel generator, batteries and a flywheel. The output is fed into the RO unit and electricity grid.

Fig. 9.30 View of the RO plant on the Island of Syros
Source: [31]

The RO unit is installed in five 12 m containers. The main container houses the purification plant and the rinsing system to flush the membranes after shut-down. It also houses the main electrical switchboard and control panels. The RO units are housed in two containers.

The number of RO modules allows the output of the plant to vary, according to the amount of energy available. Each RO module incorporates an innovative energy recovery system, which is based on a piston accumulator principle. A drinking water storage tank and a seawater storage tank are also part of the system. The seawater storage tank acts as an energy buffer too, as operation of the seawater pump is restricted to periods when high wind power is available. A similar principle applies to the drinking water storage tank: freshwater is only produced during periods of high wind power availability. Since the maximum power consumption of the desalination plant is 200 kW, and the nominal power of the wind generator is 500 kW, any energy surplus is fed into the island's electricity grid.

In recent years, following on from the operation of the above prototype plants, there has been a trend towards the development of larger plants, to cover basic water needs of remote locations, islands, etc. It appears that RES desalination technologies are now starting to gain ground, and are becoming more mature. Additionally, the market has already provided compact Wind/RO solutions on standard scales.

The Renewable Energy Development Centre (CDER) in Morocco has announced the installation of a Wind/RO plant in Akhfennir village [32]. Akhfennir is an Atlantic coastal village of about 4,000 inhabitants, with sufficient tourist and industrial potential. The area has significant wind potential, an arid climate, no water resources and no electricity grid. Drinking water is currently supplied by tankers from Tan Tan City, which is located about 100 km away. The cost of water is about 12 US$/m^3 while the demand for water is expected to be 850 m^3/day by the year 2010, and 1,430 m^3/day by 2020. The project is planning the development of a 650 kW Wind Turbine to drive a desalination unit of 850 m^3/d water capacity. A diesel generator will be also included. There are also plans for a new desalination RES project in Tan Tan city.

9.3.3 Wind/RO Market and Economics

The RES market has been steadily increasing over the last 20 years. Wind technology has been commercially accepted all over the world, and in a wide variety of applications, for some years now. In parallel, the seawater desalination market is well established with many potential applications worldwide. In recent years, a few manufacturers/suppliers, the majority of which are from the wind turbine industry, have been providing the market with compact Wind/RO solutions for electricity and water production. For instance, two German companies [33, 34] supply units, with capacities from 175 to 2,000 m^3/day, packed in containers [35]. The systems can operate on a standalone basis or connected to an electricity grid. A Danish company

[36] also provides turnkey solutions for freshwater production using wind turbines. The RO units are designed as containerised modules with production capacities from 10 to 3,800 m^3/day. Furthermore, several large Wind/RO projects, involving installations with a capacity of 1,200 m^3/day and more in Morocco, the United Arab Emirates and in Aqaba, Jordan, have been announced. The foreseeable increase in the RES desalination market is expected to raise competitiveness and reduce the cost of such systems.

The cost of water produced from RES-driven RO systems varies considerably, as it depends heavily on the RES potential of the plant's location. Furthermore, the overall system design (size of the system, use of energy recovery devices, type of materials used, etc.) significantly affects the final cost.

An economic analysis of wind-powered desalination systems, performed by the La Laguna University in Tenerife, Spain [37] examines the influence of different parameters on the levelised cost of the water produced. These parameters are: plant capacity, climatic conditions, energy requirements, economic and financial parameters. However, for applications already in existence, most of which are pilot and demonstration projects, the cost ranges from 2.5 up to 10 € /m^3 [28]. In Mallorca this is comparable with the cost of water from other sources; during times of drought, water imported by boat reaches a cost of 2 €/m^3, while the transfer cost of water in the Greek islands ranges from 2.2 to 7 €/m^3. The cost of water from wells in Aqaba, Jordan, approaches 1.7 €/m^3 [30].

9.4 Wave/Reverse Osmosis Technologies Combination

The use of wave energy for desalination is mainly realised via power generation using pump-turbine systems, where power can be continuously produced using alternators.

9.4.1 Wave/RO Technology Description

A system operating using the above concept was installed in India [14] and consists of a turbine, alternator, inverter, battery bank and an RO unit for seawater desalination. Similar to other RO units driven by RES, a storage mechanism is necessary to smooth fluctuations of the RE power, as well as to provide some degree of autonomy.

An example of wave energy driven desalination is the DelBuoy RO concept, which was developed about 20 years ago. DelBuoy is an example of a wave-powered pumping system, which uses oscillating buoys to drive piston pumps anchored to the seabed. These pumps feed seawater to submerged RO modules [38]. The unit produced around 1.1 m^3/day of potable water under specific site conditions. [3] During

[3]Specific site conditions: average wave height: 0.6–1.5 m, average wave period: 3–8 s and water depth: 15–20 m.

the 1980s, several prototypes were tested in order to confirm laboratory results and evaluate the performance of system components under actual operating conditions. A commercial demonstration project on the southwest coast of Puerto Rico took place, however the project was not continued [9]. According to published literature, the system was capable of producing water at a cost of 5.25 $/m^3$ [39].

The McCabe wave pump is another example using a wave-powered pumping system to drive RO units. The water production capacity of the system is 750 m^3/day, at an operating pressure of 70 bar and with a water recovery ratio of 35% [9]. For wave heights of 1.5 m, the output is 260 m^3/day of fresh water and 30 kW of electricity. The final cost of the produced water is estimated at 1.85 $/m^3$. A prototype device [9] was constructed in Shannon, Ireland. The device was in operation for 4 years, during which time it survived a number of storms. These trials led to the optimisation of the geometry of the full-size device, which will be about 20% longer than the prototype.

Another concept of wave-powered RO technology is known as the *water hammer* effect or unsteady incompressible duct flow. [4] This effect occurs when the velocity of a fluid flowing in a pipe is changed, for instance, by the rapid closing of a valve. The concept was investigated by Sawyer and Maratos [40] who showed, by theoretical estimation, that it is feasible to use the water hammer effect to provide sufficient pressure to drive an RO unit [41]. Maratos also investigated the possibility of using the hydro-ram (typically used to pump small amounts of water) to pump large quantities of seawater required for desalination [42]. The hydro-ram is a pumping device based on the water hammer effect that produce as much as 160 times the supply head; it has been used to pump water to heights of at least 180 m. This hydrostatic head can then be used to drive an RO unit.

A similar concept is the Tapchan device, which is simply a tapered channel that increases the height of the waves, channelling the flow into a raised reservoir. This provides a static pressure head that can be used to create a one-way flow of seawater. The reservoir is in turn used to drive a hydro-ram.

9.4.2 Wave /RO Applications

There are only few installations of wave/RO applications. As mentioned in Sect. 9.4.1, DelBuoy was the first reported wave RO installation, installed in Puerto Rico during the 1980s. An example of a wave energy-driven RO desalination plant actually under operation, is the one in Vizhinjam in India. The plant was constructed in 1990 to demonstrate the conversion of wave energy into electricity using an OWC device [14]. In 2004, an RO plant of 10 m^3/day product water capacity was commissioned at this site, in order to produce fresh water using wave energy. Figures

[4]The effect that occurs when the velocity of a fluid flowing in a pipe is changed, for instance by the rapid closing of a valve. This rapid change in velocity creates a pressure wave, which theoretically can be used to drive useful work.

Fig. 9.31 The Impulse turbine, Vizhinjam plant

9.31, 9.32 and 9.33 show the turbine, the alternator and the RO unit of the Vizhinjam plant. A schematic of the wave/RO plant in Vizhinjam is shown in Fig. 9.34 [14].

The wave power rotates a turbine connected to an 18 kW variable speed, permanent magnet, brushless alternator. The alternator produces AC current, which via a rectifier, is converted to a constant DC voltage to charge the battery bank (120 V/300 Ah, VRLA type).[5] An inverter is used to convert the DC voltage from the batteries

Fig. 9.32 View of the alternator, Vizhinjam plant

[5]VRLA: Valve Regulated Lead Acid.

Fig. 9.33 The RO unit at Vizhinjam

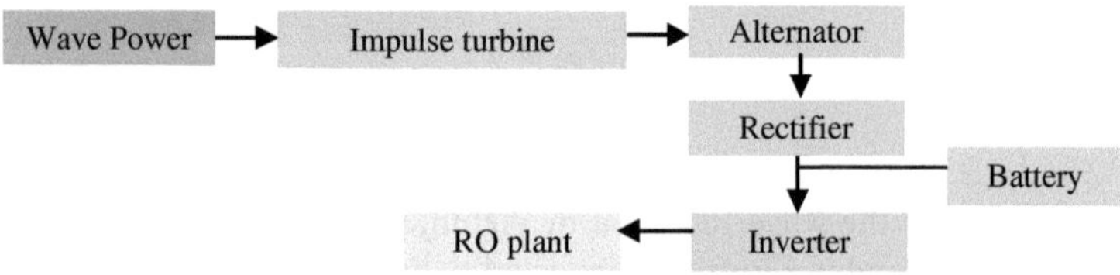

Fig. 9.34 Scheme of the wave/RO plant in Vizhinjam, India

to AC for the RO unit. The feedwater to the RO unit has a salinity of 35,000 mg/l, while that of the product water is less than 500 mg/l. The system has been reported to be operating satisfactorily [14].

In 2005, a Greek company announced the installation of a Wave/RO prototype. The company developed an innovative wave energy device based on a patented point absorber. The wave energy converter is directly connected to an RO unit, eliminating the need for a high-pressure pump, which is the main load of the unit [15].

9.4.3 Wave/RO Market and Economics

At present the market for RO desalination driven by wave energy systems is modest, mainly due to the rather limited development of wave technology. It is also noted that although most of the manufacturers/suppliers of wave converters support the use of their products in conjunction with RO desalination units, there is very little, if any, detailed information provided on the coupling of these two technologies.

There are very few sources of economic data regarding Wave/RO desalination. As mentioned previously (Sect. 9.4.1), the cost of water produced by wave/RO units ranges from 1.85 to 5.25 $\$/m^3$.

However, the cost of water from Wave/RO plants is expected to decrease, depending on the progress of wave energy technology.

9.5 Final Remarks

Renewable energies still only play a small role in desalination. However, significant water needs exist in areas with high renewable energy potential and with limited availability of conventional energy. In combination with this, the continuously rising environmental concerns and the dramatic increase of fossil fuel prices, means that water desalination around the world should be increasingly powered by wind, solar and other clean energy resources. Such environmentally-friendly systems are now becoming available at a reasonable cost. The lessons learnt from installed Wind/RO units have hopefully been passed on and incorporated into plants currently being built and tested. Wave energy technologies, although not yet economically competitive compared with other mature renewable energy technologies, in the medium-term could become significant providers of electricity, in areas adjacent to the resource. In the longer-term, wave energy could become a much more important part of the world's energy portfolio.

Both types of renewable energy sources, wind and wave, have huge potential to be exploited and can be increasingly used to drive desalination systems to meet the growing demand for freshwater in many parts of the world.

Acknowledgments The author would like to acknowledge the contribution of Dr K. Perrakis from the Regulatory Authority for Energy (RAE), Greece.

Abbreviations

AC	alternating current
AWS	Archimedes wave swing
BP	booster pump
DC	direct current
DG	diesel generator
HAWT	horizontal axis wind turbine
HPP	high-pressure pump
ITC	Instituto Tecnologico de Canarias, Spain
IEA	International Energy Agency, Paris, France
LIMPET	land installed marine energy transformer
OSW	oscillating water column
ppm	parts per million
PV	photovoltaic
R&D	research and development
RES	renewable energy sources
RO	reverse osmosis
TDS	total dissolved solids
VAWT	vertical axis wind turbine
VRLA	valve regulated lead acid
W/T	wind turbine
CRES	Centre for Renewable Energy Sources, Pikermi, Greece

EERE Energy Efficiency and Renewable Energy, US Department of Energy
NREL National Renewable Energy Laboratory, Colorado, USA

References

1. www.riso.dk
2. Hernandez-Gonzalez C., Victor Olmos Garcia et al., Basic Aspects for Application of Wind Energy, IDEA, THERMIE Programme, (1994)
3. Energy Efficiency and Renewable Energy, Wind and Hydropower Energy Technologies Program, US Department of Energy, www.eere.energy.gov
4. American Wind Energy Association, Wind Power Today, AWEA, (2007)
5. IEA Wind Energy Annual Report, ISBN 0-9786383-1-X, (2006)
6. Nambudripad G., Riding, The Waves: A Look at the European Marine Energy Sector, Frost and Sullivan
7. Ocean Energy Conversion in Europe, Recent Advancements and Prospects, Centre for Renewable Energy Sources, Centre for Renewable Energy Sources, CRES, Greece, (2006)
8. Thorpe T.W., An Overview of Wave Energy Technologies: Status, Performance and Costs, Wave Power: Moving towards Commercial Viability, Broadway House, Westminster, London, ETSU, (1999)
9. Davies P.A., Wave-Powered desalination: resource assessment and review of technology, Desalination **186**, 97–109, (2005)
10. Anibal T. de Almeida, Desalination with Wind and Wave Power, Desalination for the 21st Century, pp. 305–325, Springer (2007)
11. WaveNet, European Community, ERK5-CT-1999-20001, (2003)
12. www.ist.utl.pt
13. Raju V.S., Ravindran M., Wave energy: potential and programme in India, Renewable Energy, **10**(2/3), 339–345, (1997)
14. Sharmila N., Jalihal P., et al., Wave powered desalination system, Energy **29**, 1659–1672, (2004)
15. www.wavedragon.net
16. www.oceanpd.com
17. Carcas M., Ocean Power Delivery Ltd, The Pelamis Wave Energy Converter, www.oceanpd.com
18. http://en.wikipedia.org
19. Options for the Development of Wave Energy in Ireland A Public Consultation Document, Marine Institute- Sustainable Energy Ireland, NPD, (2002)
20. Thorpe T., A Brief Overview of Wave & Tidal Energy, (2003)
21. www.oceanpowertechnologies.com
22. Technology White Paper On Wave Energy Potential on the U.S. Outer Continental Shelf, Minerals Management Service Renewable Energy and Alternate Use Program U.S. Department of the Interior Available for Downloading at http://ocsenergy.anl.gov, (2006)
23. Robinson M., Renewable Energy Technologies for Use on the Outer Continental Shelf, Ph.D. National Renewable Energy Lab, NREL, (2006)
24. www.wave-energy.gr
25. www.pelamiswave.com
26. Folley M., Penate Suarez B., Whitteker T., An autonomous wave-powered desalination system, Desalination **220**, 412–421, (2008)
27. Baltas P., Tzen E., Perrakis K., et al., Desalination Guide Using Renewable Energy Sources, CRES, Greece, ISBN 960-90557-5-3, (1998)
28. ADU RES Project, INCO Programme, MPC-1-50-9093, Autonomous Desalination Units Using RES, WP2 Report, (2005)

29. Papapetrou M., Epp C., Tzen E., Autonomous Desalination Units based on RE Systems-A Review of Representative Installations Worldwide, Solar Desalination for the 21st Century, pp. 343–353, Springer (2007)
30. Tzen E., Theofilloyianakos D., Kologios Z., Autonomous reverse osmosis units driven by RE sources experiences and lessons learned, Desalination **221**, 29–36, (2008)
31. Morris R., Renewable Energy Powered Desalination Systems in the Mediterranean Region, United Nations Educational, Scientific and Cultural Organization, UNESCO, (1999)
32. Enzili M., Wind Energy in Morocco, Potential and Projects, CDER, presented in EWEC Conference, Athens, Greece, (2006)
33. www.enercon.de
34. www.synlyftsystems.de
35. Paulsen K., Hensel F., Design of an autarkic water and energy supply driven by RE using commercially available components, Desalination **203**, 455–462, (2007)
36. www.danvest.com
37. Garcia-Rodriguez L., Romero-Ternero V., Gomez-Camazo C., Economic analysis of wind-power desalination, Desalination **137**, 259–265, (2001)
38. Hicks D., Pleass Ch., et al., DELBUOY: Ocean wave-powered seawater reverse osmosis desalination System, Desalination **73**, 81–94, (1989)
39. Miller J., Review of Water Resources and Desalination Technologies, SAND 2003-0800, (2003)
40. Sawyer R.A, Maratos D.F, Wavepower fro seawater desalination using unsteady incompressible duct flow (water hammer), Mediterranean Conference on Policies and Strategies for Desalination and Renewable Energies, Santorini, Greece, (2000)
41. Sawyer R.A., Maratos D.F., An investigation into the economic feasibility of unsteady incompressible duct flow (water hammer) to create hydrostatic pressure for seawater desalination using reverse osmosis, Desalination **138**, 307–317, (2001)
42. Maratos D.F., Technical Feasibility of wavepower for seawater desalination using the hydroram (Hydram), Desalination **153**, 287–293, 2002

Chapter 10
Operating RE/Desalination Units

Michael Papapetrou, Essam Sh. Mohamed, Dimitris Manolakos, George Papadakis, Vicente J. Subiela, and Baltasar Peñate

Abstract This chapter presents 10 small standalone RE (renewable energy)/ desalination systems operating around the world, employing different technologies, such as PV/RO, solar/MED, etc. The examples show that several technological combinations are well enough developed to provide potable water under harsh conditions in isolated sites. However, even established technologies face problems and limitations. Continuous R&D in combination with wide scale implementation is needed to improve their reliability.

The cost of the produced water is still quite high, but the decreasing cost of RE equipment, and experience from RE/desalination implementation is driving the cost down. At the same time, the cost of conventional water supply is increasing, especially in isolated sites where water is transported by ships or trucks, and thus the cost is directly related to the oil price. Therefore RE/desalination is becoming competitive at more sites and the different technologies are competing to be ready for this potential market.

10.1 Introduction

This chapter is dedicated to the presentation of selected desalination systems powered by renewable energy (RE), installed and operating in various locations around the world. At least one example of each technological combination described in previous chapters of this book has been included. Concrete examples, which highlight the maturity of each technology and allow a better overview of the available options, are given, comparing their strengths and weaknesses.

All systems presented here are standalone installations, producing potable water via desalination of brackish water or seawater, independently of the electric grid

M. Papapetrou (✉)
WIP-Renewable Energies, Sylvensteinstr. 2, 81369, Munich, Germany
e-mail: michael@papape.com

A. Cipollina et al. (eds.), *Seawater Desalination*, Green Energy and Technology,
DOI 10.1007/978-3-642-01150-4_10, © Springer-Verlag Berlin Heidelberg 2009

and powered only by renewable energy sources. Most installations are small, with a production capacity from a few cubic meters up to a few hundred cubic meters per day. Some of the examples are from experimental projects and consist of units operating with artificial seawater within research institutes, their performance being continuously monitored. Others are operating in remote locations, actually providing local populations with water for their everyday needs.

Overall, the involvement of research institutes, in all systems presented in this chapter, is indicative of the status of the technology. RE/desalination is still taking its first steps in the market, with systems striving to demonstrate that the technology is well enough developed to be used in the harsh conditions of remote and isolated sites, where it is needed the most.

In Sect. 10.2 selected examples are presented, grouped under the following technology combinations:

- Solar/Multi-effect Distillation (MED)
- Solar/Humidification-Dehumidification
- Organic Rankine Cycle (ORC)/Reverse Osmosis (RO)
- Solar/Membrane Distillation (MD)
- PV/ Reverse Osmosis
- Wind/ Reverse Osmosis
- Hybrid PV and Wind/ Reverse Osmosis
- Wave/Reverse Osmosis

Section 10.3 is dedicated to the analysis of relevant costs and the factors that affect them, while Sect. 10.4 gives some general data about operating RE/desalination systems around the world and trends in the market. Final remarks and conclusions are summarised in Sect. 10.5.

10.2 Selected Operating Systems Worldwide

10.2.1 Solar Thermal Multi-Effect Distillation

The system was installed in 2004, as part of the AQUASOL project (FP5-EVK1-CT2001-00102), in the CIEMAT-Plataforma Solar de Almería premises in Spain. CIEMAT-Plataforma Solar de Almería is responsible for the installation. It consists of the following main elements [1]:

- A multi-effect distillation plant with 14 cells in vertical arrangement
- A stationary Compound Parabolic Concentrator solar collector field
- A thermal water-based storage system
- A double effect (LiBr–H_2O) absorption heat pump
- A smoke-tube gas boiler

The system is a pilot plant with a nominal daily fresh water production of 3 m^3/h, a thermal energy consumption of 57.5–70.4 kWh/m^3 and a performance ratio of around 11.

The operating principle is the following: water is heated as it circulates through the solar collectors and is then directed to the thermal storage system. From the storage tank the hot water is taken to provide the MED system with the required heat. When the solar energy is insufficient, the gas boiler provides heat to drive the system. As opposed to conventional MED plants, this system operates with hot liquid water as the heat transfer medium, rather than steam, within the first effect.

The MED part of the system has actually been operating since 1988, but connected to a different energy system; the current installation is an improved version of the previous configuration. Table 10.1 shows the main specifications of the MED system [1].

The energy system can be seen in Figure 10.1 and consists of a solar field with 252 AoSol 1.12 × CPC (Compound Parabolic Concentrator) collectors, with a total area of 500 m^2. The technical characteristics of each collector are given in Table 10.2. There are four rows, each with 63 collectors and the total flow rate of 14.97 m^3/h is equally distributed between the four rows.

The system can operate with solar energy as the only source of thermal energy and this operating mode has been tested and demonstrated. However, there is also a double effect heat pump, powered by a gas boiler, which can be used either continuously, with a 30% minimum contribution, or only when there is not enough solar radiation to operate the system.

For the solar-only operating mode, the average daily water production ranges from 20 to 30 l for every square meter of solar collector surface. So the system, using only solar energy, produces on average 10–15 m^3/day of distillate water. It can produce up to 72 m^3/day of water in hybrid operating mode [2].

This MED system is operating as an experimental unit and provides valuable experience on the use of solar energy in thermal desalination. It also demonstrates

Table 10.1 Overview of the multi-effect distillation system

Distillate production	3 m^3/h
Output salinity	500 ppm TDS
Number of effects	14
Heat source energy consumption	190 kW
Vacuum system	Hydro ejectors (seawater at 3 bar)
Maximum brine temperature	70°C

Table 10.2 Solar thermal supply of the multi-effect distillation system

Dimensions	2,012 × 1,108 × 107 mm
Aperture area	1.98 m^2
Operating pressure	6 bar
Optical efficiency	0.70–0.71
Thermal loss factor	3.4 W/m^2K

Fig.10.1 The solar thermal field of the solar MED plant in Almeria

that there are economies of scale to be exploited in this technology and that hybrid systems using natural gas can substantially improve the economics. Some North African countries have expressed interest in installing thermal desalination systems powered by concentrated solar power and natural gas.

10.2.2 Solar Thermal Multi-Effect Humidification System

Within the framework of the ADIRA project (www.adira.gr), a demonstration plant based on the multi-effect humidification (MEH) technology supplied by solar thermal energy, was installed in the Municipality of Geroskipou (Paphos district, Cyprus). The system was installed by the Solar & Other Energy Systems Laboratory (NCSR "DEMOKRITOS") in the new municipal sports centre, to demonstrate the desalination concept and to gain experience from its implementation and operation under real conditions.

The system consists of a desalination unit, a thermal solar system and auxiliary parts for seawater supply and freshwater post-treatment processes. The desalination unit was provided by the German/Austrian water management company TiNOX-MAGE.

The main operating principle is shown in Figure 10.2 and can be described as follows: The seawater is heated through highly corrosion-resistant heat exchangers that transfer the heat from the sun. The seawater then enters a corrosion-free evaporation chamber. In this evaporation chamber, it evaporates from efficient anti-bacterial fleece surfaces. The produced steam is transported to the condenser without the need for any additional energy. During condensation, the majority of the energy used for evaporation is recovered via the application of materials with extremely low heat transfer resistance.

The collectors are on the roof of the main swimming pool building. They are placed with a 45° inclination using appropriate mountings in parallel rows, oriented south-west. The solar collectors have a total surface area of 96 m^2, a selective absorber surface and an efficiency of at least 45%, for solar radiation of 1,000 W/m^2. The difference between the fluid temperature in the collector and the ambient temepratue is about 50°C.

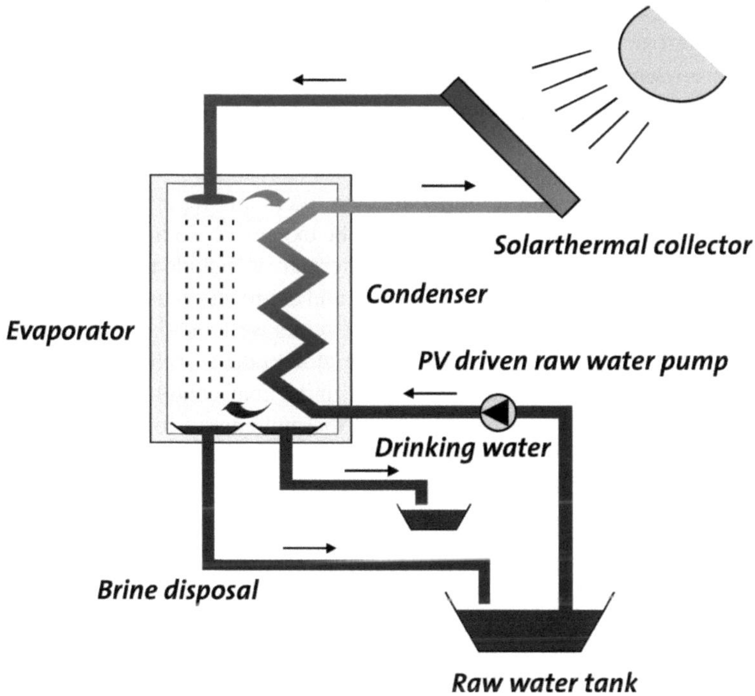

Fig.10.2 The main principle of the MEH technology
Source: TiNOX-MAGE

The tank and the heat exchanger are situated next to the existing electro-mechanical installation room. The cylindrical hot water storage tank is installed vertically through its main axis and has a capacity of 5 m^3. It is thermally insulated with a total heat loss coefficient lower than 12 W/K and appropriate anti-corrosion protection that allows use of water at temperatures of up to 90°C.

The heat exchanger (plate type), used in large systems for disconnecting the solar collector circuit from the seawater circuit, provides thermal power of at least 60 kW and a primary-secondary circuit flow of about 3.6 m^3/h. The piping is appropriately insulated to ensure that the thermal loss coefficient factor (per length unit) is less than 0.3 W/K/m. The insulation is also protected from external elements (e.g. rain, ultraviolet radiation).

A distinguishing feature of this system is its seasonal hybrid character. During the hot period (when there is high water demand and low thermal load) priority is given to desalination, while during winter months (when there is low water demand and high thermal load), the operator can give priority to heating. There is, of course, the possibility of simultaneous operation during spring and autumn. Thus the system demonstrates optimum exploitation of available RES potential, leading to high techno-economical performance. Moreover, the connection of the system to the swimming pool heating system enables the durability of the desalination unit. For material resistance reasons, the temperature in the evaporator should not exceed 87°C.

All components of the unit that come into contact with salt water are manufactured with corrosion-free materials. Condenser/evaporator devices are made of taint-free polypropylene material, compatible for use in drinking water installations. The casings of the humidification chamber and collection basins are made of high-grade stainless steel.

For post-treatment of water, common solutions include the use of a dosing pump to add acid, passing through re-mineralisation filters, pH regulation and UV disinfection. For small units this is considered to be expensive, so it was decided to test the feasibility of alternative solutions. The alternatives include re-mineralisation by mixing with seawater (using a dosing pump) and disinfection using a UV lamp.

Experience has shown that the technology works well and there were only a few minor problems, mainly of a practical nature, due to the fact that it was one of the first times that the product had been tested under real conditions. The produced water is relatively expensive, but the use of thermal energy for other purposes, in periods of low water demand, improves the economics of the plant [3].

10.2.3 Solar Rankine Reverse Osmosis System

The novel idea of using solar thermal energy to drive the high pressure pump of a reverse osmosis desalination unit through a organic Rankine cycle was developed at the Agricultural University of Athens. The concept was successfully tested in a pilot system installed in Greece.

This system was installed and operated as part of an R&D project. It was a pilot plant that proved the main concept of the technology: RO desalination operation, using only heat from solar energy, and without the need for electricity. The system was installed in 2005 and is still operative at the date of publication of this book, however various improvements are planned that will maximise its efficiency and optimise its performance. The cost of the product water, when using only solar power, is around 12 €/m^3 for the first generation system and expected to reduce further in subsequent improved versions. Thermal energy exploitation can operate equally well with other sources, such as geothermal and waste heat. If the right conditions are met, the cost in these cases becomes much lower, approaching 2.5 €/m^3 [4].

The key components of the system are identified below and illustrated in Fig. 10.3, which presents the layout of the system.

1. Vacuum tube solar collector array, consisting of 54 collectors (*Thermomax model TDS 300*) of a total gross area of 216 m^2
2. Circulator
3. Preheater and Evaporator, 35 and 73 kW, respectively
4. Condenser, 100 kW
5. Expander, 2 kW
6. HFC-134a pump, 2,000 kg/h controlled by a frequency regulator

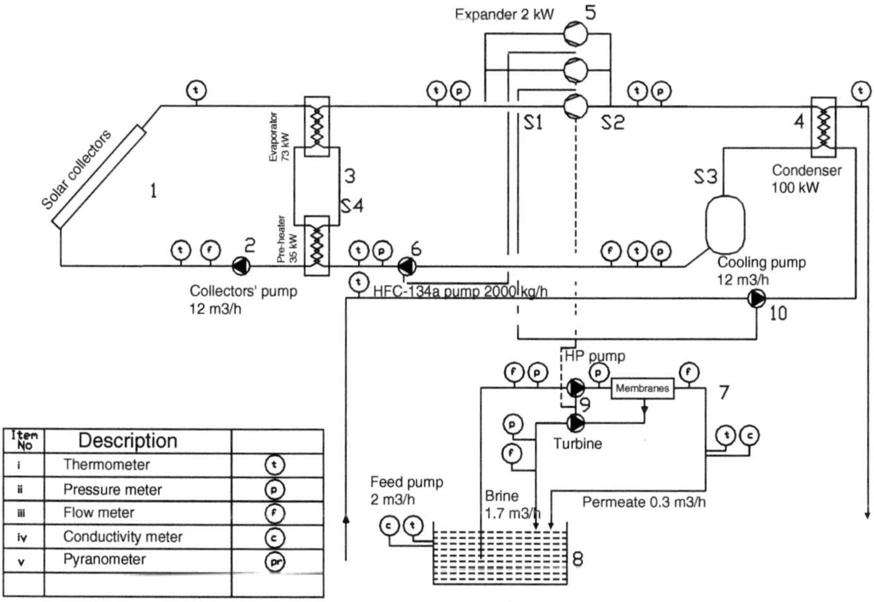

Fig. 10.3 Layout of a solar organic rankine cycle (SORC) for RO desalination

7. RO unit, 0.3 m^3/h fresh water production
8. Water reservoir, 1 m^3
9. RO energy recovery system, consisting of Axial Piston Pumps

The operating principle is briefly described below:

The thermal energy produced by the solar collector array evaporates the refrigerant HFC-134a on the preheater-evaporator surfaces of the Rankine engine (Fig. 10.3). The superheated vapour is then driven to the expander, where the generated mechanical work produced by expansion, drives the high-pressure pump of the RO desalination unit. The superheated vapour at the expander's outlet is directed to the condenser where it condenses. Finally, the saturated liquid at the condenser outlet is pressurised by a positive displacement pump and the thermodynamic cycle is repeated. A special energy recovery system of Axial Pistons Pumps (APP) has been integrated into the RO unit to minimise the specific energy consumption [5].

In essence, the process that takes place is the transformation of the heat, produced by the solar collector field, to mechanical energy, via the ORC process.

10.2.4 Solar Membrane Distillation System

The system consists of two solar autonomous membrane distillation units: a compact unit with a nominal capacity of 100 l/day and a larger unit with a nominal capacity of 1 m^3/day. These were installed and tested at the Canary Islands Institute of Technology (ITC) facilities in Pozo Izquierdo (Grand Canary Island, Spain).

The operation of these units are the main outcome of an EU project (MEMDIS project, Contract Number NN5-2001-00819), coordinated by the Fraunhofer Institute for Solar Energy Systems, ISE.

The operation of the systems is based on highly innovative distillation technology which uses hydrophobic membranes. This type of membrane has the characteristic of allowing the passage of water vapour, but not liquid water.

The operating principles of these systems are fully described in Chap. 7 and will therefore not be described here.

Both systems operate on solar heat from high efficiency collectors. In Figure 10.4 can be seen the solar thermal system powering the larger unit installed in the Grand Canary Island. A small PV field is used to power the circulating pumps.

In the compact system, seawater is fed directly into the collectors, while the larger system has two loops (seawater and freshwater). Freshwater is circulated through the collector field to collect heat, which is transferred to the feed seawater through a heat exchanger.

The compact system was designed to operate 6–10 h per day, as there is no energy storage system. However, the larger system is capable of operating up to 22 h daily. It includes a set of 6 membranes, and a 4 m^3 hot water storage tank.

The two solar MD systems have been under test in Pozo Izquierdo for more than 4 years. The operation of the systems has demonstrated that this combination has the following advantages:

- Low temperature operation, very suitable for high efficiency solar collectors
- Scaling does not occur, so no chemical pretreatment is required

Fig. 10.4 The solar collector system of the 1 m^3/day MD plant

- No problems in intermittent operating mode
- Process is at atmospheric pressure, there is no need for vacuum generation
- Salinity of the feed has almost no effect on process efficiency
- Very high quality water is obtained
- System is modular, so a large range of different water capacity units can be implemented
- Good adaptation to variable solar power

Other main advantages are the ease of operation and the low maintenance requirements. Nevertheless, it has to be taken into account that this is a pre-commercial, technical concept model and certain improvements are still to be made.

10.2.5 Photovoltaic/Reverse Osmosis Systems

10.2.5.1 PV/RO System in Tunisia

The system, based on the ITC patent PCT n° ES 2004/000568, is a stand-alone 50 m^3/day RO desalination unit, powered by a 10.5 kWp PV field with battery energy accumulation. It is located in a remote inland Tunisian village. The unit has been operative since May 2006 and produces freshwater from the brackish water well, located in the nearby oasis (salinity: 3,500 mg/l). The whole system is controlled automatically and 2.1 m^3/h freshwater is produced and distributed to the town through five public standpipes.

The village of Ksar Ghilène has a population of 300 inhabitants dedicated to agriculture and the keeping of livestock. The drinking water supply used to depend on tankers coming from a well located 60 km away. There was no possibility of electrical grid connection, with the nearest point at a distance of 150 km. The annual average daily solar irradiation is 5.6 kWh/m^2, with a mean ambient temperature of 26°C (temperature varies from 0 to 45°C).

The project was structured with the following phases: design of the desalination unit and solar PV generator, study of infrastructure, hydraulic and civil engineering works, equipment transportation to the village, installation and start-up of the whole system, complemented by practical training of local technicians, and follow-up and evaluation of the project. A general view of the system is presented in Fig. 10.5.

The PV solar generator provides electricity to the desalination unit through a 10 kW inverter and batteries with a capacity of 600 Ah at 120 V DC. The inverter feeds the power for the loads of the RO desalination unit, which includes a 1 kW feedwater pump, a prefiltration system and a 3 kW high pressure pump. The configuration of the RO unit is one pressure vessel with 3 membranes of 8 inches diameter (20 cm). The main operating parameters of the desalination system are:

- Recovery: 70%.
- Freshwater flow: 2.1 m^3/h
- Freshwater salinity: less than 500 ppm.

Fig. 10.5 The building housing the RO system, the PV and the water storage tank of the PV/RO system in Tunisia

To avoid the effect of high temperatures on the equipment, a passive cooling solution was implemented. The building containing the desalination unit and the power control equipment is half-buried, and uses the shade produced by the solar PV modules, located on the building roof, to prevent overheating.

During the first 2.5 years, the desalination plant successfully produced more than 6,500 m^3 of freshwater in more than 3,500 h of operation. The system, in this period, produced an average of 7.5 m^3/day with a daily average energy consumption of 15 kWh.

10.2.5.2 PV/RO System in Turkey

This system was installed in 2007, within the framework of the ADIRA project (www.adira.gr), to provide drinking water for a hotel in the Fethiye area, in Turkey. It operates autonomously of the electricity network and is powered by photovoltaic panels, backed-up by batteries. The Istanbul Technical University was responsible for the installation which desalinates brackish water, through reverse osmosis, with a nominal capacity of 2 m^3/day, a specific energy consumption of 15 kWh/m^3 and a water recovery of 15%.

The energy production system consists of 36 polycrystalline PV solar modules, with a total power of 6 kWp, which produce the DC electricity. The DC electricity then passes through a charge controller that protects the battery from over-charging or deep discharge, and then to the battery bank. The DC electricity stored in the battery is converted into regulated AC electricity, suitable for driving the pumps of the RO system. The technical characteristics are given in Table 10.3 and a picture of the system can be seen in Figure 10.6.

Fig. 10.6 The PV/RO system in Turkey

The reverse osmosis unit consists of two thin film composite polyamide membranes (diameter 2.50 inches, length 40.00 inches, i.e. about 6 and 102 cm, respectively) and a high pressure pump. The pretreatment system consists of a spiral wound cartridge filter, a sediment cartridge filter, an activated carbon cartridge filter and the sodium metabisulphite dosing system. The post-treatment system consists of a calcite filter and UV system. The system produces between 0.96 and 1.6 m^3/day, depending on irradiation. Table 10.4 presents the main specifications of the desalination system.

The produced water from the ADS (autonomous desalination system) is distributed to the restaurants and bars of the hotel complex from the freshwater storage

Table 10.3 Energy supply characteristics for the PV/RO system in Turkey

PV module type	Polycrystalline
Peak power (P_{mpp})	80 W
Peak power voltage (V_{mpp})	16.9 V
Open circuit voltage (V_{oc})	21.5 V
Short circuit current (I_{sc})	4.97 A
Operating temperature range	–40 to 80°C
Total number of panels	36
Battery bank	
Battery capacity	200 Ah
Total battery voltage	24 V
Inverter	
Instant max. power	6 kW
Rated power	3 kW
Input voltage range	21–30 V

Table 10.4 Desalination unit specifications of the PV/RO system in Turkey

Sediment filter	Cartridge
Max. operating pressure	1.0 MPa
Filtration material	Spiral wound fibre filament
Activated carbon filter	Cartridge
Max. Operating pressure	1.0 MPa
Fill material	Granular activated carbon
High pressure pump	Vertical shaft gradual centrifuge pump
Flow	0.6 m^3/h
Revolutions/Power	2,900 rpm/0.75 kW
Desalination membranes	Thin film composite polyamide
Diameter/Length	2.50 inches/40.00 inches
Calcite filter	Cartridge type
Max. operating pressure	1.0 MPa
Fill material	Calcite
Length	50.8 cm
UV system	0.5 m^3/h
Power supply	220 V/50 Hz
Min UV dosage	30,000 J
Lamp life	9,000 h

tank. The customer demand for high quality water is expected to be around 3 m^3/day and the ADS plant satisfies a third to a half of this demand. Customers generally prefer bottled water for drinking, however the water produced from the plant is safe, and perfectly suitable for the preparation of soft drinks.

The system has been operating without any technical difficulties, demonstrating a "real world" PV/RO application. The cost of the produced water has been estimated at 18 €/m^3, taking into consideration both the capital and O&M costs. This is relatively high, mainly because it is a small pilot system, the costs included project development and consultancy fees and the PV panels in Turkey cost more than the average world market price, at the time of installation.

10.2.6 Wind/Reverse Osmosis Desalination System

The system, installed at the ITC facilities in Pozo Izquierdo (Gran Canaria Island, Spain), is a standalone 18 m^3/day seawater RO desalination unit, powered by a 15 kW wind generator, with support of batteries. The commissioning was in 2004 and has been tested under different wind conditions. The whole system is controlled and monitored automatically. Table 10.5 outlines the main characteristics of the RO unit and Figure 10.7 shows pictures of the wind generator and the RO unit.

The power supply system consists of a wind turbine with a rated nominal power of 15 kW, a three-phase self-exciting induction generator for a static condenser battery, a charger and a three-phase sine wave inverter, both micro-processed. It also has battery storage, with autonomy of more than 1 h. The main technical characteristics are given in Table 10.6.

Table 10.5 Desalination unit specifications of the Wind/RO system in Gran Canaria

Country	Spain
Organisation	Instituto Tecnológico de Canarias (ITC)
Year	2004
Nominal freshwater production	0.75 m^3/h
Freshwater quality	<500 ppm
Specific energy consumption	9.3 kWh/m^3
Water recovery	24%
Type of unit	Pilot plant
Expected product water cost	3–5 €/m^3
Energy system	Wind with battery bank storage
Energy recovery	Not available

The nominal pressure of operation of the RO unit is 5.5 MPa and the recovery ratio is 24%. The high pressure pump, which is the main load of the RO unit, has a nominal power consumption of 7 kW. The pump motor is a three-phase motor. The energy consumed by the RO unit is 9.3 kWh/m^3.

The pretreatment system consists of a sand filter and a cartridge filter; no chemicals are added. A freshwater storage tank of 250 m^3 is also available. The product water has a salinity of less than 500 ppm TDS. The RO operates when there is adequate wind, with the option to keep the system running for up to an hour using batteries. On average there are 15 h per day of operation in the winter and 20 h per day in the summer.

The control and data acquisition systems receive signals from the sensors in the plant and make adjustments relating to the start/stop configuration of the installation. Two microprocessors are exclusively used to control and manage the available energy in the electricity generating system. The battery bank guarantees that the flushing circuit is always full of freshwater.

Because of the intermittent operation, some scaling problems have been observed, mainly with $CaCO_3$.

Table 10.6 Energy supply characteristics of the Wind/RO system in Gran Canaria

Wind turbine	VERGNET GEV 10–15/15 kW
Number of blades	2
Generator type	Asynchronous
Start-up wind speed	4.5 m/s
Maximal wind speed (tower up)	60 m/s
Control system	Charger CB120/15 kW
Charger characteristics	CB120/15 kW, 380 Vac/120 Vcc, 125 A
Battery capacity	190 Ah
Type of battery	Lead-acid batteries
Max battery discharge (%)	70%
Battery efficiency (%)	85%
Number of cells	60
Inverter type	Sinus 3 phase 120 V/15 KVA/400 Vac-50 Hz
Inlet voltage	90–160 V_{cc}
Output voltage	400 V_{ac}

Fig. 10.7 The ITC wind/RO system in the Canary Islands

The wind turbine costs € 67,000, the RO system costs € 41,000 and other equipment and works cost in total € 15,000, bringing the overall system cost to € 123,000 and the cost of water in the range 3–5 €/m^3 [6].

10.2.7 Wind/PV Reverse Osmosis Desalination Systems

10.2.7.1 Hybrid System at the Agricultural University of Athens

This reverse osmosis system is powered by a hybrid energy system consisting of a small wind turbine and a PV installation. It was installed and is operating as part of a research project. The continuous operation of the pilot system has been closely monitored over several years, offering unique long-term experience of a small PV/wind/RO operating system, including energy recovery. The feedwater has seawater salinity, but is artificially created onsite as needed by the system. Table 10.7 below gives an overview of the main characteristics of the installation.

The hybrid wind/PV system consists of 18 *Arco solar (Shell SM50-H)* modules, with a total rated peak power of 846 Wp and a *Whisper* 1 kW H80-24 wind generator. The hybrid configuration gives the system the possibility of operating at night and at times of low solar radiation, e.g. in winter, which minimises the battery bank capacity requirement; this is one of the most problematic elements of standalone systems, especially in hot climates. The direct current produced by the PV modules charges the battery bank via the *Tarom 235* charge controller. The three-phase current produced by the wind turbine passes throw the *EZ Wire* rectifier-charge controller, which transforms it to regulated DC current, for charging the battery bank in parallel to the PV modules. The wind turbine controller allows for the connection of

Table 10.7 Main characteristics of the Wind/PV/RO system in Athens

Country	Greece
Organisation	Agricultural University of Athens
Year	2004
Nominal daily fresh water production	2.2 m^3/day
Freshwater quality	<500 ppm
Specific energy consumption	3.3–5.2 kWh/m^3
Water recovery	10%
Type of unit	Pilot plant
Expected product water cost	7–9 €/m^3
Energy system	PV/Wind or Hybrid with battery
Energy recovery	Clark pump

up to 700 Wp PV modules [7]. Table 10.8 includes the technical characteristics of the energy supply system and Figure 10.8 includes pictures of the energy system.

The reverse osmosis unit consists of two spiral wound seawater *Filmtec* membranes with a maximum potable water production capacity of 90 l/h at 25°C. A feedwater positive displacement rotary vane pump pressurises the NaCl solution (50 mS/cm), from the main mixing tank to one of the two cylinders of the Clark pump (the energy recovery unit). The high pressure brine enters the second Clark pump cylinder and exchanges its hydraulic energy with the medium pressure feedwater; the result of these actions is to increase the feedwater pressure to the required membrane pressure. The rotary pump is directly connected to a permanent magnet brushless DC motor with a maximum power of 510 W. The details of the RO system are listed in Table 10.9.

The system may be operated via the battery thus resulting in constant speed and constant pressure on the membranes; the energy being supplied by the PV only, the wind turbine only or by both sources simultaneously. Alternatively the system

Fig. 10.8 The hybrid energy system that powers the RO at the Agricultural University of Athens

Table 10.8 Energy supply characteristics of the Wind/PV/RO system in Athens

PV module type	Arco solar
Peak power (P_{mpp})	47 W
Total number of panels	18
Charge controller	Tarom 235/24 V
Wind turbine type	Whisper H-80
Rotor diameter	3 m
Start up wind speed	3.1 m/s
Peak power	1,000 W at 10.5 m/s
Voltage	24 V
Controller type	EZ WIRE 120 A
Tower height	6 m
Battery bank	FIAMM TMHD 425/3
Capacity at C5	315 Ah
Total battery voltage	24 V

may be operated directly from the PV, by-passing the batteries, with the RO pump operating at variable speed and variable pressure on the membranes.

Despite the intermittent operation of the system, the average freshwater quality was kept within acceptable limits. There has been no need to change the membrane throughout the 3 years the system has been operating. The energy recovery device was the main cause of system breakdown, due to internal pressure losses in the pistons. The manufacturer of the energy recovery device has been constantly improving

Table 10.9 Desalination unit characteristics of the Wind/PV/RO system in Athens

DC motor	
Type	Drive systems LV74.9
Rated power	510 W
Rated voltage	24 V
Maximum current	25 A
Rated RPM	1,500 rpm
Rotary vane pump	
Type	Fluid-o-tech PO700
Maximum pressure	1.6 MPa
Rated flow rate at 1,450 rpm	0.8 m^3/h
Energy recovery unit	
Type	Eco systems Clark pump
Feed flow low limit	0.9 m^3/h
Pressure limit	7.0 MPa
Membrane	
Type	Spiral wound thin film composite
Model	Filmtec SW30-2540
Maximum operating pressure	6.9 MPa
Maximum operating temperature	45°C
Maximum feed flow rate	1.4 m^3/h
Product water flow rate	0.083 m^3/h
Minimum salt rejection	99.2%
Single element recovery	8%

the quality of materials used and the device is now working well, with only three breakdowns in the 3 years of operation. The use of low pressure membranes could enhance system performance by using the low energy produced by the RE system during the day.

10.2.7.2 Hybrid System in Lavrio, Greece

In 2003 the Centre for Renewable Energy Sources (CRES) installed one of the first standalone reverse osmosis units powered by a hybrid wind/PV energy supply system. It has been in practically continuous operation since then and its monitoring has provided valuable experience. Table 10.10 gives and overview of the main characteristics of the installation and some pictures are given in Figure 10.9.

The PV array consists of three sub-arrays of 12 modules each. Each array is connected to a charge controller (3 controllers in total). The solar chargers are of 45 A each. The PV array has an adjustable tilt angle during the year (2 positions) according to solar altitude. The 900 W wind generator (W/G) has its own charger, which controls the voltage from the generator and prevents battery overcharging and deep discharge. A uniquely designed resistor bank is used as a dump load. A circuit breaker to stop the wind generator, when necessary, is also included. The energy produced from both the PVs and the W/G drives the RO unit. A battery bank of 1,850 Ah/100 h is used as an energy buffer and to provide constant power to the RO unit. Two inverters are used in order to convert the DC voltage from the battery bank to AC for the main load (RO). An inverter (I) of 1.5 kW drives the booster pump and is also able to provide electricity for other auxiliary loads. A second inverter (II) of 3 kW drives the RO unit high pressure pump. The reason for the use of two inverters is mainly to increase the reliability of the autonomously operating system. The RO unit should be able to operate, at least for flushing processes, at times when no energy is available from the hybrid system [8]. Table 10.11 provides the details of the hybrid energy system.

The RO unit operates in a closed water circuit as no seawater is available at the site where the system is installed. The seawater for the unit is produced by mixing freshwater, from the water distribution network, with salt, in a water storage tank

Table 10.10 Main characteristics of the Wind/PV/RO system in Lavrio

Country	Greece
Organisation	Centre for renewable energy sources
Year	2003
Nominal daily fresh water production	3.12 m^3/day
Freshwater quality	<500 ppm
Specific energy consumption	15 kWh/m^3
Water recovery	20%
Type of unit	Pilot plant
Expected product water cost	20–23 €/m^3
Energy system	Hybrid Wind/PV with battery bank storage
Energy recovery	Not available

Table 10.11 Energy supply characteristics of the Wind/PV/RO system in Lavrio

Equipment	Technical characteristics
PV array	4 kWp Siemens SM110 36 modules in parallel, adjustable tilt
Solar charger	3 × 45 A, Steca/Tarom PWM
Wind generator	900 W, Whisper H40, 24 V EZ-Wire Universal Charger
Battery bank	1,850 Ah/100 h, Fulmen Solar 2 V, 12 cells in series
Inverter I	1,500 W, Siemens ESW 3,024 24 V DC, 230 V AC, (3,000 VA–30 min), 1 phase
Inverter II	3,000 W, Respect 24 V DC, 230 V AC, 1 phase
RO membranes	2 × SW30-254 FilmTec
Operating pressure	5.8 MPa
Feedwater salinity	36,000 ppm TDS
Product water salinity	~230 ppm TDS
Booster pump	0.45 kW
High pressure pump	2.2 kW

with a capacity of 2 m^3. A booster pump of 0.45 kW (1-phase motor) drives the feedwater, with a pressure of around 0.15 MPa, to the pretreatment system. This is supplied by the 1.5 kW inverter (I) at 24 V. The pretreatment system for this particular application is simple and consists of a carbon filter for water de-chlorination, and a cartridge filter for polishing-filtration, for the removal of only very small amounts of materials.

A positive displacement high pressure pump, of 2.2 kW (3-phase motor), pressurises the feedwater to a pressure of around 8 MPa to the RO membranes. The electricity for the high pressure pump is supplied by the 3 kW inverter (II) at 24 V. For the RO module system, 5 cm *FilmTec, SW30-2540* spiral wound membranes are used. Two pressure vessels connected in series, each with one 5 cm diameter membrane module, are used to desalinate 36,000 ppm TDS seawater to freshwater of around 230 ppm TDS salinity. The hourly product water capacity is around 0.13 m^3. The high pressure brine exits the membrane, passes through a throttling valve and then is transferred to the water storage tank. No energy recovery system is used in this application.

10.2.8 Wave Reverse Osmosis Desalination System in India

As interest in wave energy is growing and the technology is developing, the combination with desalination is becoming an attractive idea [9]. An early experimentation with this concept was performed in India. The main characteristics of the system are presented in Table 10.12.

The wave energy plant was constructed in 1990 in Vizhinjam Kerala, India. It demonstrates electricity generation from wave energy, using the Oscillating Water Column (OWC) principle. A column of water rises and falls, compressing and depressurising a coloumn of air above it. The trapped air flows through a turbine and thus electricity is generated.

Fig. 10.9 The hybrid RO system installed in Lavrio by the centre for renewable energy sources

Table 10.12 Overview of the Wave/RO system in Lavrio

Country	India
Organisation	National Institute of Ocean Technology
Year	2003
Nominal freshwater production	0.6 m^3/h
Freshwater quality	<500 ppm
Water recovery	33%
Type of unit	Pilot plant
Energy system	Wave energy (OWC) with battery bank storage

After experimenting with different configurations, the system was optimised to a twin 1 m diameter, horizontal axis Wells turbine coupled to an impulse turbine, designed, fabricated and erected for that purpose.

The desalination system has a capacity of 0.6 m^3/h and a 33% water recovery. Pretreatment involves filtration and the water is then fed to the RO system by the high pressure pumps. The pH level is brought within the 7–7.5 range through neutralisation after the water has passed through the RO process. The produced water, which is of very good quality, is also chlorinated against biological contamination.

The impulse turbine drives a variable speed 18 kW permanent magnet brushless alternator that ensures constant DC voltage under varying wave conditions. A set of 120 V, 300 Ah valve regulated lead acid batteries keep the plant running when wave power is low. An inverter produces the AC voltage for running the RO plant.

Overall the system has been operating as planned without any major technical difficulties. It is a promising technology that needs further testing and development and holds the potential to contribute to the water supply in dry areas both in India and around the world.

10.3 Cost of Water from RE/Desalination

The high specific cost of freshwater from an RE/desalination system is one of the main barriers to the introduction of these systems on the water market.

There are significant differences between the cost of one cubic meter of desalted water from conventional desalination plants and from RE powered desalination systems. Thorough and detailed studies on the cost of conventionally desalinated water have already been made [10].

The following figures give typical ranges of costs for the main conventional desalination technologies:

- RO (seawater): 0.35–1.5 €/m^3
- RO (brackish water): 0.15–0.55 €/m^3
- MED: 0.5–1.5 €/m^3
- MSF: 0.5–1.5 €/m^3

These values are, for the moment, much lower than the costs of RE/desalination, which are in the range of 1–20 €/m^3, depending on the technology used, the salinity of the feedwater and other site-specific factors such as renewable energy potential. While examples of operating costs for the different technologies are given in the previous paragraphs or in other chapters, this section provides a short general cost analysis and some recommendations to help project developers achieve water costs closer to the lower end of the range for RE/desalination systems.

10.3.1 Cost Components and Limiting Factors

For calculating the specific cost of water produced by RE/desalination systems, the following main cost elements need to be considered:

- Capital costs:
  - Engineering
  - Feedwater system (in some cases, well drilling and pumping)
  - Freshwater distribution system (storage and pipes)
  - Brine disposal system (particularly relevant for inland places)
  - Desalination and renewable energy system equipment
  - Auxiliary equipment (hydraulic and electrical elements)
  - Control and monitoring systems

- Other costs such as construction, land and transport
- Local management (pre-visits, contacts, meetings, training)

- Maintenance and operating costs:
  - Personnel
  - Consumables (chemical products, distilled water for batteries)
  - Replacement costs (change of membranes, filters and other maintenance requirements)
  - Subcontracting (those derived from external technical support)

Other factors that affect the cost of the system and of the produced water include: the accessibility of the site, the renewable energy resources in the area, the salinity and other characteristics of the feedwater, and the demand for freshwater.

The low commercial availability of this kind of system and the lack of practical experience are limiting factors, increasing further the cost and reducing the acceptability from an end-user point of view. Currently most of the systems are small pilot units, installed for R&D purposes [11]. For more information, a review of current technologies by Mathioulakis, Bellessiotis, and Delyannis can be consulted [12].

10.3.2 Cost Reduction for PV/RO Systems

Photovoltaic systems and reverse osmosis plants are mature technologies for electricity supply and water production, respectively. This is one of the reasons for the extensive use of PV/RO combinations in autonomous desalination. In cases of inland remote areas with small demand (under 100 m^3/day), the PV/RO concept is currently considered as one of the most suitable options.

The following recommendations could contribute to minimising the cost of water produced by autonomous PV/RO systems, according to an analysis performed under the ADU-RES project [13, 14]:

- *Use of energy recovery systems (seawater installations)*: These systems are always recommended as long as the size of the desalination plant allows them. The cost of the desalination unit is increased, but the specific energy consumption is significantly reduced, requiring less PV power. Since the PV equipment is the most expensive part of the whole system, the specific water cost could be decreased by more than 10%; this decrease is particularly notable in the case of battery-less systems.
- *Use of batteries (brackish water and seawater installations)*: According to current capital costs, the use of batteries is recommended for autonomous systems. It is a cost effective way of maximising daily operating hours, and consequently daily water production. There is some experience of battery-less systems, but complex control software is needed to regulate the differences between generated and consumed power, at any given moment. The battery-less option offers a

very attractive potential, as the elimination of batteries would simplify the system and would imply a significant saving in maintenance and investment cost.

- *Appropriate pretreatment system*: Use of beach wells for the intake of seawater reduces pretreatment requirements. The physicochemical characteristics of the feedwater help to determine the optimal selection of pretreatment system and the required chemical dosing, so prolonging the life of the system and reducing overall costs.
- *Selection of high efficiency pumps*: The use of electronic speed control for variable power consumption, and materials that are suitable and which have guaranteed performance with seawater contact, is recommended. Monitoring and predictive maintenance may also be beneficial.

10.3.3 Cost Reduction for Wind/RO Systems

Wind powered reverse osmosis systems are currently the best option for medium sized systems (from 100 to 2,000 m^3/day) if the site has adequate wind resource. Wind energy is an established technology with a lower unit cost of produced electricity compared to PV.

The inclusion of energy storage systems is unavoidable in autonomous wind desalination plants, as the sudden variations in wind speed and produced power cannot be directly taken up by the loads of the system. Two kinds of back-up systems are recommended: batteries for long term storage (several hours) and flywheels for very short periods (seconds).

A detailed description on the performance and operation of RO units driven by off-grid wind farms, based on the ITC (Canary Islands Institute of Technology) tests in Pozo Izquierdo, indicates that this is a very promising option from a technical and economic point of view [15–17]. An analysis of a modular RO system, (units of 100 m^3/day) coupled to an off-grid wind farm (2×230 kW), concludes that the theoretical minimum cost of water could be less than 0.83 €/m^3, in an area with high wind speed potential, which is the case of the southeast of Grand Canary Island [18].

10.3.4 Cost Reduction for Solar Thermal Systems

The technological concept of solar thermal driven desalination systems includes different options: solar stills, solar/MED, solar/MD and MEH, solar/MSF; and offers a wide range of water production capacities. They are not as common as other autonomous desalination concepts (see Sect. 10.4) and thus there is very little available information about economic data available. Nevertheless, some projects have operated successfully and a few recommendations on cost reduction can be given.

- *Solar/MED*: According to the experience of CIEMAT – PSA (see Sect. 10.2.1) the main actions to reduce water production costs are the following: the effi-

cient incorporation of a static solar collector system, of CPC type, to supply heat at medium temperature (70–100°C); the development of a new Double Effect Absorption Heat Pump; and the reduction to zero of any discharge by recovering brine flow with a solar dryer [19]. Costs of solar/MED are in the range 0.67–9.9 $/m^3$, depending on the cost of equipment, capacity and the type of solar system; the lowest values are for the units supplied by solar ponds [20].
- *Solar/MSF*: According to the review by S.Al-Hallaj et al. [20], the unit costs of a complete solar-based MSF system is \$2.84/m^3. A wider range (0.7–5.36) is given by Miller [10].
- *Solar stills*: According to the review by S.Al-Hallaj et al. [20], costs of solar stills can be from even under 3 $/m^3$(for conventional solar stills) and more than 28 $/m^3$ (for multi-effect stills). The most significant factor effecting this cost is the cost of investment of the still. There are no significant economies of scale.
- *Solar/MD*: From to the experience of the Fraunhofer Institute (see Sect. 10.2.4) the costs of the first prototypes of this technology are fairly high (25 €/m^3 for a unit of 150 l/d). However, they estimate a future cost of about 10–15 €/m^3. Cost reduction can be achieved by increasing the quantity of units produced, further research into efficiency, and the implementation of an effective control system [21].

10.4 Overall Considerations and Conclusions

There are many RE/desalination technologies in various stages of development. As a result, a potential end-user has a variety of options to choose from and can adapt the system to suit their specific requirements. The choice depends on various parameters, the main ones being:

- Salinity of raw water
- Required quantity and quality of product water
- Availability of solar, wind, wave or other renewable energy resource
- Cost of land

An overview of the most commonly used RE/desalination technologies, is given in the review by Papapetrou et al. [22]. In total, 91 plants were identified, with the following characteristics:

- capacity of 50 m^3/day or less;
- autonomous plants (i.e. not connected to the central electricity grid); and
- powered by renewable energy sources.

Of these, 31 were distillation plants, 56 were membrane desalination units and 4 used hybrid or other technologies. The chart below (Fig. 10.10) shows the tech-

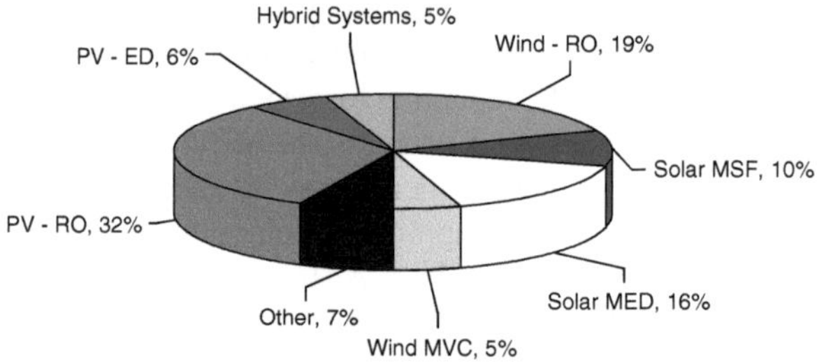

Fig. 10.10 Technology combinations of the plants identified in the ADU-RES study
Source: [22].

nology combinations used in the identified installations. The most popular combination is PV/RO, followed by wind/RO and then solar distillation technologies such as solar/MED.

There are many technological combinations, such as RO powered by wind or PV and solar/MED, that have been successfully tested on various scales and in various locations. Other technologies, such as MD and MEH, are in development phase, while many other concepts, such as the Organic Rankine Cycle RO, are just entering experimental phases with scale models. It is not expected that one single technology will dominate the market; each has its own strengths and weaknesses and each targets end users with differing requirements. For example, thermal plants can be much more competitive in applications where adequate waste heat is available, whereas RO becomes more attractive for the desalination of brackish water where less energy is required.

Future R&D and cost trends may be decisive in determining which technology combination will gain prominent market place. A breakthrough in PV panel technology, that would reduce their cost drastically, would strengthen the trend of using them for RO and ED (electrodialysis) desalination. On the other hand, improvements in the energy efficiency of wind/MVC (mechanical vapour compression) and a solution to the scaling problem could propel these into the spotlight.

The problems that even established technologies such as RO and distillation are facing, when applied in combination with renewable energy in small standalone systems, should not be underestimated.

Reverse Osmosis has to cope with sensitivity of membranes to fouling and scaling, because of the start-stop cycles and partial load operation during periods of oscillating power supply [23] typical of wind turbines or PVs. The solution tends to be a large energy storage system such as batteries, which increases the capital cost. As an alternative, battery-less operation has been tested in combination with frequent replacement of the membranes. However, more R&D is required before this option can be offered as an off-the-shelf product with high quality water production and easy operation.

Solar/MED plants, on the other hand, tend to suffer from corrosion because of the high temperature of operation. They are characterised by high requirements for thermal energy and therefore become more cost-effective as size increases, but only when land is available at a reasonable price. Solutions to the corrosion problem have been demonstrated, either by using corrosion-free materials or by processing the feedwater and exchange surfaces, however the costs of these approaches are relatively high.

Many RE/desalination combinations are technically mature enough to be used to supply water in certain areas. There are many niche markets where water is being transported at similar or higher costs, and the installation of autonomous desalination units would make commercial sense. In other communities, without access to electricity or water networks, local populations are very poor and could never meet the costs of financing such units by themselves. In these cases, innovative financing mechanisms and subsidies from development organisations are needed. The research community has demonstrated that autonomous desalination, powered by renewable energy, is technically possible and their efforts are focused on further technical improvement and cost reduction. The remaining problems can only be overcome with the impetus and resources that free market applications can bring.

Acknowledgments Section 10.2.2 was written based on information and text provided by those responsible for installation of the system: Emmanouil Mathioulakis, Solar & other Energy Systems Laboratory – NCSR "DEMOKRITOS", e-mail: sollab@ipta.demokritos.gr. Section 10.2.5.2 was written based on information and text provided by those responsible for installation of the system: Prof. Seval Sözen and Senem Teksoy from the Istanbul Technical University (ITU), Faculty of Civil Engineering, Turkey, e-mail: senem.teksoy@itu.edu.tr, sozens@itu.edu.tr. Section 10.4 was developed with the kind help and suggestions of Prof. Gonzalo Piernavieja, Canary Islands Institute of Technology, e-mail: gpiernavieja@itccanarias.org

Abbreviations

ADS	autonomous desalination system
CPC	compound parabolic concentrator
MD	membrane distillation
MED	multi-effect distillation
MEH	multi-effect humidification
MSF	multi stage flash
O&M	operation and maintenance
ORC	organic rankine cycle
PV	photovoltaic
R&D	research and development
RE	renewable energy
RO	reverse osmosis
UV	ultra violet

References

1. D. Alarcón, J. Blanco, S. Malato, M.I. Maldonado, P. Fernández, 'Design and setup of a hybrid solar seawater desalination system: The AQUASOL project', ISES Solar World Congress 2005, Orlando, USA, Solar Energy: Bringing Water to the World, (August 8–12, 2005)
2. J. Blanco, D. Alarcón, W. Gernjak and E. Guillén (2008) The AQUASOL System: Solar Collector Field Efficiency and Solar-Only Mode Performance. In: 2008 14th Biennial CSP SolarPACES Symposium, March 4-7, 2008, Las Vegas, NV. CD-550-42709

3. E. Mathioulakis, V. Belessiotis, G. Panaras, "Experience gained by the operation of a pilot solar thermal desalination plant in Cyprus", 9th National Conference for Renewable Energy Sources, Geroskipou-Cyprus, (March 2009)
4. D. Manolakos, S. Kyritsis, G. Papadakis, J. Karagiannis, P. Soldatos, 'Cost analysis and economic evaluation of an autonomous low-temperature solar rankine cycle system for reverse osmosis desalination – comparison with a simple solar still desalination system'. WSEAS Transactions on Environment and Development, Issue 2, Vol. 1, November 2005, ISSN: 1790-5079 205
5. D. Manolakos, G. Papadakis, Essam Sh. Mohamed, S. Kyritsis, K. Bouzianas, Design of an autonomous low-temperature solar Rankine cycle system for reverse osmosis desalination, Desalination 183 (2005) 589–596
6. E. Tzen et al. "Autonomous Desalination Units using RES", ADU-RES project (www.adu-res.org) deliverable 2.1, (October 2005)
7. Essam Sh. Mohamed, G. Papadakis, E. Mathioulakis, V. Belessiotis, An experimental comparative study of the technical and economic performance of a small reverse osmosis desalination system equipped with an hydraulic energy recovery unit, Desalination, 194 (2006) 239–250
8. E. Tzen, D. Theofilloyianakos, Z. Kologios, 'Autonomous reverse osmosis units driven by RE sources experiences and lessons learned', Desalination 221 (2008) 29–36
9. N. Sharmila, P. Jalihal, A.K. Swamy, M. Ravindran, 'Wave powered desalination system', Energy 29 (2004) 1659–1672
10. J.E. Miller, "Review of water resources and desalination technologies". SAND 2003-0800. Printed March 2003 (Available in: http://www.sandia.gov/water/docs/MillerSAND20030800.pdf)
11. L. García, "Seawater desalination driven by renewable energies: a review". Desalination 143 (2002) 103–113
12. E. Mathioulakis, V. Bellessiotis, E. Delyannis, "Desalination by using alternative energy: Review and state-of-the-art". Desalination 203 (2007) 346–365
13. Several authors. Deliverable 4.1. "Report on techno-economic performance and cost reduction potential". ADU RES EU co funded coordination action INCO-CT-2004-509093 (www.adu-res.org)
14. V. Subiela, B. Peñate, A. Sánchez, K. Soulis, "Cost reduction strategies for autonomous desalination units based on renewable energies". International Desalination Association. Proceedings of the World Congress on Desalination and Water Reuse. Maspalomas, Gran Canaria Island, Spain. (October 21–26, 2007)
15. J.A. Carta, J. González, V. Subiela, "The SDAWES project: an ambitious R&D prototype in wind-powered desalination". Desalination 161 (2004) 33–47
16. J.A. Carta, J. González, V. Subiela, "Operational analysis of an innovative wind powered reverse osmosis system installed in the Canary Islands". Solar Energy 75 (2003) 153
17. V. Subiela, J.A. Carta, J. González, "The SDAWES project: lessons learnt from an innovative project". Desalination 168 (2004) 39–47
18. Water Department, ITC, Seawater desalination plants connected to an off-grid wind farm. Internal Report. (2005)
19. J. Blanco, D. Alarcón, E. Zarza, S. Malato, J. León, "Advanced solar distillation: A Feasible Technology to the Mediterranean Area". *4th ISES – Europe Solar Congress*. Bologne, Italy. (June 23–26, 2002)
20. S. Al-Hallaj, S. Parekh, M.M. Farid, J.R. Selman, "Solar desalination with humidification-dehumidification cycle: Review of ecocomics". Desalination 195 (2006)
21. M. Wieghaus, J. Koschikowski, M. Rommel, Fraunfofer Institute for Solar Energy Systems, ISE, Germany. "Solar membrane distillation ideal for remote areas". Desalination & Water Reuse 18(3) (November/December 2008)
22. M. Papapetrou et al. "Autonomous Desalination Units Based On Renewable Energy Systems – A Review of Representative Installations Worldwide", Solar Desalination for the 21st Century (ISBN 978-1-4020-5506-5), Springer Netherlands (2007) 343–353
23. R. Morris, "Renewable Energy Powered Desalination Systems in the Mediterranean Region", UNESCO Workshop, (1999)

Chapter 11
Protecting the Marine Environment

Sabine Lattemann

Abstract As the use of desalination increases in many parts of the world, the problem changes from water scarcity to energy consumption and moves from overused, polluted freshwater bodies to the marine environment. The main environmental concerns of desalination activity revolve around the emissions of greenhouse gases and air pollutants, the concentrate and chemical discharges into the sea, the use of large quantities of seawater for cooling purposes and as feed water, causing the impingement and entrainment of marine organisms, and construction-related impacts on coastal and nearshore habitats. This chapter gives a synopsis of the key concerns of desalination plant impacts on the marine environment. An overview of seawater desalination capacities by sea region is given, followed by a more in-depth discussion of the concerns associated with desalination activity in three sea regions, the Gulf, the Red Sea and the Mediterranean Sea, which together account for 70% of global seawater desalination capacity.

11.1 Worldwide Installed Capacity

Today, many countries of the world, particularly in the Middle East, depend heavily on desalinated water, while many more turn to desalination in order to develop and diversify their water supply options in the face of economic development, demographic growth and climate change. The worldwide installed seawater desalination capacity is increasing at a rapid pace. The latest figures from the 20th IDA Worldwide Desalting Plant Inventory indicate that about 28 million m^3/day are presently produced from the world's oceans and regional seas [1], which is comparable to the average discharge flow rate of the River Seine in Paris (328 m^3/s).

In terms of sea areas, the largest number of seawater desalination plants are found in the Gulf with a total desalination capacity of approximately 12.1 million

S. Lattemann (✉)
Institute for Chemistry and Biology of the Marine Environment, Carl von Ossietzky University of Oldenburg, Postfach 2503, 26111 Oldenburg, Germany
e-mail: sabine.lattemann@icbm.de

A. Cipollina et al. (eds.), *Seawater Desalination*, Green Energy and Technology, DOI 10.1007/978-3-642-01150-4_11, © Springer-Verlag Berlin Heidelberg 2009

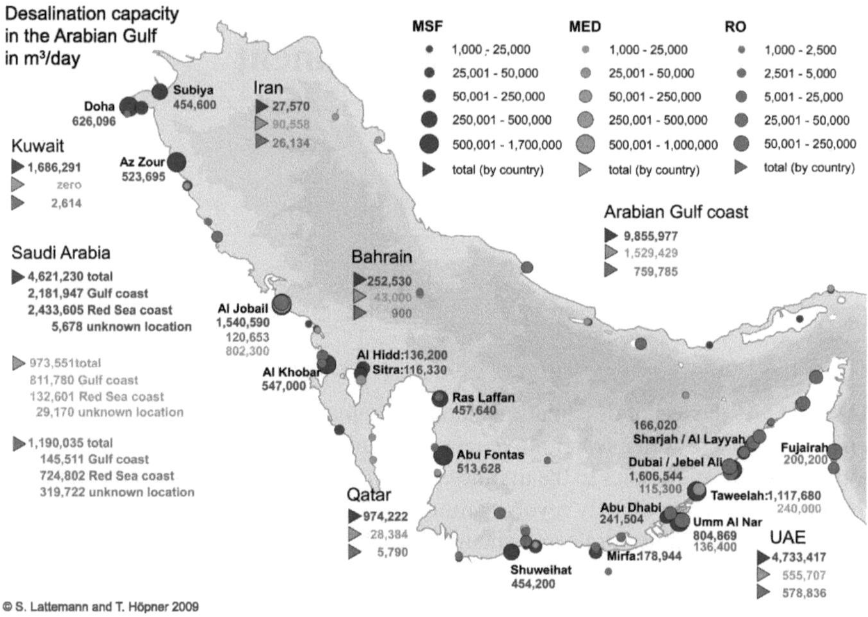

Fig. 11.1 Seawater desalination capacity in the Gulf. The map shows all sites with cumulative capacities > 1,000 m^3/day. The total capacity of each riparian state is given, as is the capacity installed in the sea region (first published in [14], updated with primary data from [1])

m^3/day – or a little less than half (44%) of the worldwide daily production (Fig. 11.1). The main producers in the Gulf (and worldwide) are Saudi Arabia (25% of the worldwide seawater desalination capacity, of which 11% is in the Gulf, 12% is in the Red Sea, and 2% is unaccounted for), the United Arab Emirates (23%) and Kuwait (6%). Thermal desalination processes dominate in the Gulf region (about 94% of all production), as water and electricity are often generated by large co-generation plants that use steam from power plant turbines as a heat source for desalination. Most of the water (81%) in the Gulf is produced by the process of multi-stage flash distillation (MSF). Minor processes are multi-effect distillation (MED) and reverse osmosis (RO), which account for 13 and 6% of the production, respectively (primary data from [1]).

In the Red Sea region, desalination plants have a combined production capacity of 3.6 million m^3/day (13% of worldwide capacity, Fig. 11.2). Similar to the Gulf, most of the water is produced by large co-generation plants, mainly on the Saudi Arabian coast in the locations of Yanbu, Rabigh, Jeddah, Assir and Shoaiba, where the world's largest desalination complex with a capacity of 1.6 million m^3/day is located. Saudi Arabia accounts for more than 92% of the desalinated water production from the Red Sea, with 2.6 million m^3/day (78%) produced by thermal plants. Egypt, the second largest producer of desalinated water in the region, accounts for only 7% of the production from the Red Sea, with 90% (0.2 million m^3/day) coming

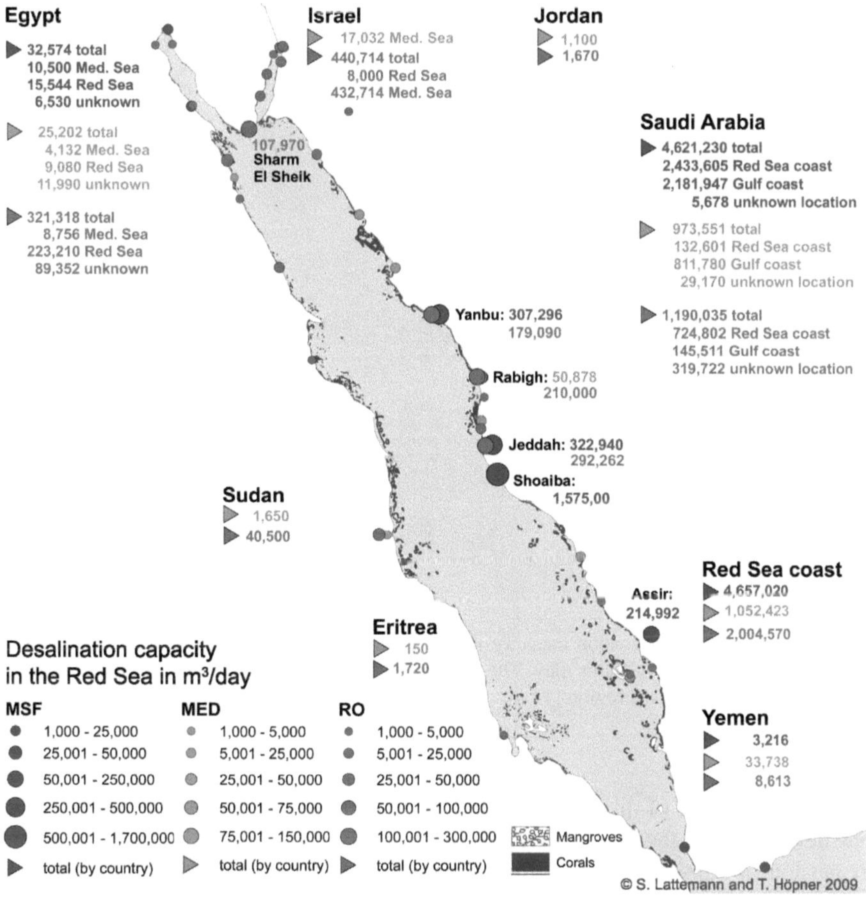

Fig. 11.2 Seawater desalination capacity in the Red Sea. The map shows all sites with cumulative capacities > 1,000 m^3/day. The total capacity of each riparian state is given, as is the capacity installed in the sea region [14, 1]

from smaller RO plants on the Sinai Peninsula and in the tourist resorts along the Red Sea coast.

In the Mediterranean Sea, the total water production from seawater is about 4.0 million m^3/day (14% of worldwide capacity, Fig. 11.3). Spain, with about 8% of the worldwide desalination capacity, is the largest producer of desalinated water in the region with an installed capacity of 2.2 million m^3/day. About 65% (1.4 million m^3/day) of the Spanish installations are located on the Mediterranean coast and the Balearic Islands, and 25% on the Canary Islands. The Spanish "Agua program" will further augment water supply along the Mediterranean coast by increasing the desalination capacity to over 2.7 million m^3/day by 2010. While thermal processes dominate in the Gulf and Red Sea, 70% of the Mediterranean and 99% of the Spanish production on the Mediterranean coast is produced by seawater reverse osmosis (SWRO).

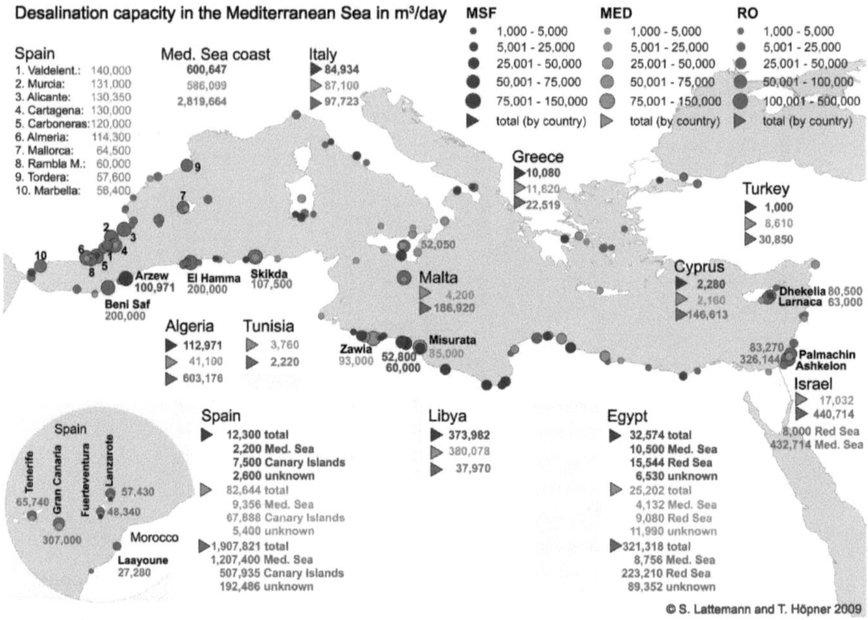

Fig. 11.3 Seawater desalination capacity in the Mediterranean. The map shows all sites with cumulative capacities > 1,000 m^3/day. The total capacity of each riparian state is given, as is the capacity installed in the sea region [14, 1]

Larger numbers of distillation plants are found only in Libya, Algeria and Italy, but new plants in these countries are increasingly also RO plants. A tremendous expansion of capacity is currently taking place in Algeria, North Africa's fastest growing desalination market, where the first sizeable SWRO plant (200,000 m^3/day) was opened in February 2008. It is the first in a series of other projects with capacities between 50,000 and 500,000 m^3/day, which will increase the country's desalination capacity to 4 million m^3/day by 2020 [2]. In Israel, two large SWRO are currently in operation, the Ashkelon plant with a capacity of 330,000 m^3/day – the world's largest SWRO project to date – and the Palmachin plant. Desalination presently accounts for approximately 8% of Israel's water supply, and will increase to 30% by 2020 (1.8 million m^3/day) [3].

While seawater desalination is already a well-established technology in the above mentioned sea regions, the age of large-scale desalination projects is about to begin in other parts of the world.

In California, a potential 15–20 new desalination projects are expected between now and 2030 with a combined production of 1.7 million m^3/day. The two largest and most advanced projects are the 200,000 m^3/day facilities in Carlsbad and Huntington Beach [4]. In Australia, the first large SWRO plant with a capacity of 144,000 m^3/day became operational in Perth in 2006. Another project currently under construction is the Sydney plant with an initial capacity of 250,000 m^3/day, which can be increased to 500,000 m^3/day. Further projects include the Melbourne, Brisbane and South East Queensland plants, with projected capacities of up to 400,000 m^3/day each, and projects in Adelaide, the Upper Spencer Gulf and a second plant near Perth, with capacities of between 120,000 and 140,000 m^3/day each. A third impressive example is China, where desalination capacity could increase from 366,000 m^3/day by a factor of 100 until 2020 [2].

Most desalinated water today is produced in industrial-sized facilities. These include large thermal plants in the Middle East with production capacities of up to 1.6 million m^3/day. Outside the Middle East region, SWRO is the dominant process. The majority of SWRO plants (86%) are small (< 5,000 m^3/day) and account for only 18% of worldwide production of 9.4 million m^3/day, while less than 50 large facilities ($\geq$50,000 m^3/day) account for almost half of the worldwide production. The largest plant currently produces 330,000 m^3/day and several projects up to 500,000 m^3/day are being planned. Due to growing desalination activity in many sea regions and the growing number of large facilities, concerns over potentially negative impacts of the technology on the environment are being raised.

11.2 Marine Environmental Impacts of Desalination Plants

The list of potential environmental impacts of desalination plants is long and in some aspects, such as land use, analogous to other development projects. A full analysis of the potential impacts was carried out as part of the European Research project "MEDINA" (Membrane based desalination – an integrated approach, 6th Research Framework) [5]. The main concerns for the *marine environment* are construction-related impacts which may cause habitat destruction, the impingement and entrainment of organisms with the intake water and, most significantly, the concentrate and chemical discharges into the sea. These may have adverse effects on water and sediment quality and may impair marine life and coastal ecosystems if not well designed and managed.

11.2.1 Intake and Outfall Structures

Intake structures can generally be subdivided into open intakes and below ground intakes. Below ground intakes are completely embedded in the seafloor, either in beach sediments onshore, such as beach wells or infiltration galleries, or in offshore marine sediments, such as horizontally drilled drains (HDD). Open intakes are the most commonly used intake system for large desalination plants, although HDDs are reportedly being used successfully in some larger SWRO plants [6]. Screens, such as fine-meshed, travelling or drum screens, are usually placed in front of the open intakes. In some cases, a breakwater basin is constructed around the intake area.

The most commonly used method of concentrate disposal is surface water discharge, either via a single open outfall or via a diffuser system. Options for co-discharge exist with power plant cooling water. Large distillation plants are typically co-located with power plants, but co-location is also an option under consideration for SWRO projects in California [7, 8], Florida [9] and Israel [10, 11]. The use of existing intake and outfall structures reduces construction-related impacts; the cooling water serves as feedwater for the desalination plant and provides dilution of the concentrate before discharge [12].

The construction of the intake and outfall structures may cause a disturbance to, or compaction of, sediments. If placed above the seafloor, intake and outfall structures and pipelines can act as an artificial breakwater, which may cause wave refractions, change current patterns and interfere with dynamic sediment processes, such as erosion or deposition. The suspension of sediment material into the water column may cause a temporarily increased turbidity near the construction site, which could affect water quality by increasing levels of nutrients or pollutants, or by reducing oxygen levels, depending on sediment properties. During operation, the intake of large quantities of seawater may affect water circulation, especially in areas characterized by weak natural currents.

Increased turbidity during construction may have short term indirect effects on marine species, such as filter-feeding organisms, and the resettling of sediments may cause the burial of benthic flora and fauna. More severe is the destruction of benthic habitats along the construction corridor, which often take between one and several years to recover from disturbances. Construction activities may also disturb sensitive wildlife, such as marine mammals or seabirds. Dredging and drilling activities, particularly, produce low frequency noise emissions below water, including structure-borne sound emissions and vibrations, which can travel considerable distances. Impacts depend on the sound level, distance from the source and the hearing ability of the organisms.

Structures above the seafloor provide hard-bottom substrates to which living organisms can attach, such as algae, anemones or mussels. The prolific growth of these artificial reefs often attracts other invertebrate species for food or shelter, such as echinoderms (e.g. starfish), crustaceans (e.g. lobsters) or marine snails (e.g. abalone), and often results in an increased density of fish species. This "reef effect" may alter the existing community structure and local predator-prey relationships.

Open seawater intakes usually result in the loss of eggs and larvae from fish and invertebrate species, spores from algae and seagrasses, phytoplankton and zooplankton, as well as smaller marine organisms that are drawn into the plant (entrainment). The survival rate is minimal, but the question is whether entrainment mortality represents a significant additional source of mortality for the affected species, which then negatively affects the ability of species to sustain their population. Plankton organisms are generally prevalent in coastal surface waters and have rapid reproductive cycles, while most fish and invertebrate species, as part of their reproduction strategy, produce large numbers of eggs and larvae to compensate for a high natural mortality rate. In most cases, entrainment is not considered to be significant. This perception may change where cumulative sources of mortality (e.g. multiple power or desalination plants) are involved and where endangered species, species of commercial interest or marine protected areas, are affected. While it is relatively simple to estimate the levels of entrainment for a specific project, it is difficult to evaluate the indirect impacts on the ecosystem, especially in places where several projects are involved.

Death of larger marine organisms at intake screens may be caused by suffocation, starvation or exhaustion from being pinned up against the intake screens (impingement) [13]. Impingement may be a significant source of mortality for endangered or protected marine species, such as sea turtles or sea snakes, or species of commercial interest. If the intake velocity of the feedwater is reduced to velocities of about 0.1 m/s, which is comparable to ocean background currents, it could be expected that mobile organisms would be able to swim away from the intake area.

11.2.2 Reject Streams

The waste water stream mainly contains the natural constituents of the intake seawater in concentrated form. Depending on the process, environmental concerns may arise due to the high concentration of inorganic salts or due to the increased temperature of the waste stream, which may increase ambient salinity and temperature in the discharge site and may negatively affect local ecosystems. Furthermore, the pretreatment of the intake feedwater involves chemical additives (Figs. 11.4 and 11.5), some of which are discharged along with the waste water. As seawater is a highly corrosive medium, the waste stream may also contain small amounts of metals that pass into solution on corrosion of metallic parts inside the plant. Although the following review of waste water properties (based on [5, 14–18]) is formally subdivided into concerns related to the physical properties and concerns related to the chemical additives, the likely synergetic effect of thermal and osmotic stress and effects, caused by exposure to residual chemicals, must be taken into account.

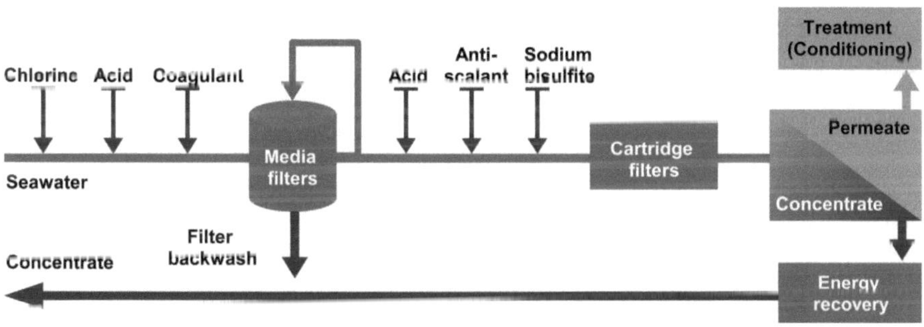

Fig. 11.4 Flow-chart of an SWRO system showing the conventional pretreatment and chemical dosage steps and the different waste and side streams (adapted from [14])

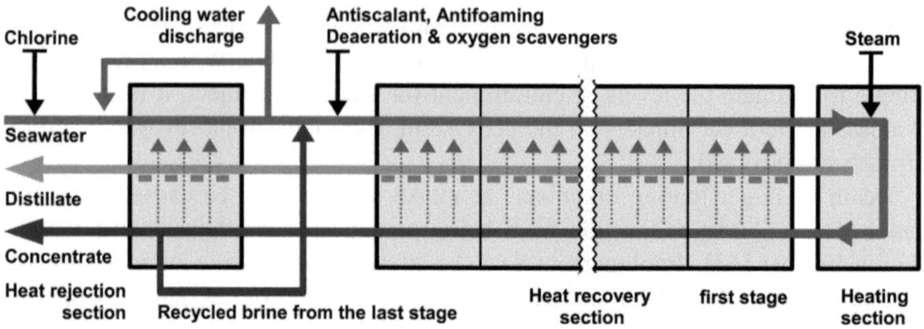

Fig. 11.5 Flow-chart of an MSF distillation plant showing the conventional pretreatment and chemical dosage steps and the different waste and side streams (adapted from [14])

11.2.2.1 Salinity, Temperature and Density

The *salinity* of the concentrate is largely a product of the plant recovery rate, which in turn depends on the salinity of the source water and the process configuration. SWRO plants have higher recovery rates than distillation plants and therefore higher reject stream salinities,[1] which typically range between 65 and 85. Although the brine blow-down from the last stage in MSF distillation plants may have a salinity of almost 70, the brine is effectively diluted with a threefold amount of cooling water.[2] Dilution results in a salt concentration that is rarely more than 15% higher than the salinity of the receiving water, while the RO brine may contain twice the seawater concentration. The brine and cooling water discharges of thermal plants are between 5 and 15°C warmer than ambient seawater, whereas the temperature of the RO concentrate is similar to ambient values.

The concentrate discharge may lead to an increase in salinity in the discharge zone. The salinity increase can be controlled by pre-dilution with other waste streams such as cooling water, dissipation by a multi-port diffuser system, or discharge into a mixing zone that can effectively dissipate the salinity load due to strong wave action and currents.

In addition, the concentrate and cooling water discharge from distillation plants may cause "thermal pollution" in the discharge site. The World Bank guidelines recommend that the discharge water temperature of thermal power plants does not result in an increase greater than 3°C of ambient temperature at the edge of the

[1]The UNESCO definition of Practical Salinity Units (psu) is used, which is the conductivity ratio of a seawater sample to a standard potassium chloride (KCl) solution. Salinity is given as a dimensionless quantity. A salinity of 35 corresponds to 35 g of salt per 1,000 g of seawater, or 35,000 mg/l.

[2]Thermal plants use cooling water for temperature control. The seawater flow rate into thermal plants is 3–4 times higher than the feed into RO plants, for the same amount of product water extraction. The cooling water is discharged along with the concentrate, so that both reject streams mix before disposal.

mixing zone. For the mixing zone, the guidelines recommend to use 100 m from the point of discharge when sensitive aquatic ecosystems are absent [19].

The density difference between waste water and ambient seawater is a controlling factor for mixing and spreading of the discharge plume in the receiving water body. In surface water, density is a function of salinity and temperature. Due to the high salt content, the RO reject stream has a higher density than the ambient seawater. For example, ambient salinity levels of 36–40 and seasonal temperature variations of 15–30°C (e.g. typical Mediterranean surface water) result in density variations of 1,023–1,030 kg/m^3. An SWRO plant with a feedwater salinity of 36, operating at 50% recovery, would produce a concentrate with a salinity of 72. At 20°C, the density of the concentrate would be 1,053 kg/m^3, which is negatively buoyant compared to an ambient density of 1,025 kg/m^3. Unless adequately dissipated, the plume would sink to the seafloor, forming a mass of high salinity water which would spread over the seafloor in the vicinity of the outfall pipe, and likely diffuse into the sediment pore water.

As increased salinity and temperature have opposing effects on density, the reject streams of distillation plants can either be positively, neutrally or negatively buoyant. They are usually positively buoyant due to the influence of large amounts of cooling water discharge. For example, a seawater salinity of 45 and a temperature of 33°C, is characteristic of Gulf seawater. The reject water of an MSF distillation plant would be negatively buoyant compared to the ambient density (1,028 kg/m^3) at a salinity of 50 and a temperature increase of 5°C (1,030 kg/m^3), and positively buoyant at a temperature increase of 10°C (1,027 kg/m^3).

The reject streams of SWRO and distillation plants generally affect different realms of the marine environment: SWRO concentrate which spreads over the sea floor may affect benthic communities, whereas the reject streams of distillation plants may affect the pelagic (open water) community. However, it must be pointed out that mixing and dispersal processes are largely influenced by site-specific oceanographic conditions. In order to analyze plume spreading at a specific project site, the existing hydrological conditions need to be investigated and accompanied by modeling studies and density calculations.

Salinity is a vital environmental parameter for marine life. Similar to "thermal pollution", increased salt concentrations can be harmful and even lethal to marine life. In general, toxicity depends on the sensitivity to increased salinity of the species, the natural salinity variations of their habitats, and their life cycle stage. For instance, field and laboratory studies on the Mediterranean seagrass *Posidonia oceanica* showed that a salinity of around 45 caused about 50% mortality in 15 days and growth rates were reduced by 50% at a salinity of 43 [20]. In contrast, two seagrass species common in Western Australia, *P. australis* and *P. amphibolis*, seem to be more adapted to naturally higher salinities. Denser covers of *Posidonia* meadows are being observed at salinities of between 40 and 50 [21, 22]. Available studies suggest that some seagrass species are more tolerant to hypersaline conditions than others. Also, some macrofauna taxa such as echinoderms (e.g. sea urchins, starfish), which are strictly marine, seem to be more sensitive to salinity variations than, for example, organisms found in estuaries, which are able to adapt to a wide range of salinities including fresh, brackish and saltwater environments. Furthermore, organisms in young life cycle stages, such as sea urchin embryos, are often considered to be more sensitive than adults.

Most marine organisms can adapt to minor deviations in salinity and can recover from extreme, short term exposures to increased salinities. For example, *P. oceanica* plants that survived in a salinity of 43 over 15 days were able to recover when returned to normal conditions [20]. However, only a few species would be tolerant to high salt concentrations over extended periods of time. Marine organisms normally thrive in those environments to which they are adapted and which provide favourable environmental conditions. Natural salinity values vary between 30 and 37 in the Atlantic Ocean, between 36 and 40 in the Mediterranean Sea, between 37 and 43 in the Red Sea, and can range up to 60 in naturally saline environments of the Gulf. Salt concentrations that considerably and continuously exceed the ambient levels to which the native species are adapted may result in osmotic stress. This will drive mobile animals away from the discharge site and can cause a die-off of the sessile flora and fauna. For example, salinity increases near the outfall of the Dhekelia SWRO in Cyprus were reported to be responsible for a decline of macroalgae forests, and echinoderm species vanished from the discharge site [23]. Observations on the distribution of marine species from naturally hypersaline environments in the Gulf, indicate that salinities above 45 considerably alter the benthic community [24]. This emphasises the importance of salinity as a controlling environmental factor.

Similarly, thermal discharges may have an effect on species distribution by changing the annual temperature profiles in the discharge site. This could enhance biological processes by increasing seawater temperatures to favourable conditions in winter, but could result in thermal stress when critical values are exceeded in summer. Marine organisms could be attracted or repelled by the warm water, and species more adapted to the higher temperatures and seasonal pattern may eventually dominate in the distillation plant discharge site. In extreme cases, the thermal discharge may cause a die-off of sessile marine species.

Salinity and temperature thresholds must be established for local conditions, taking the sensitivity of species, natural salinity and temperature variations into account. In Spain, extensive field and laboratory studies have been carried out to investigate the effects of increased salinity on *Posidonia* meadows. It has been recommended to avoid discharges of desalination concentrate into meadows, or to dilute the discharge salinity so that it exceeds a value of 38.5 for less than 25% of the time and a value of 40 for less than 5% of the time [25], compared to ambient salinities in the western Mediterranean of 37–38.

In Western Australia, guidelines for fresh and marine waters specify that the median increase in salinity is to be less than 5% compared to background levels. The criteria for concentrate discharge set for the Perth SWRO plant require that salinity is within 1.2 units of ambient levels at a distance of 50 m and within 0.8 units at a distance of 1,000 m from the discharge point [21]. In the U.S.A., the Environmental Protection Agency (EPA) recommends that the salinity variation from natural levels should not exceed 4 units, in areas permanently occupied by food and habitat forming plants, and when natural salinity is between 13.5 and 35 [7]. For an SWRO plant in Okinawa, Japan, a maximum salinity of 38 in the mixing zone and a maximum increase of 1 unit, where the plume meets the seafloor, were established [26].

The strict discharge thresholds that have been established in some places underline the fact that even minor salinity increases may be harmful to marine ecosystems, and should thus be avoided by adopting an advanced discharge design that effectively dilutes the salinity load. The salt load, however, is not a concern for sea areas such as the Gulf, the Mediterranean or the Red Sea as a whole, where natural evaporation exceeds the effects from desalination by several orders of magnitude. The key to avoiding impacts is to sufficiently dilute and disperse the salinity load to ambient concentrations. The same argument, however, does not hold for chemical additives. The salt is of natural origin, whereas the additives are of anthropogenic origin, and some have a tendency to accumulate in the environment. Dilution is therefore a questionable means of impact mitigation for most of these compounds.

11.2.2.2 Chemicals for the Control of Biofouling

In most desalination plants, chlorine is added to the intake water to control bio fouling. The initial chlorine concentration is quickly reduced inside the plant due to the oxidant demand of seawater, mainly caused by reactions with organic seawater constituents. In RO plants, the water is usually dechlorinated with sodium bisulfite before it makes contact with the RO membranes, which are sensitive to oxidizing chemicals. Chlorine concentrations will therefore be very low to non-detectable in the reject streams of RO plants, while distillation plants discharge residual chlorine to surface waters in varying concentrations.

The chlorination practice in distillation plants is generally similar to coastal power plants using once-through cooling systems. Low-level chlorination is routinely used in power plants in doses of 0.5–1.0 mg/l and resulting oxidant levels of 0.1–0.2 mg/l in the cooling water [27]. In distillation plants, doses of between 0.4 and 4.0 mg/l and resulting levels of between 0.1 and 0.5 mg/l at the point of discharge have been reported [14]. Environmental concentrations are scarce, with levels of up to 0.02 mg/l in the dispersing plume near power plant outlets [27], and levels of between 0.03 and 0.5 mg/l in the vicinity of distillation plant outlets [28, 29]. The latter value of 0.5 mg/l should decrease to 0.05 mg/l within a 1 km distance [30].

The European reference document on the application of best available techniques (BAT) of industrial cooling systems recommends that the average daily (24 h) value of free residual oxidant emissions, from a once-through cooling system, should be $\leq$0.2 mg/l at the outlet, for continuous seawater chlorination, and the average hourly value within one day should be $\leq$0.5 mg/l for intermittent and shock seawater chlorination [31]. The World Bank Pollution Prevention and Abatement Handbook also recommends a total residual chlorine concentration of $\leq$0.2 mg/l for effluents from thermal power plants [19]. BAT and World Bank standards are not binding, as discharge limits are normally controlled by local regulations. While some countries may choose to accept higher levels, regulations may be stricter in other places. In Qatar, the EU's seawater chlorination BAT is challenged by new stricter standards. The state environmental regulator has reduced, incrementally over a number of years, the maximum chlorine concentration permitted in discharged cooling seawater, from 0.2 to 0.05 mg/l [32]. In Europe, the regulatory authorities are increasingly concerned about the formation of chlorinated by-products: discharge limits may be tightened in the future there too. For instance, chlorination has already been banned in the Venetian Lagoon [33].

The practice of low-level chlorination of 0.5–1.0 mg/l deliberately applies a chronic, but not acute, toxicity to the sessile species within the cooling water flow and potentially also in the receiving water in close proximity to the discharge. It causes, however, the mortality of a proportion of the entrained planktonic organisms, depending on species sensitivity, life stage and the thermal regime involved. Following discharge, the effluent plume mixes with fresh seawater, and the sequential oxidant demand rapidly negates the remaining toxicity of the water. This is why almost no sign of measurable residual oxidant can be found beyond the mixing zone of coastal power plants [34].

Despite the fact that residual chlorine levels are quickly decreased following discharge, the potential for adverse effects still exists in the mixing zone of the plume. The high toxicity of chlorine has been confirmed by many studies. Based on toxicological data from a wide spectrum of marine species, the US EPA recommends a long-term water quality criterion for chlorine in seawater of 7.5 μg/l and a short-term criterion of 13 μg/l [35]. These are estimates of the highest concentration of a material in surface water to which an aquatic community can be exposed, briefly or indefinitely, without resulting in an unacceptable effect. The actual toxicity depends very much on species sensitivity and life cycle stage. The EU environmental risk assessment for hypochlorite has determined a PNEC (predicted no effect concentration) for saltwater species of 0.04 μg/l free available chlorine, based on fish, invertebrate and algae toxicity data. Furthermore, the EU risk assessment notes that the synergetic effects of thermal stress and exposure to residual chlorine should be taken into account, as demonstrated in many studies, e.g. on the discharge of power plant cooling effluents [36].

To conclude, the discharge and environmental levels reported for distillation plants may be harmful to the marine life in the plume mixing zones. It is therefore necessary to establish effluent standards and mixing zone regulations for desalination plants. The former encourages source control principles, such as effluent treatment, while the latter is associated with the concept of a mixing zone, in which gradual mixing takes place and where the numerical water quality standards could otherwise be exceeded [37]. Mixing zones can extend over considerable areas of the water body, depending on the effluent volume and the hydrology of the water body. The boundaries of the mixing area are therefore often regulated. In order to meet mixing zone regulations, properly sited outfalls with optimized high efficiency mixing design are usually needed [38].

Further potential concerns arise from the formation of halogenated organic by-products: mainly by-products of trihalomethanes (THMs) such as bromoform.

Increased THM levels of 9.5 μg/l [28] and of up to 83 μg/l [39] have been reported near distillation plants. Concentrations of other halogenated organics, such as haloacetic acids, are usually considerably lower. Oil contamination of the seawater may give rise to compounds like chlorophenols or chlorobenzenes [27, 39–41]. The THM levels reported near distillation plants are similar to mean bromoform concentrations of between 3.5 and 25.1 μg/l in the effluents of various different coastal power plants using chlorination [34].

While toxicity of the applied oxidant is known to decline rapidly with dilution due to compounding seawater demand, the same cannot necessarily be said of the more chemically stable by-products. For THMs like bromoform, the main route of loss from water is through volatilisation, with a half-life of 1–2 days in shallow water, and reported aerobic biodegradation half-lives of around 1 month. The half-lives of other chlorination by-products in seawater are reported to be in the range of several days to several weeks [34].

The concentrations encountered in desalination and power plant effluents were far below reported *acute* toxicity data. However, recorded data in current literature is limited and long term *chronic* exposure studies have yet to be published [27]. Some by-products, such as chlorinated hydrocarbons, are persistent and some components tend to accumulate in the fat of aquatic organisms and cause chronic mutagenic and carcinogenic toxicity [31].

Due to environmental and health issues raised as a result of residual chlorine and disinfection by-products, several alternative pretreatment methods have already been considered. None have gained acceptance over chlorine use, however chlorine dioxide is becoming an alternative to chlorine dosing in certain areas of the Gulf [18] and is also used in the Tampa Bay SWRO plant in Florida [9]. Chlorine dioxide – like chlorine – is a strong oxidant, but is thought to form less THMs if added in small quantities. Therefore, environmental impacts are considered to be relatively lower than for chlorine, but like all biocides, chlorine dioxide may affect non-target organisms if discharged into surface waters. A minimal impact is to be expected for *intermittent* discharge of chlorine dioxide into marine waters which undergo rapid mixing, based on toxicity tests for giant kelp, purple sea urchins and kelp bass [42]. The quoted study found chlorine dioxide to be markedly less toxic, with no observed effect concentrations (NOEC) a thousand times higher than for total residual chlorine. However, the study concludes that the results probably underestimate the effects of continuous exposure, citing 96 h LC_{50} values for fish species in the range between 20 and 170 μg/l.

11.2.2.3 Removal of Suspended Matter (RO Plants Only)

Conventional pretreatment relies on a combination of chemical treatment for coagulation-flocculation and subsequent media filtration in order to remove the suspended material from the intake water. Ferric chloride ($FeCl_3$) and ferric sulfate ($FeSO_4$) salts are typically used for coagulation. Dosing of sulfuric acid for pH adjustment and the addition of polyelectrolytes, which have similar properties to both polymers (high molecular weight compounds) and electrolytes (salts), can enhance the coagulation process. The dosage of coagulants and polyelectrolytes is normally correlated to the amount of suspended material in the intake water. It can range between < 1 and 30 mg/l for coagulants and between 0.2 and 4 mg/l for polyelectrolytes [14]. The particulate material is retained as the seawater passes through the filter beds. The filter backwash water, which contains natural suspended material and the coagulant chemicals, can either be discharged into the sea or can be dewatered and the sludge disposed on land. The clarified backwash water, which contains about 1% of the particulate material retained in the pretreatment filter, is normally discharged into the sea.

Large SWRO plants with a sludge separation step are, for example, the Perth and Sydney (Australia), Carlsbad (California) and Barcelona (Spain) projects. Examples where the sludge is discharged into the sea are the Ashkelon plant in Israel and the Hamma plant in Algeria, with capacities of 320,000 and 200,000 m^3/day, respectively. However, the Ashkelon plant and the new SWRO plants in Israel (e.g. Hadera) plan to collect the backwash water in a storage tank and then to discharge it continuously in order to avoid turbidity peaks. The use of ultra- and micro-filtration membranes (UF/MF) prior to RO is an emerging area in SWRO applications. It may reduce or eliminate the need for chemicals such as coagulants during pretreatment, but UF/MF membrane pretreatment systems require frequent chemically enhanced backwashing and periodic cleaning.

The filter backwash may significantly increase the amount of suspended matter in the discharge site, which may be an aesthetic problem as ferric salts can turn the mixing zone of the backwash plume into a deep red-brown colour. The filter backwash can thereby increase turbidity and decrease light penetration in the water column. Coagulant chemicals are commonly used in water treatment systems and are, in general, non toxic to aquatic life. Iron is also not considered a priority pollutant, as it is a common natural element in seawater. The discharge of large sludge volumes, however, may cause physical effects that can have negative impacts on marine life. For instance, lower light penetration could affect the productivity of benthic macroalgae, seagrasses or corals if present in the discharge site, and sedimentation of the material may blanket benthic plants and animals.

11.2.2.4 Chemicals for the Control of Scaling

Calcium carbonate is the main scale forming species in desalination plants. In thermal desalination plants, magnesium hydroxide and sulphate scales are also formed due to the high operating temperatures. Scale formation is usually controlled either by the addition of sulphuric or hydrochloric acid, the addition of a special scale inhibitor ("antiscalant"), or a combination thereof.

Acids must be added in relatively high concentrations of 20–100 mg/l to the feed stream due to a stoichiometric reaction with calcium carbonate. Target pH values are usually between 6 and 7, compared to a natural seawater pH of about 8.3. A pH effect on the receiving water is unlikely, due to the good buffering capacity of seawater, which will neutralize surplus acidity quickly following discharge.

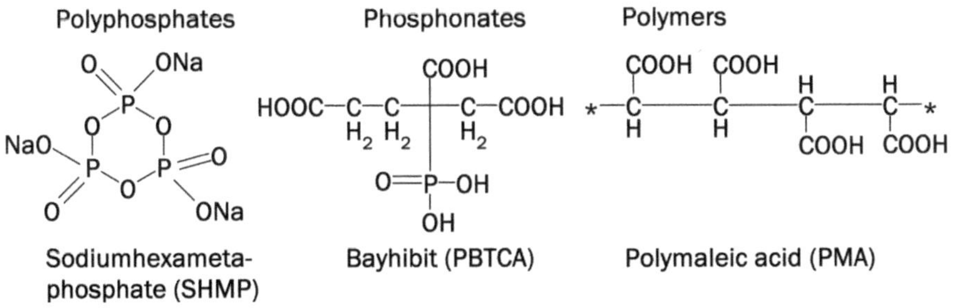

Fig. 11.6 Chemical structures of a phosphate antiscalant, a phosphonate antiscalant and a polymeric, polycarboxylic antiscalant (adapted from [14])

Antiscalants prevent scale formation in non-stoichiometric doses of 1–2 mg/l by retarding the nucleation process and impairing crystal growth. The main types of antiscalants are organic polymers (mainly polyacrylic acid and polymaleic acid), phosphonates and polyphosphates [14] (Fig. 11.6). Antiscalants are not harmful to invertebrate and fish species as the dosing levels are considerably lower than the concentrations at which acutely toxic effects can be observed. However, antiscalants may be harmful to algae at higher concentrations of around 20 mg/l. Material data sheets for some commercial antiscalant products state that observed inhibition of algae growth is likely to be due to the product's complexing ability and not to its toxicity, as such. Antiscalants prevent scale formation by dispersing and complexing divalent cations, such as calcium and magnesium, which are also needed for algae growth.

Antiscalants generally have a slow to moderate rate of elimination from the environment via abiotic and biological degradation processes. Most antiscalants are classified as "inherently biodegradable". As they exhibit complexing properties it seems plausible that antiscalants may also interfere with the natural processes of dissolved metals in seawater following discharge. This could be of concern in areas of high desalination activity, especially in combination with slow removal from the environment. Based on a typical dosage of 2 mg/l, it is estimated that 23 t of antiscalants could be discharged into the Mediterranean Sea and almost 65 t could be discharged into the Gulf every day from desalination plants. No field investigations concerning the actual environmental fate and interactions of antiscalants in areas of high desalination activity have been carried out to date [43].

Problems of eutrophication have been observed near the outlets of desalination plants in the Gulf where polyphosphates were used [29, 44]. These are easily hydrolysed to orthophosphates, which are an essential nutrient for primary producers. However, polyphosphates are only used on a limited scale at present.

11.2.2.5 Corrosion

Copper-nickel alloys are common heat exchanger materials in distillation plants. Corrosion of these materials typically causes contamination of reject streams with copper and nickel. Copper concentrations in reject streams of 15–100 μg/l have been reported [14]. The concentrate from RO plants may contain traces of iron, nickel, chromium and molybdenum, but metal contamination is generally below critical level, as non-metal equipment and corrosion-resistant stainless steels are commonly used in RO desalination plants.

The presence of copper does not necessarily mean that it will adversely affect the environment. Natural concentrations of copper range from an oceanic background level of 0.1–100 μg/l in estuaries [45]. It is therefore difficult to distinguish between natural copper levels and anthropogenic effects, e.g. caused by industrial outfalls or oil pollution. The discharge levels of thermal plants, however, are within a range that could affect natural copper concentrations. The US EPA recommends a maximum copper concentration of 4.8 μg/l in seawater for brief exposure and 3.1 μg/l for long-term exposure [35]. Values of the same order of magnitude were determined for European saltwater environments, with a PNEC of 5.6 μg/l [46] and a water quality objective for the Mediterranean Sea of 8 μg/l [47]. Copper, like most metals, is transported and accumulated in sediments, which is a major concern for point discharges, as this could lead to increased sediment concentrations in the discharge sites. This stresses the importance of estimating and evaluating total loads, in addition to concentrations. Metals in sediments can be assimilated by benthic organisms, which often form the basis of the marine food chain, leading to bioaccumulation and biomagnification.

11.2.2.6 Cleaning

The cleaning procedure depends on the type of fouling. In RO plants, alkaline solutions (pH 11–12) are typically used to remove silt deposits and biofilms from membranes, while acidic solutions (pH 2–3) are applied to dissolve metal oxides or scales. These solutions often contain additional chemicals to improve the cleaning process, such as detergents (e.g. dodecylsulfate, dodecylbenzene sulfonate) or oxidants (e.g. sodium perborate, sodium hypochlorite). After cleaning, or prior to storage, membranes are typically disinfected. For this purpose, either oxidizing biocides (such as chlorine and hydrogen peroxide) or non-oxidizing biocides (such as formaldehyde) are applied. The cleaning of distillation plants is comparatively simple: the plants are usually washed with warm acidic seawater, to which corrosion inhibitors (e.g. benzotriazole derivates) may be added, to remove alkaline scales from heat exchanger surfaces [14].

After the cleaning process is complete and the cleaning agents have been circulated through the membranes, the membranes are rinsed with fresh water (taken from the product water). In many cases, the residual membrane cleaning solution and also the first rinse, which contains most of the constituents from cleaning, are neutralized and diverted to a sanitary sewer for processing. The ensuing rinses are typically disposed of with the brine. Discharge into sewers may not be standard practice in all locations; little is known on current practice of waste disposal of most SWRO plants. It is possible that cleaning wastes are either discharged by direct blow-down immediately after cleaning, or by storage and continuous blending into the waste stream [14]. For instance, practices on the Canary Islands indicate that reject products such as chemical additives, pretreatment and membrane cleaning solutions, and other waste waters are usually discharged into the sea [48].

The accidental or deliberate discharge of cleaning solutions into surface waters may be harmful to marine life near the outlet, due to very high or low pH values and due to the presence of hazardous chemicals. For example, detergents like dodecylbenzene sulfonate have surface active properties, i.e. they have one lipophilic and one hydrophylic residue and are therefore soluble both in water and in organic solvents. This might, for example, disturb the intracellular membrane system of organisms. If complexing agents such as Ethylene Diamine Tetraacetic Acid (EDTA) are released into the sea, they could interact with dissolved metal ions and interfere with the natural processes of these elements. EDTA was furthermore found to be poorly degradable and persistent in the environment. Oxidizing or non-oxidizing biocides (e.g. chlorine or formaldehyde) used for disinfection, are particularly hazardous as they are effective biocides that may be harmful to marine life if released into surface water [14].

11.3 Cumulative Impacts on Sea Regions

11.3.1 The Gulf

Due to their waste discharges, desalination plants were identified as a major source of land-based marine pollution in the Gulf [49]. They are mainly large MSF distillation plants located in the very shallow, southern part of the Gulf. It is estimated that the combined discharge of all MSF plants amounts to a waste water flow of more than 1,000 m^3/s, which is the equivalent of a large river such as the Shatt Al-Arab. This waste water is characterised by a higher temperature and increased salinity and residual additives, including chlorine and antiscalants, as well as corrosion products such as copper. Daily discharge loads of these compounds from desalination plants into the Gulf are estimated to be at about 23.7 t of chlorine, 64.9 t of antiscalants and 296 kg of copper (updated from [43]).[3] Residual chlorine and chlorination by-products such as trihalomethanes, chlorophenols and chlorobenzenes are detectable near desalination plants, and chlorine pollution has been reported to be affecting two mud flat areas in the Bay of Kuwait [50], which is probably caused by the Doha power and desalination plant. There is no information on the environmental fate and effects of the antiscalant discharges, and signs of copper contamination attributed to desalination activity in water, sediment and organisms are also missing, possibly

[3] Assuming a chlorine concentration of 250 μg/l in the brine and cooling water discharges of MSF and MED plants, copper levels of 15 μg/l in the brine of MSF plants, and a dosing rate of 2 mg/l of antiscalants to the feed water of MSF plants operated at 10% recovery, and a dosing rate of 2 mg/l of antiscalants to the feed water of RO plants operated at a limited (33%) recovery due to the high salinity in the Gulf.

due to a lack of monitoring. To date, no impact assessment studies have been published, either based on field investigations or which comprehensively investigate the single or cumulative impacts of desalination plants on the Gulf's ecosystem.

The marine environment of the Gulf has been degraded over recent years as a result of a wide range of land-based pollution sources and anthropogenic activities [49]. Considering that the Gulf is a very shallow, semi-enclosed sea, with an average depth of 35 m and a narrow opening to the open ocean of only 56 km, the question is how it has been able to withstand the manifold pressures so far. This may be due to the good mixing of the water column caused by wind and tidal action. The general circulation is driven by density gradients and is characterized by a cyclonic, anti-clockwise pattern. Bottom water with high salinity flows along the southern shoreline, out of the Gulf, and is compensated by oceanic surface water through the Strait of Hormuz which flows along the Iranian coast to the north. With an anticipated turnover time of about 3–5 years, waterborne pollutants are eventually flushed out of the Gulf. For other substances, such as metals which do not degrade and tend to be transported into the sediments, the Gulf may act as a sink with the risk of long-term accumulation of these substances.

11.3.2 The Red Sea

Similar to the Gulf, the Red Sea is a semi-enclosed body of water with limited exchange with the open ocean through the narrow strait of Bab el Mandeb (29 km wide, 130 m deep). However, the Red Sea is considerably larger and deeper than the Gulf with an average depth of 490 m. A net inflow of water into the Red Sea occurs through the Bab El Mandeb which is partially balanced by an outward bottom current. These flows correspond to moderate turnover times of 6 years for the surface layer and 200 years for the whole water body [51]. A pycnocline separates the surface and deep water layers at a depth of about 250–300 m. The Red Sea is known for its outstanding and fragile biological habitats. The southern parts are influenced by oceanic water and dominated by fine grained sediments and associated mangroves, seaweeds and calcareous algae, while the nutrient-deprived northern parts are characterized by fringing coral reefs [52].

Similar to the Gulf, desalination plants are considered a major source of land-based pollution in the Red Sea [53]. Although the total installed capacity is lower than in the Gulf region, the world's largest MSF plant and other large plants are found in the Red Sea area. The daily discharges of chemicals from desalination plants into the Red Sea can be estimated to amount to 5.6 t of chlorine, 20.7 t of antiscalants and 74 kg of copper (updated after [52]).[4] So far, no scientific model exists that allows a conclusion on the actual impacts of these discharges on the Red Sea's ecosystem. The stratification of the water column, the existence of sills and the long turnover times of the deep water layer bear the risk that pollutants have either relatively long residence times or remain in the Red Sea indefinitely.

[4]Based on the same assumptions as for the Gulf region.

In combination with fragile and ecologically important ecosystems, it is probable that the Red Sea is very susceptible to disturbances by harmful materials. The rapid expansion of urban centres in Saudi Arabia has been achieved through the extensive use of desalinated water to meet the demands of the population and industry [53]. A similar development will probably take place in the Gulf of Aqaba. The transition from the Gulf of Aqaba to the Red Sea, separated by the 250 m deep Strait of Tiran, is a smaller version of the transition from the Red Sea to the ocean. The double semi-enclosed nature of the Gulf of Aqaba makes it one of the places most susceptible to pollution.

11.3.3 The Mediterranean Sea

The complex geomorphology of the Mediterranean basin is reflected in a complex surface water circulation, which is characterized by the formation of ring-shaped currents in most of the Mediterranean regional seas. As tidal currents are generally weak and therefore have little influence on the dispersal and dilution of pollutants, surface circulation is the primary factor controlling the transport of contaminants within the Mediterranean Sea. Transport between basins and passage out of the Mediterranean is limited by narrow and shallow straits. Vertically, the water column is not well-mixed due to the deep average depth of the Mediterranean, of almost 1,500 m. Three different water masses can be distinguished within this stratification: The *surface water* consists of inflowing Atlantic water from the west, which moves in an anti-clockwise direction along the Algerian coast to the east. High evaporation rates in the eastern basin cause an increase in salinity. In combination with winter cooling, the surface water increases in density and sinks, forming the *Levantine Intermediate Water* (LIW). The LIW flows towards the west at a depth of 200–500 m and enters the Adriatic and Balearic Seas, where strong cooling events cause a further increase in density and lead to the formation of the *East and West Mediterranean Deep Water* (at a depth below 600 m). Mixing of the two deep waters is largely restricted by the 400 m deep sill that forms the Strait of Sicily. Above this deep layer, the LIW circulates through both basins and eventually exits the Mediterranean via Gibraltar. The circulation takes place slowly and the turnover time, from entry as Atlantic surface water until its return to the ocean, is about 80 years [54, 55]. Some deep water bodies may be much older, in the order of 100–300 years, whereas some of the water may exit in only a few decades. Therefore, marine pollution problems in the two basins are to a large extent independent of each other and the long turnover times allow for a rapid accumulation of substances [56].

A recent report of the European Environment Agency and UNEP identified priority issues in the Mediterranean environment, including land-based pollution by sewage, urban run-off and industrial effluents [57]. The report does not list or discuss desalination activity in any form, however, an earlier report addressed sea water desalination in the Mediterranean specifically, including environmental concerns [16, 58]. The document pointed out that seawater desalination is an industrial process and a growing industry in the Mediterranean, which may have adverse effects on the coastal environment if not well designed and managed.

Hot spots of desalination in the Mediterranean are along the Spanish, Algerian, Libyan and Israeli coasts and on some larger islands including Mallorca, Malta, Sicily and Cyprus. The main process used is RO. The concentrate from RO plants is primarily characterized by high salinity, and typically contains antiscalants from pretreatment. The daily discharge of antiscalants from desalination plants into the Mediterranean Sea may amount to 23 t plus 1.9 t of chlorine and 18 kg of copper.[5] Side streams, such as backwash waters from media filters, are often discharged into the sea, and the fate of cleaning solutions could be similar.

11.3.4 Discussion

Naturally, enclosed seas have a very limited exchange of water with the open ocean which favours long residence times of pollutants. The Gulf has the world's highest density of desalination plants. If anywhere, the impact of desalination activity should be visible in the Gulf. However, a holistic study investigating the cumulative impact of desalination plants on the Gulf's marine environment is still to be done. If there are effects, they may not be immediately apparent for two reasons: firstly, favourable mixing and flushing may disperse the pollutants, and secondly, impacts from desalination activities may be overshadowed by other sources of pollution or activity, such as the permanent oil burden or land reclamation. While no conclusive evidence can be found concerning the Gulf *as a whole*, the risk of damage to the ecosystems in close proximity to plants does exist. It is possible to link environmental effects to desalination activity on some occasions, but even here the scientific data is incomplete. The few available studies are typically short-term, limited in scope, and without ecologic baseline or effects monitoring. They fall short of recognizing the potentially synergetic effects of single waste components on marine organisms and the complexity of potential responses by the ecosystems.

The Mediterranean and Red Sea have a lower density of desalination plants than the Gulf, although some parts show increased desalination activity. Due to their longer coastlines, greater water depth and lower total desalination capacity, cumulative impacts are less likely and are of secondary importance behind other issues of higher priority, such as sewage and industrial discharges or eutrophication. In the Mediterranean, where RO is the dominating process, the problem of chemical discharges is "reduced" to the antiscalant loads and intermittent backwash or cleaning wastes. Chemical impacts from desalination plants are therefore still limited and loads are fairly well distributed. Localized impacts, however, may be

[5] Assuming a dosing rate of 2 mg/l of antiscalants to the feed water of RO plants operated at 33% recovery (for comparability). For MSF plants, the same assumptions as for the Gulf and Red Sea have been used.

significant, especially when important ecosystems are affected. These include the seagrass meadows in the Mediterranean, which have been classified as a priority habitat by the European Habitats Directive[6] and have been found to be very sensitive to salinity increases, as well as, for instance, the coral reefs and mangroves of the Red Sea that are of global importance. The Red Sea is a unique ecological treasure which is vulnerable to ecological damage. With regard to the expected future demand for seawater desalination in both the Mediterranean and Red Sea region, it is important to regulate the development of new plants and to adopt a precautionary approach. It is therefore of paramount importance to investigate and mitigate all potential impacts on a project- and site-specific basis.

11.4 Protecting the Marine Environment from Harmful Effects

A widely recognized approach for investigating and mitigating the impact of development projects on the environment is environmental impact assessment (EIA). To this day, only a handful of EIA studies on desalination plants have been carried out and made publicly available. There is still a surprising paucity of useful experimental data, both from laboratory tests and from field monitoring [59]. In some cases, the investigations were carried out under tight time constraints. For instance, only 4 months were given for an EIA study of a large SWRO plant in Algeria [60]. This shows that environmental concerns can be of secondary importance when a ready supply of freshwater is urgently needed. However, the opposite is also true: comprehensive environmental studies are currently being carried out on the major SWRO projects in Australia, and environmental concerns are the major hurdle in obtaining planning permission for new projects in California.

In the EU, the EIA Directive[7] regulates which project categories have to be subject to an EIA by member states. The directive covers groundwater abstraction schemes, dams and works for the transfer of water resources between river basins, but does not list desalination plants. This may be due to the fact that desalination plants were small and only used at a minor scale in Southern Europe at the time when the directive was first introduced in 1985, and later amended in 1997. As EIAs are mandatory for other large water supply projects, it would be consistent to include desalination projects in the directive as well. Furthermore, desalination projects should be an integral part of water resources management planning which not only considers the development of new or existing water supplies, but also the economic use and reuse of water, where possible. According to EU regulations, a strategic environmental assessment (SEA Directive)[8] is mandatory for plans and programmes in the field of water management.

---

[6]Directive 92/43/EEC.

[7]Directive 97/11/EC amending Directive 85/337/EEC.

[8]Directive 2001/42/EC.

A central element of all EIA studies is the comparison of possible alternatives, such as alternative project sites or alternative technologies in order to identify the option with the least environmental impact. For example, the selection of a suitable site for a new desalination plant can be a very effective way of preventing later impacts. In general, ecosystems or habitats which are worth protecting on a regional or global scale, which are inhabited by protected, endangered or rare species, which are important in terms of their productivity or biodiversity, or which play an important role as feeding or reproductive areas in a certain region, should be avoided. The site for the desalination plant should furthermore provide sufficient capacity to dilute and disperse the salt concentrate and to dilute, disperse and degrade any residual chemicals. The load and transport capacity of a site will primarily depend on water circulation and exchange rate as a function of currents, tides, surf, water depth and shoreline morphology. In general, exposed rocky or sandy shorelines with strong currents and surf would be preferred to shallow, sheltered sites with little water exchange, as proposed in a sensitivity scale of coastal ecosystems to desalination discharges [61].

Several different technical approaches can be implemented to mitigate the environmental effects of waste discharges, including pre-dilution with cooling water or the use of multi-port diffuser systems to dilute and disperse the salt load. Advanced diffuser systems can achieve a maximal dilution with a minimum salinity increase of 1 unit above background levels, outside the mixing zone [62]. To avoid impacts from high temperature, the outfall should achieve maximum heat dissipation, from the waste stream to the atmosphere, before entering the water body, e.g. by elongated outfall channels, reservoirs or cooling towers. To analyze plume spreading in a specific project site, the environmental and operational conditions should be investigated using hydrodynamic modelling studies, in conjunction with field investigations, before and during operation of the desalination plant [18].

Negative impacts from chemicals can be minimized by treatment before discharge, by substitution of hazardous substances and by implementing alternative, non-chemical treatment options wherever possible. In particular, biocides such as chlorine, which may be harmful to non-target organisms in the discharge site, should be replaced or treated prior to discharge. Chlorine can be effectively removed by various chemicals, such as sodium bisulfite as used in RO plants. For the treatment of cooling waters of coastal power plants and ballast water from ships, products based on peracetic acid have been approved by environmental authorities. Filter backwash water should be treated by dewatering and land-deposition where possible, whilst cleaning solutions should be treated on-site in special treatment facilities or discharged into a sanitary sewage system [18].

The use of alternative pretreatment methods should be considered where feasible. Prefiltration with Ultra Filtration (UF) or Micro Filtration (MF) membranes may eliminate or reduce the need for chemical pretreatment, if well designed. UF/MF pretreatment often uses in-line coagulation and shock chlorination, and usually requires chemically enhanced backwash and periodic cleaning [15] of the membranes and is therefore not entirely "chemical free". However, the advantage of intermittent backwashing and cleaning over continuous pretreatment is that waste waters are produced in smaller volumes and can be treated separately. A non-chemical treatment option is irradiation for disinfection of the intake water with UV-light at a wavelength of 200–300 nm. UV-light damages the DNA structure of microorganisms and forms highly reactive and short-lived active substances in seawater (i.e. free radicals). However, to date UV-irradiation has not been found to be an effective pretreatment for larger desalination plants.

In order to mitigate the impacts of open intakes, a combination of varied meshed screens and a low intake velocity should be considered. This can minimize the impingement and entrainment of larger organisms, such as fish or turtles, while the entrainment of smaller plankton organisms, eggs and larvae can be minimized by locating intakes away from productive areas, e.g. into deeper waters, offshore or underground (e.g. horizontal drains or beach wells). As the intake water quality is often better in these locations than in nearshore and surface waters, lower chemical pretreatment may be required. However, the initial soil disturbance during construction of below ground intakes or long pipelines may be higher, especially when this involves drilling or excavation activities. Co-location of desalination and power plants should also be considered where feasible, as the power plant cooling water can serve as feedwater to the desalination plant, which minimizes the impact of entrainment and impingement, the usage of chemicals and construction and land use impacts [18].

11.5 Final Remarks

A standard chemical pretreatment is still preferred over non-conventional methods, as outlined above, for most new desalination projects. Although many of the measures above have reportedly produced good results in smaller projects, their performances still need to be validated on a larger scale, i.e. in large plants and over longer periods of time. Many useful ideas have been put forward in recent literature to minimize the environmental footprint of desalination. The best practicable environmental option, however, can only be identified through project- and site- specific environmental impact assessment studies. A catalogue of best available techniques (BAT) and best environmental practices (BEP) may be useful in guiding practitioners, consultants and decision-makers in their choices.

All these measures will probably increase the cost of water production. For example, over US$ 15 million were spent on planning permission costs for the proposed Carlsbad SWRO project and a Climate Action Plan will be added, of an estimated US$ 76 million, to the gross capital cost of the plant over its 30-year life. For two Australian SWRO plants currently under construction, the advanced seawater intake and concentrate outfalls cost more than the entire capital cost of the Ashkelon plant [63]. Sustainable desalination is not a utopia, and requires a commitment to providing water at a reasonable price, which not only includes the usual construction and operating costs, but also the costs that are necessary to reduce the environmental impact, including the costs of environmental studies, advanced technology and compensation measures.

For a long time, the earth's atmosphere was regarded as a bottomless pit with an endless capacity to absorb emissions. Today, as many of the world's large rivers are running dry, it is often proclaimed that the sea provides an unlimited source of drinking-water. This is true insofar as 97% of the world's water lies in the oceans, and desalination removes only an infinitesimally small amount of water compared to natural evaporation. The problem lies rather in the waste products, the concentrate and chemical discharges, and the impact on local ecosystems. Many coastal areas and regional seas are already under stress from manifold anthropogenic activities, including land reclamation and habitat degradation, eutrophication and land-based pollution, fishing and maritime shipping. As desalination plants depend on intact marine ecosystems which provide a good raw water quality, marine environmental protection is also in the inherent interest of the industry.

Acknowledgments The idea for this chapter emerged during a two day seminar hosted by Miriam Balaban and the European Desalination Society in L'Aquila, entitled "Desalination and the Marine Environment", where the author met one of the editors of this book. The results presented stem from previous and current research projects in which the author participated. Special acknowledgement must be given to the research project MEDINA of the Sixth European Research programme [5].

Abbreviations

BAT	best available techniques
BEP	best environmental practices
DNA	deoxyribonucleic acid
EDTA	ethylene diamine tetraacetic acid
EIA	environmental impact assessment
EPA	environmental protection agency
HDD	horizontally drilled drains
IDA	international desalination association
LIW	levantine intermediate water
LC50	lethal concentration with 50% mortality of test organisms
MED	multi-effect distillation
MEDINA	membrane-based desalination – an integrated approach
MF	microfiltration
MSF	multi-stage flash distillation
NOEC	no observed effect concentration
PNEC	predicted no effect concentration
RO	reverse osmosis
SEA	strategic environmental assessment
SWRO	seawater reverse osmosis
THMs	trihalomethanes
UF	ultrafiltration
UNEP	United Nations Environment Programme
UV	ultraviolet light

References

1. IDA, Media Analytics Ltd. (2007) The 20th IDA worldwide desalting plant inventory. MS Excel format, Media Analytics Ltd., Oxford, UK
2. GWI (2008) DesalData. http://desaldata.com. Accessed 15 May 2008
3. Dreizin Y, Tenne A, Hoffman D (2008) Integrating large scale seawater desalination plants within Israel's water supply system. Desalination 220:132–149
4. Voutchkov N (2007) Advances and challenges of seawater desalination in California. Proceedings of the IDA world congress on desalination and water reuse, Gran Canaria
5. Lattemann S (2007) Overview and description of the relevant impacts of membrane desalination plants, with emphasis on concentrate and chemical discharges. Project report D 9.2.1, Membrane-based desalination – an integrated approach (MEDINA), European sixth framework programme, thematic priority global change and ecosystems, specific targeted research project, project no 036997
6. Peters T, Pinto D, Pinto E (2007) Improved seawater intake and pre-treatment system based on Neodren technology. Desalination 203:134–140
7. City of Carlsbad, Poseidon Resources (2005) Environmental impact report for the Carlsbad seawater desalination facility. http://www.carlsbad-desal.com/EIR.asp. Accessed 22 June 2008
8. City of Huntington Beach, Poseidon Resources (2005) Environmental impact report for the Huntington Beach seawater desalination facility. http://www.surfcity-hb.org/Government/Departments/Planning/major/poseidon.cfm. Accessed 22 June 2008
9. Tampa Bay Water, http://www.tampabaywater.org/watersupply/tbdesalprotect.aspx. Accessed 22 June 2008
10. Einav R, Lokiec F (2003) Environmental aspects of a desalination plant in Ashkelon. Desalination 156:79–85
11. Sinapsa Ltd. (2007) Summary of the environmental document for the desalination plant in Hadera, Israel
12. Alspach B (2007) The evolution of SWRO co-location in the US: Analysis and alternatives. Proceedings of the IDA world congress on desalination and water reuse, Gran Canaria
13. Damitz B, Furukawa D, Toal J (2006) Desalination feasibility study for the Monterey Bay region. Final report prepared for the Association of Monterey Bay Area Governments (AMBAG). http://www.ambag.org/publications/reports/Desal2006/AMBAGFINALDesal Study.pdf. Accessed 22 June 2008
14. Lattemann S, Höpner T (2003) Seawater desalination – impacts of brine and chemical discharges on the marine environment. Desalination publications, L'Aquila, Italy
15. Hodgkiess T, Oldfield JW, Höpner T et al. (2003) Assessment of the composition of desalination plant disposal brines. Middle East Desalination Research Center (MEDRC), Project No 98-AS-026
16. Lattemann S (2003) Guidelines for the environmental sound management of seawater desalination plants in the Mediterranean region. In: UNEP/MAP/MEDPOL (ed.) Sea water desalination in the Mediterranean: Assessment and guidelines. MAP technical reports series no 139, UNEP/MAP, Athens
17. WHO (2008) Desalination for safe water supply, guidance for the health and environmental aspects applicable to desalination. World Health Organization (WHO), Geneva, Eastern Mediterranean Regional Office (EMRO), Cairo
18. Lattemann S, Höpner T (2008) Environmental impact and impact assessment of seawater desalination. Desalination 220:1–15
19. World Bank Group (1998) Thermal Power: Guidelines for new plants, pollution prevention and abatement handbook, http://www.ifc.org/ifcext/enviro.nsf/AttachmentsByTitle/gui thermnewWB/FILE/thermnewPPAH.pdf. Accessed 22 June 2008
20. Latorre M (2005) Environmental impact of brine disposal on *Posidonia* seagrasses. Desalination 182:517–524
21. Crisp G, Rhodes M, Procter D et al. (2007) Perth Seawater Desalination Plant ñ Blazing a

Sustainability Trail Proceedings of the IDA World Congress on Desalination and Water Reuse, Maspalomas, Gran Canaria
22. Walker DI (1989) Regional studies – seagrass in Shark Bay, the foundations of an ecosystem. In: Larkum AWD, McComb AJ, Sheperd SA (eds.) Biology of seagrasses – a treatise on the biology of seagrasses with special reference to the Australian region, Elsevier, Amsterdam
23. Argyrou M (1999) Impact of desalination plant on marine macrobenthos in the coastal waters of Dhekelia Bay, Cyprus. Department of Fisheries, Ministry of Agriculture, Natural Resources and Environment, Nicosia, Cyprus
24. Coles SL, McCain JC (1990) Environmental factors affecting benthic infaunal communities of the western Arabian Gulf. Mar Env Res 29:289–315
25. Buceta JL, Gacia E, Mas J et al. (2003) Estudio de los efectos de incrementos de salinidad sobre la fanerogama marina *Posidonia*oceanica y su ecosistema, con el fin de prever y minimizar los impactos que pudieran causar los vertidos de aguas de rechazo de plantas desaladoras. Ingenieria Civil 132
26. Environmental impact assessment report for the seawater desalination project in Okinawa, Japan, executive summary. Okinawa Bureau for Enterprises
27. Jenner HA, Taylor CJL, van Donk M et al. (1997) Chlorination by-products in chlorinated cooling water in some European coastal power stations. Mar Env Res 43:279–293
28. Ali MY, Riley JP (1986) The distribution of halomethanes in the coastal waters of Kuwait. Mar Pollut Bull 17:409–414
29. Abdel-Jawad M, Al-Tabtabaei M (1999) Impact of current power generation and water desalination activities on Kuwaiti marine environment. Proceedings of the IDA world congress on desalination and water reuse, USA
30. Al-Ghadban AN, Al-Ajmi A (1993) Environmental impact assessment: Integrated methodology – a case study of Kuwait, Arabian Gulf. Coast Man 21:271–298
31. European Commission (2001) Reference document on best available techniques in the industrial cooling systems. European integrated pollution prevention and control (IPPC) bureau. http://eippcb.jrc.es. Accessed 22 June 2008
32. KEMA (2006) Quatargas trials EU's seawater chlorination best available technique, meeting the strict new standard. KEMA information leaflet, The Netherlands
33. Cristiani P (2005) Solutions to fouling in power station condensers. Appl Therm Eng 25:2630–2640
34. Taylor CJL (2006) The effects of biological fouling control at coastal and estuarine power stations. Mar Pollut Bull 53:30–48
35. U.S. EPA (2006) National recommended water quality criteria, National Center for Environmental Publications and Information, Cincinnati. http://www.epa.gov/waterscience/criteria/wqctable/nrwqc-2006.pdf. Accessed 22 June 2008
36. ECB (2007) Draft risk assessment report for sodium hypochlorite, European Chemicals Bureau (ECB), http://ecb.jrc.it/documents/Existing-Chemicals/RISKASSESSMENT/DRAFT/R045071lenvhh.pdf. Accessed 22 June 2008
37. Jirka GH, Bleninger T, Burrows R et al. (2004) Environmental quality standards in the EC-Water Framework Directive: Consequences for water pollution control for point sources. Eur Water Man Online, European Water Association, http://www.ewaonline.de/journal/pdf. Accessed 22 June 2008
38. Bleninger T, Jirka GH, Lattemann S (2008) Environmental planning, prediction and management of brine discharges from desalination plants, Middle East Desalination Research Center (MEDRC), project no 07-AS-003
39. Saeed T, Khordagui H, Al-Hashash H (1999) Contribution of power/desalination plants to the levels of halogenated volatile liquid hydrocarbons in the coastal areas of Kuwait. Desalination 121:49–63
40. Allonier AS, Khalanski M, Camel V et al. (1999) Characterization of chlorination by-products in cooling effluents of coastal nuclear power stations. Mar Pollut Bull 38:1232–1241

41. Abarnou A, Miossec L (1992) Chlorinated waters discharged to the marine environment, chemistry and environmental impact. Sci Tot Env 126:173–197
42. Hose JE, Fiore DD, Parker HS et al. (1989) Toxicity of chlorine dioxide to early life cycle stages of marine organisms. Bull Env Contam Toxicol 42:315–319
43. Lattemann S, Höpner T (2008) Impacts of seawater desalination plants on the marine environment of the Gulf. In: Abuzinada AH, Barth HJ, Krupp F et al. (eds.) Protecting the Gulf's marine ecosystems from pollution, Birkhäuser, Basel
44. Shams El Din AM, Aziz S, Makkawi B (1994) Electricity and water production in the Emirate of Abu Dhabi and its impact on the environment. Desalination 97:373–388
45. Kennish MJ (1997) Practical handbook of estuarine and marine pollution, 2nd ed. CRC Press, Boca Raton.
46. Hall LW, Anderson RD (1999) A deterministic ecological risk assessment for copper in European saltwater environments. Mar Poll Bull 38:207–218
47. UNEP/WHO (1996) Guidelines for authorizations for the discharge of liquid wastes into the Mediterranean Sea. MAP technical reports series no. 107, UNEP, Athens
48. Sadhwani JJ, Veza JM, Santana C (2005) Case studies on environmental impact of seawater desalination. Desalination 185:1–8
49. UNEP (1999) Overview on land-based sources and activities affecting the marine environment in the ROPME sea area. UNEP regional seas reports and studies no. 168
50. Scott DA (1995) A directory of wetlands in the Middle East. IUCN, Switzerland, and IWRB, U.K.
51. Sheppard C (2000) The Red Sea. In: Sheppard C (ed.) Seas at the millennium, Pergamon, Amsterdam
52. Hoepner T, Lattemann S (2003) Chemical impacts from seawater desalination plants–a case study of the northern Red Sea. Desalination 152:133–140
53. UNEP (1997) Assessment of land-based sources and activities affecting the marine environment in the Red Sea and Gulf of Aden. UNEP regional seas reports and studies no. 166.
54. Tomczak M, Godfrey JS (1994) Regional oceanography: An introduction. Pergamon, Amsterdam
55. Robinson AR, Leslie WG, Theocharis A et al. (2001) Ocean currents: Mediterranean Sea circulation, Academic Press, London
56. UNEP (1996) State of the marine and coastal environment in the Mediterranean region. MAP technical report series no. 100, UNEP, Athens
57. EEA (2006) Priority issues in the Mediterranean environment. European Environment Agency, Copenhagen, United Nations Environment Programme, Athens
58. UNEP/MAP/MED POL (2003) Sea water desalination in the Mediterranean: Assessment and guidelines. MAP technical reports series no. 139, Athens
59. National Research Council (2008) Desalination: A national perspective, committee on advancing desalination technology, water science and technology board, Division on Earth and Life Studies, National Research Council of the National Academies. http://www.nap.edu/catalog.php?recordid=12184. Accessed on 22 June 2008
60. Mooij C (2007) Hamma water desalination plant: Planning and funding. Desalination 203:107–118
61. Höpner T, Windelberg J (1996) Elements of environmental impact studies on coastal desalination plants. Desalination 108:11–18
62. Melbourne Water, GHD (2007) Melbourne augmentation program seawater desalination feasibility study
63. Pankratz T (2008) SWRO economics oil's long reach. Water desalination report 44, 13

Index

MIX
Papier aus verantwortungsvollen Quellen
Paper from responsible sources
FSC® C105338

If you have any concerns about our products,
you can contact us on
ProductSafety@springernature.com

In case Publisher is established outside the EU,
the EU authorized representative is:
Springer Nature Customer Service Center GmbH
Europaplatz 3, 69115 Heidelberg, Germany

Printed by Libri Plureos GmbH
in Hamburg, Germany